Heterostructure Epitaxy and Devices

NATO ASI Series

Advanced Science Institutes Series

A Series presenting the results of activities sponsored by the NATO Science Committee, which aims at the dissemination of advanced scientific and technological knowledge, with a view to strengthening links between scientific communities.

The Series is published by an international board of publishers in conjunction with the NATO Scientific Affairs Division

A	**Life Sciences**	Plenum Publishing Corporation
B	**Physics**	London and New York
C	**Mathematical and Physical Sciences**	Kluwer Academic Publishers
D	**Behavioural and Social Sciences**	Dordrecht, Boston and London
E	**Applied Sciences**	
F	**Computer and Systems Sciences**	Springer-Verlag
G	**Ecological Sciences**	Berlin, Heidelberg, New York, London,
H	**Cell Biology**	Paris and Tokyo
I	**Global Environmental Change**	

PARTNERSHIP SUB-SERIES

1.	**Disarmament Technologies**	Kluwer Academic Publishers
2.	**Environment**	Springer-Verlag / Kluwer Academic Publishers
3.	**High Technology**	Kluwer Academic Publishers
4.	**Science and Technology Policy**	Kluwer Academic Publishers
5.	**Computer Networking**	Kluwer Academic Publishers

The Partnership Sub-Series incorporates activities undertaken in collaboration with NATO's Cooperation Partners, the countries of the CIS and Central and Eastern Europe, in Priority Areas of concern to those countries.

NATO-PCO-DATA BASE

The electronic index to the NATO ASI Series provides full bibliographical references (with keywords and/or abstracts) to more than 50000 contributions from international scientists published in all sections of the NATO ASI Series.
Access to the NATO-PCO-DATA BASE is possible in two ways:

– via online FILE 128 (NATO-PCO-DATA BASE) hosted by ESRIN,
Via Galileo Galilei, I-00044 Frascati, Italy.

– via CD-ROM "NATO-PCO-DATA BASE" with user-friendly retrieval software in English, French and German (© WTV GmbH and DATAWARE Technologies Inc. 1989).

The CD-ROM can be ordered through any member of the Board of Publishers or through NATO-PCO, Overijse, Belgium.

Series 3: High Technology – Vol. 11

Heterostructure Epitaxy and Devices

edited by

Jozef Novák

Institute of Electrical Engineering,
Slovak Academy of Sciences,
Bratislava, Slovakia

and

Andreas Schlachetzki

Institute for Semiconductor Technology,
Technical University Braunschweig,
Braunschweig, Germany

Springer-Science+Business Media, B.V.

Proceedings of the NATO Advanced Research Workshop on
Heterostructure Epitaxy and Devices
Smolenice Castle, Slovakia
October 15–19, 1995

A C.I.P. Catalogue record for this book is available from the Library of Congress.

ISBN 978-94-010-6593-1   ISBN 978-94-009-0245-9 (eBook)
DOI 10.1007/ 978-94-009-0245-9

Printed on acid-free paper

 NATO Advanced Research Workshop

H E A D '95

2nd international workshop on

HETEROSTRUCTURE EPITAXY AND DEVICES

OCTOBER 15 - 19, 1995

HOUSE OF SCIENTISTS, SMOLENICE CASTLE, SLOVAKIA

Organised by

Institute of Electrical Engineering
Slovak Academy of Sciences
Bratislava

Institute for Semiconductor Technology
Technical University of Braunschweig
Braunschweig

CONTENTS

SECTION II. HETEROSTRUCTURES

SECTION III. COMPOSITE SYSTEMS

SECTION IV:CHARACTERIZATION

SECTION V. DEVICES

PREFACE

From October 15 to 19, 1995 a Workshop on Hetero-structure Epitaxy and Devices was held at Smolenice Castle near Slovakia's capital Bratislava. The intention of this Workshop was to establish and strengthen ties between scientists of the formerly Socialist East and Middle-European states with their colleagues from the Western countries. With this aim the Workshop found the financial support by NATO which tremendously helped to facilitate organizing the meeting

That the Workshop was also a scientific success is evidenced by the present volume comprising a selection of the contributed papers. We are confident that the reader of these Proceedings can convince himself of the high quality of the work whose results are presented here. We hope that this and the numerous discussions between the participants of the Workshop will promote cooperations among scientists from the countries represented at the meeting.

It is a pleasure to express our gratitude to NATO and, as representatives of the institutions involved in the organization, to Lubomir Malacký (Institute of Electrical Engineering, Slovak Academy of Sciences) and Hergo-Heinrich Wehmann (Institute for Semiconductor Technology, Technical University Braun-schweig) whose dedicated work was most essential for the Workshop.

J. Novák   A. Schlachetzki

November 1995

SIMULATION OF III-V LAYER GROWTH

Y. ARIMA
Department of Physics, Gakushuin University
1-5-1 Mejiro, Toshima-ku, Tokyo 171, Japan

AND

T. IRISAWA
Computer Center, Gakushuin University
1-5-1 Mejiro, Toshima-ku, Tokyo 171, Japan

1. Introduction

Since it was reported [1] that the intensities of RHEED for the growing surface of a *GaAs* crystal in the process of MBE oscillate with a period corresponding to the completion of a monolayer, this phenomenon has been applied to the thin layer growth of man-made superlattices. It can be said that MBE method is the process of the crystal growth under the extreme conditions that all incident atoms are incorporated into the lattice because of the low enough substrate temperature to neglect the evaporation from the surface. Recently, as the surface structure is studied in detail by using STM, one can see not only the monolayer steps but also a few atom clusters.

We carried out the Monte Carlo study to understand the structural feature of surface in MBE growth, because we can not apply the usual vapor growth theory [2]. However, we must devise the simulation method because of the slow growth rate under MBE growth conditions.

In the next section, we argue the simulation of one-component system. At first, we clarify the simulation methods to be used in this paper. We define the characteristic length for MBE growth with kinematical consideration, and show the periodic changes and flatness in the structure of a surface growing under MBE conditions [3, 4]. Two growth modes (layer by layer and step flow) of vicinal surface under MBE conditions are distinguished by the condition whether the nucleation occurs or not. Moreover, we discuss the origin of anisotropy of·the surface structure. In section 3, we extend our simulation model to the *A-B* two-component system. We find

1

J. Novák and A. Schlachetzki (eds.), Heterostructure Epitaxy and Devices, 1–10.
© *1996 Kluwer Academic Publishers.*

some feature which are not seen in one-component system. We discuss the orderliness of crystal in the stoichiometric A-B system, especially.

2. Simulation of One Component System

2.1. SIMULATION MODEL

At first, we show the simulation models and the algorithm to be used in this paper. We simulate the growth of the (001) face of a simple cubic crystal. Events to be treated are adsorption, evaporation and surface diffusion of the atoms. The anisotropy of bond energy (ϕ_x, ϕ_y and ϕ_z) and that of surface diffusion are taken into account. When we set $\phi = \phi_x = \phi_y = \phi_z$, this model is equivalent to Gilmer and Bennema's model [5].

The ratio of evaporation rate K_{ij}^{-} and adsorption rate K^{+} is given by

$$K_{ij}^{-}/K^{+} = \exp[(1-i)\phi_x/kT + (1-j)\phi_y/kT - \Delta\mu/kT] \qquad (1)$$

and the ratio of the surface diffusion rate K_{ij}^{D} and adsorption rate is given by

$$K_{ij}^{D}/K^{+} = (\lambda_s/a)^2 K_{ij}^{-} \Big/ K^{+}, \qquad (2)$$

where $\Delta\mu$ is the chemical potential difference of the crystal and vapor, a the lattice constant, k the Boltzmann constant, T the temperature, i and j ($i, j = 0, 1, 2$) are the number of the nearest neighbor bonds in the direction x and y, respectively. Surface diffusion length λ_s is given by

$$\begin{aligned} \lambda_s^2 = D_s \tau_s &= a^2 exp[(\phi_z - E_D)/kT] \\ &= a^2 exp[(1-\varepsilon)\phi_z/kT] \end{aligned} \qquad (3)$$

using surface diffusion constant $D_s = a^2 \nu \exp[-E_D/kT]$ and the mean life time of adatoms $\tau_s = \nu^{-1} \exp[\phi_z/kT]$, where ν is the vibrational frequency and E_D activation energy of surface diffusion. And we represent E_D as $E_D = \varepsilon\phi_z$ ($0 < \varepsilon < 1$). In addition, we treat the surface diffusion anisotropy by introducing a parameter p. We define the probabilities that an adatom migrates toward the x-direction or the y-direction by p or $1-p$, respectively ($0 \le p \le 1$). The isotropic surface diffusion also can be treated by setting $p = 0.5$.

If we use the usual time constant ($\tau = 1/(K^{+} + K_{00}^{-} + K_{00}^{D})$) simulation method, the surface structure can hardly change in the case of low substrate temperature and large surface diffusion length such as in MBE growth conditions. Therefore, we use the so called waiting time method [6]. Then, we can simulate the movement of adatoms by showing them on the display. The details of the simulation procedure with the waiting time method are described as follows.

As $\tau_w = 1/\sum_{m,n} N_{mn} K_{mn}$ is the waiting time until the next event happens, the transition probability Q_{ij} that an atom at (i,j) state changes to another state is given by

$$Q_{ij} = N_{ij} K_{ij} \tau_w, \tag{4}$$

where, N_{ij} is the number of the lattice points with state (i,j), and the transition rate K_{ij} is defined by

$$K_{ij} = K^+ + K_{ij}^- + K_{ij}^D. \tag{5}$$

Then, the simulation procedure is; **step 0**: calculate Q_{ij} and pass τ_w, **step 1**: select the state (i,j) according to Q_{ij}, **step 2**: chose the lattice point of state (i,j) randomly, **step 3**: execute the event in proportion to K^+, K_{ij}^-, K_{ij}^D, **step 4**: modify the various values such as Q_{ij} and N_{ij}, then repeat the **step 0**. In this procedure, the computation time is spent in the steps 1 and 4 mainly.

2.2. PERIODIC CHANGES AND FLATNESS IN THE STRUCTURE OF A SURFACE

In the conventional vapor growth theory [2] applying for ordinary growth conditions, evaporation of adatoms on the surface is included as one of the important processes. Thus, the important parameter to decide the growth rate or the surface structure is the average waiting time τ_s of an adatom to evaporate (i.e. the mean life time on the surface) or the average diffusion length λ_s of an adatom in that time. On the other hand, MBE growth condition is the extraordinary one in which no evaporation occurs, hence, the conventional theory can not be applied. In such a condition, important parameter is the waiting time τ_c of an adatom to be captured by another adatom (i.e. the actual mean life time of an adatom) or the expected diffusion length λ_c of an adatom in that time.

The time τ_c and the length λ_c are defined [3, 4] by the kinematical consideration [7] neglecting the evaporation of the adatoms and the dissociation of two-atom clusters (so-called dimer) on the surface,

$$\tau_c = (JD_s)^{-1/2} \;\; and \;\; \lambda_c = (D_s/J)^{1/4}, \tag{6}$$

where J is the incident flux and D_s the surface diffusion coefficient. The length λ_c coincides with the linear size of the territorial area of each stable nucleus. Thus, there is only one adatom in the territory on average and no further nucleation occurs on the nucleus. The growth goes on as follows. A dimer (i.e. the stable nucleus) is created in each territory whose area is λ_c^2 on the flat surface. Then, the nuclei grow by the incorporation of all atoms which come from the vapor and are migrating on the surface. When the nuclei grow to the maximum diameter λ_c and are incorporated with each other, the growth layer is completed and the surface returns to be flat. This periodical change of the surface structure is detected experimentally by the oscillation of RHEED intensity. However, when the incident flux J is so large that the length λ_c becomes as short as the lattice constant a, the flatness of the surface will be lost because the impinging atoms crystallize without the surface diffusion. Therefore, the necessary condition for the layer growth with the flat surface and without evaporation of the adatoms is given by

$$a < \lambda_c < \lambda_s. \tag{7}$$

The results of our simulation support [3, 4] our theory described above.

2.3. GROWTH OF VICINAL SURFACE

On the vicinal surface, there is a parallel sequence of steps separated by distance λ. Thus, the advancement of those parallel steps contributes to the growth of the vicinal surface. The birth and spread of the two-dimensional nucleation on the terrace also contributes to it. If former contribution is much greater than the latter one, the surface grows continuously. Under the step flow growth, the surface structure does not change periodically and the oscillation of the RHEED intensity is no longer observed. The diffusion length λ_c of an adatom to be captured by another adatom was defined under MBE growth condition in the last section. Thus, if the characteristic length λ_c is longer than the distance λ between the neighboring steps, almost all the adatoms will be incorporated by the step before being captured by another adatom. This means that the two-dimensional nucleation on the terrace do not occur under such condition. Therefore, it can be concluded that the oscillation of RHEED intensity is not observed under the condition that

$$\lambda < \lambda_c. \tag{8}$$

It is noticed that higher temperature of the substrate or smaller incident flux gives longer diffusion length λ_c.

We simulate the growth of vicinal surface under MBE condition and decide the growth mode by calculating the space-dependent correlation function $G(r)$ of the local surface height z_i [8]. It is defined by

$$G(r) = < \Delta z_i \cdot \Delta z_j > \tag{9}$$
$$\Delta z_i = z_i - < z >,$$

where the suffix i denotes the i-th lattice site on the surface, $r = |r_i - r_j|$ is the distance between two lattice sites i and j in the direction perpendicular

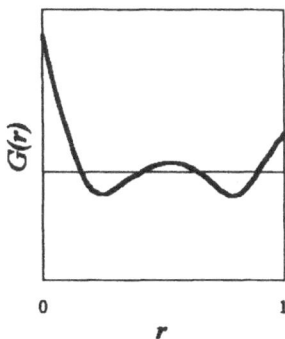

Figure 1. Correlation function of local surface height in the direction perpendicular to steps.

to steps. Δz_i is the deviation from the average height. The base of the local height z is the height of the initial surface with the straight parallel steps. Figure 1 shows the correlation function at the time when 0.5 layer have grown. The local maximum of the correlation function at the middle of the terrace signifies the occurrence of two-dimensional nucleation on the terrace. Then, we define the boundary of the two growth modes by the criterion that the local maximum of $G(r)$ equals to 0. We found that the critical value of λ_c which corresponds to the boundary of the growth mode has the linear dependence on the step distance λ. Therefore, the growth mode of the vicinal surface under MBE condition can be characterized by λ_c, which is in agreement with our theory.

2.4. ANISOTROPIC BONDING ENERGY AND SURFACE DIFFUSION

Recently, the surface structure is studied in detail by using STM, and formation of extremely anisotropic cluster on $Si(001)$ surface [9] and the difference of the step structure for the step direction on $GaAs(001)$ vicinal surface [10] are reported. It is not clear whether the main reason of this anisotropy is the effect of the anisotropic bonding or the anisotropy of the surface diffusion of adatoms. There are few suggestions [11] about that. In this section, we use Monte Carlo simulation which takes account the anisotropic effect to investigate that problem.

According to BCF theory [2], anisotropic clusters can be produced in the case that λ_s has anisotropy because the step velocity is proportional to λ_s. On the other hand, we can ignore the reevapolation of adatoms under MBE growth conditions. Therefore, we can not apply BCF theory.

Figure 2a shows the snap shot of the surface generated by simulation in the anisotropic bonding case with $\phi_x = 5kT$, $\phi_y = 10kT$ and $\phi_z = 20kT$

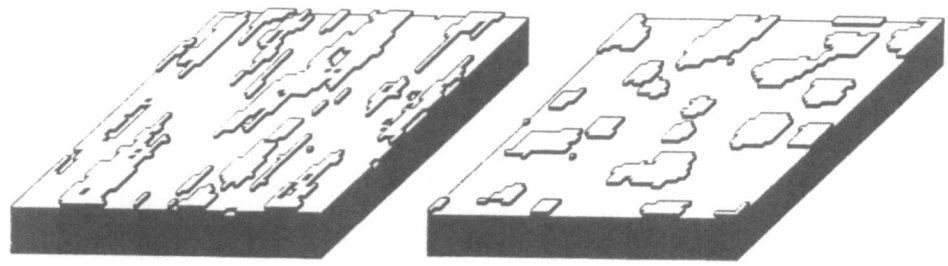

Figure 2. Snapshots of growing singular surface with a) anisotropic bonding energy ($\phi_x=5kT, \phi_y=10kT$ and $\phi_z=20kT$) and with b) anisotropic surface diffusion ($\phi_x=\phi_y=5kT$, $\phi_z=20kT$ and $\lambda_{sx} : \lambda_{sy}=1 : 99$ ($p=0.01$).

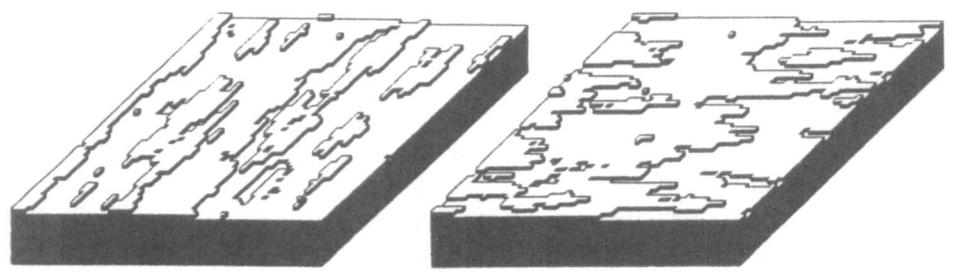

Figure 3. Snapshots of growing vicinal surface with the anisotropic bonding energy: a) $\phi_x=5kT$ and $\phi_y=10kT$ and b) $\phi_x=10kT$ and $\phi_y=5kT$.

($\lambda_s = 148a$). There is no dissociation of adatoms bonded in y-direction since the bond energy is large. On the other hand, there is a frequent dissociation of adatoms bonded in x-direction, and sometimes adatoms are crystallized at the kink site after migrating along the step. In this case, the anisotropic (long in y-direction) clusters are generated. The experimental data of the largely anisotropic clusters reported by Mo et al. [9] can be explained by this mechanism.

Figure 2b shows the snap shot of the surface in the case with $\phi_x = \phi_y = 5kT$, $\phi_z = 20kT$ and anisotropic surface diffusion length $\lambda_{sx} : \lambda_{sy} = 1 : 99$ ($p = 0.01$). In this case, any difference in the shape of clusters can hardly be seen from the case with the isotropic surface diffusion. Therefore, the anisotropy of surface diffusion does not generate anisotropic shape of clusters.

Figure 3 shows the snap shots of the vicinal surface with the anisotropic bonding in two different directions: a) $\phi_x = 5kT$ and $\phi_y = 10kT$ and b) $\phi_x = 10kT$ and $\phi_y = 5kT$. Different structures of steps are generated similar to those observed by the experiments.

3. Simulation of A-B Two-Component System

Many crystals grown by MBE are multi-component systems. However, there are still much left to be studied because the growth conditions are controlled by many parameters [12, 13].

We extend the simulation model to treat the A-B two-component system as follows. The adsorption rates of A and B atoms are determined by flux J_A and J_B as

$$K_X^+ = J_X a^2, \tag{10}$$

where suffix X is A or B. The evaporation rates of A and B atoms are given by

$$K_{Xij}^- = \nu_X \exp[(-i\phi_{XA} - j\phi_{XB})/kT], \tag{11}$$

where i and j denote the numbers of A and B atoms bonded to X ($0 < i + j < 6$), respectively, ϕ_{AA}, ϕ_{BB} and ϕ_{AB} the bond energies of A-A, B-B and A-B pairs, ν_A and ν_B the vibrational frequencies assumed to be equal. The surface diffusion is the important event in MBE growth conditions. In this paper, we assume that the surface diffusion length is constant. Then, we define the surface diffusion rates of A and B atoms as

$$K_{Xij}^D = (\lambda_{sX}/a)^2 K_{Xij}^-, \tag{12}$$

where λ_{sA} and λ_{sB} are surface diffusion lengths of A and B adatoms, respectively. In this treatment, we emphasize that the activation energy for the surface diffusion of A adatoms on A substrate atoms is not equal to that for A on B atoms, that is, A adatoms on A substrate atoms can move faster than those on B substrate atoms when $\phi_{AA} < \phi_{AB}$.

We treat the two-component system of only stoichiometric A-B crystal in this paper. It means that $\phi_{AA} = \phi_{BB} < \phi_{AB}$, $J_A = J_B$, $\lambda_{sA} = \lambda_{sB}$ and near the MBE growth conditions, that is, the growth mode is layer by layer and no evaporation of adatoms from the surface. Under this condition, there are 4 growth features realized by simulation. The crystal grows a) with large orderliness and with layer by layer mode, b) with large orderliness but with not layer by layer mode because of larger incident flux, c) with small orderliness because of smaller ratio of the bond energy ϕ_{AA} to ϕ_{AB} and with layer by layer mode because of small flux, and d) with small orderliness and with not layer by layer mode. The orderliness means the fraction of adatoms which are bonded to the different species at all the six nearest neighbors, and are normalized by the number of atoms which constitute the layer. Similar to one-component system, the characteristic lengths of the growth can be defined by

$$\lambda_{Xc} = [D_X/(J_A + J_B)]^{1/4}, \tag{13}$$

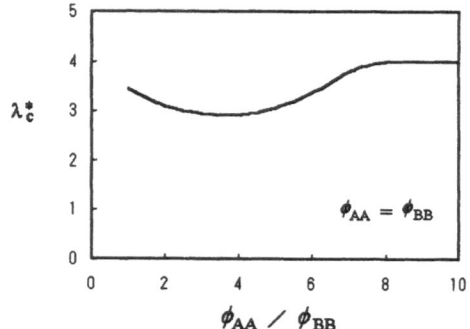

λ_c^*

$\phi_{AA} = \phi_{BB}$

ϕ_{AA} / ϕ_{BB}

Figure 4. Change of the critical condition λ_c^*.

where the suffix X is A or B. Since we treat the stoichiometric system, we omit that suffix and denote λ_c simply.

We simulated the growth with varying the fluxes J_A and J_B and found the critical value λ_c^* for the periodical changes of the surface structure. Figure 4 shows the critical length λ_c^* versus bond energy ratio $r = \phi_{AB}/\phi_{AA}$. One can see that the growth mode can not be understood by λ_c alone in the case of two-component system contrary to one-component system because the line of λ_c^* is not a straight line. That is, λ_c^* has the minimum value near $r = 4$ at $\phi_{AB}/kT = 20$. In the case of large r, the life time of an adatom is shorter when it is the same species as the substrate atom than that when it is different. That is, the territory of stable cluster has larger area than (13) because of the long diffusion distance before encountering other atoms of different species. Therefore, the surface structure becomes multi-layer, because the nucleus can be created on the larger stable cluster when r is too large. Contrary, when $r < 4$, the surface structure is stabilized, because the territory of the stable nucleus is not so large and the adatom can move easily in comparison with the case of $r = 1$, so the adatoms are caught at the edge before they encounter others.

It is clear that layer by layer grown crystal does not always have large orderliness considering the case of $r = 1$. The orderliness is changed with the ratio of bond energy ($\phi_{AA}, \phi_{BB} : \phi_{AB}$), temperature, surface diffusion length and incident atom flux.

As shown in Figure 5, the orderliness decrease with increasing incident flux. Figure 6 shows the orderliness versus the incident flux, where the incident flux is given by $\exp[\Delta\mu/kT - 3\phi_{AB}/kT]$. In the case of a small flux, as the adatom has enough time to migrate towards the stable position, the orderliness is large. On the other hand, in the case of a large flux, as the adatom collides with another adatom and crystallizes at unstable position frequently, the orderliness is small. This transition seems to be the second

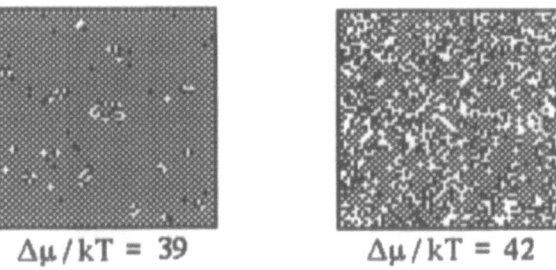

$\Delta\mu/kT = 39$ $\Delta\mu/kT = 42$

Figure 5. Snapshots of growing surface of two-component system with $\phi_{AA}=\phi_{BB}=6kT$ and $\phi_{AB}=20kT$.

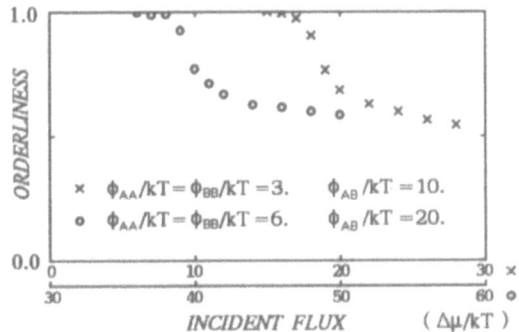

Figure 6. Plots of orderliness vs. incident flux.

order. The orderliness is a function of $(2\phi_{AB}/kT - \Delta\mu/kT)$, because the variable $(2\phi_{AB}/kT - \Delta\mu/kT)$ determines K^-_{A01}/K^+_A which is important in determining the surface structure in stoichiometric growth condition. Where, $\Delta\mu$ is not the correct chemical potential difference but a control parameter of the flux. Therefore, it can be expected that this transition exists in some region of ϕ_{AA}, $\phi_{BB} < \phi_{AB}$.

4. Concluding Remarks

Several growth features are realized by Monte Carlo simulation to understand MBE growth in atomic scale. It seems that we can observe it by using recent experimental technique. The main reason of the surface structural anisotropy is not the surface diffusion anisotropy but the anisotropic bond energy. Therefore, it is clear that we must not apply the usual vapor growth theory to analyze MBE growth. We got several knowledge about two-component systems which can be applied to many systems, though we carried out only a simple case. But there are many problems for two-component systems such as the dependence on partial pressure and non-symmetric bonding energy and so on. There is no alternative but waiting for future research and development.

Acknowledgments

We are indebted to Professor T. Nishinaga of Tokyo University. The work in this paper was supported by a Grant-in-Aid for Scientific Research by the Ministry of Education, Science and Culture, Japan, No. 03243101 and No. 04227108, and was partly supported by "Foundation for Promotion of Material Science and Technology of Japan (MST Foundation)".

References

1. Harris, J.J., Joyce, B.A. and Dobson, P.J. (1981) Oscillation in the surface structure of Sn-doped $GaAs$ during growth by MBE, *Surf. Sci.* **103**, L90-L96.
2. Burton, W.K., Cabrera, N. and Frank, F.C. (1951) The growth of crystals and the equilibrium structure of their surfaces, *Phil. Trans. Roy. Soc.* **A243**, 299-358.
3. Irisawa, T., Arima, Y. and Kuroda, T. (1990) Periodic changes in the structure of a surface growing under MBE conditions, *J. Crystal Growth* **99**, 491-495.
4. Irisawa, T., Ichimiya, A. and Kuroda, T. (1991) Periodic changes in th structure of a surface growing under MBE conditions and RHEED oscillation, *Surf. Sci.* **242**, 148-151.
5. Gilmer, G.H. and Bennema, P. (1972) Simulation of crystal growth with surface diffusion, *J. Appl. Phys.* **43**, 1347-1360.
6. Borz, A.B., Kalos, M.H. and Lebowitz, J.L. (1975) A new algorithm for Monte Carlo simulation of Ising spin systems, *Comput. Phys.* **17**, 10-18.
7. Lewis, B. and Campbell, D.S. (1967) Nucleation and initial-growth behavior of thin-film deposits, *J. Vacuum Sci. Technol.* **4**, 209-218.
8. Arima, Y. and Irisawa, T. (1991) Influence of surface diffusion on the structure of growing crystal surface, *J.Crystal Growth* **115**, 428-432.
9. Mo, Y.W, Kleiner, Y., Webb, M.B. and Lagally, M.G. (1991) Activation energy for surface diffusion of Si on Si(001): A scanning-tunneling-microscopy study, *Phys. Rev. Lett.* **66**, 1998-2001.
10. Pashley, M.D., Haberen, K.W. and Gaines, J.M. (1991) Scanning tunneling microscopy comparison of $GaAs$(001) vicinal surfaces grown by molecular beam epitaxy, *Appl. Phys. Lett.* **58**, 406-408.
11. Lu, Y.-T. and Metiu, H. (1991) Growth kinetics simulation of the Al-Ga self-organization on $GaAs$(100) stepped surfaces, *Surf. Sci.* **245**, 150-172.
12. Takata, M. and Ookawa, A. (1974) On the growth mechanism of an A-B crystal, *J. Crystal Growth* **24/25**, 515-518.
13. Takata, M. and Ookawa, A. (1974) On the growth mechanism of an A-B crystal -II Effect of dissociative adsorption of B_2 molecules-, *J. Crystal Growth* **42**, 35-40.

REAL TIME MONITORING OF EPITAXIAL GROWTH

Wolfgang Richter, Kerstin Knorr, Thomas Zettler and Martin Zorn
Institut für Festkörperphysik, TU Berlin
PN 6-1, Hardenbergstr. 36
10623 Berlin, Germany

1. Introduction

Real time monitoring in the UHV based growth methods, MBE and MOMBE by RHEED has been performed already for a long time [1]. Thus, MBE has been always ahead of its gasphase counterpart MOVPE where such a direct control was not possible since electrons cannot penetrate the gasphase. However, electromagnetic waves can penetrate gaseous environments in certain spectral regions and thus are in principle suitable probes to study the epitaxial surfaces during growth. Experimental techniques are available for example in the visible spectral range and in the x-ray region. In the latter case, similar as with RHEED, diffraction can be performed under grazing incidence (grazing incidence x-ray scattering: GIXS) and yields similar information [2]. However, for intensity reasons a synchroton has to be used as x-ray source and therefore this technique, being very useful for basic studies, cannot really be applied for growth monitoring on epitaxial equipment. A number of optical techniques operating in the visible spectral range on the other hand have been developed in the last years to such a stage that they can be utilised in extracting information from a growing surface. The two necessary requirements surface sensitivity and reasonable speed in order to monitor the changes during growth are fulfilled for example by non-linear optical techniques like Second Harmonic Generation (SHG) or by the linear techniques Spectroscopic Ellipsometry (SE), Surface Photoabsorpion (SPA) or Reflectance Anisotropy Spectroscopy (RAS/RDS) [3]. Because of the large peak powers involved (which might influence the growth process) and the still limited spectral capability non-linear techniques at present seem not to be the first choice for growth monitoring. The linear techniques in contrast can all be operated at low powers with standard white light sources thus avoiding perturbations of the growth process. On the other hand in connection with monochromators they provide spectral capabilities. The latter aspect turns out to be essential for extracting chemical information from the surface.

In the following chapter we will discuss the three linear optical techniques with respect to monitoring the surface status. The next two chapters will then deal with the application of RAS to group V element stabilised surfaces obtained under pregrowth (or after growth) conditions (chapter 3) and to surfaces during growth (chapter 4). In

11

J. Novák and A. Schlachetzki (eds.), Heterostructure Epitaxy and Devices, 11–20.
© 1996 *Kluwer Academic Publishers.*

chapter 5 finally aspects related especially to the needs of epitaxial growth will be discussed. Most of the data published so far for growth monitoring with RAS have been obtained from III-V-semiconductors and we will discuss only these. Since the optical methods can be applied to all growth techniques (vacuum as well as gasphase) we will present results from Molecular Beam Epitaxy (MBE), Metalorganic - MBE (MOMBE) and Metalorganic Vapour Phase Epitaxy (MOVPE).

2. Linear optical techniques for growth monitoring

The surface sensitivity of the linear optical techniques derives from the anisotropic optical properties of the surfaces in contrast to the isotropic bulk dielectric function of the cubic semiconductors discussed here. Because of energetic reasons solid surfaces in general reconstruct, i.e. atoms at the surface are differently bonded than in the bulk. By this nearly always the symmetry within the surface is lowered and the dielectric tensor components taken along the surface eigenvectors become different. For the growth relevant (001)-surfaces of zincblende structure III-V-semiconductors these eigenvectors are x=(110) and y=(-110). In principle any reflectance experiment performed with polarised light can measure the anisotropies of the corresponding tensor components ε_{xx} and ε_{yy} (Fig.1).

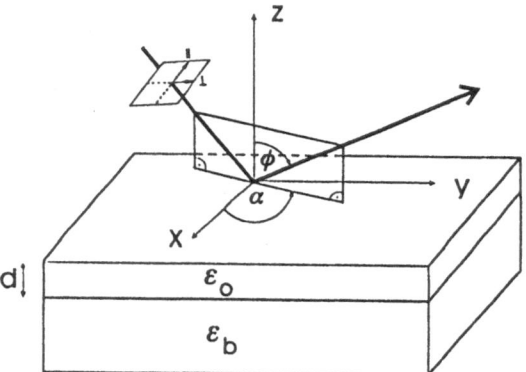

Figure 1. Schematic diagram of an optical surface monitoring experiment with polarised light. The surface layer has thickness d (a few monolayers), anisotropic eigenvectors x,y and dielectric function ε_o. The substrate dielectric function is denoted by ε_b.

The various linear optical techniques differ by the angle of incidence and the way of analysing the reflected light. Fig. 2 gives a schematic overview of the most commonly used techniques: reflectance anisotropy spectroscopy (RAS/RDS), surface photo absorption (SPA) and spectroscopic ellipsometry (SE). The operating point with respect to Fresnels formula is indicated on the left. The right side shows how the technique can be performed in a differential manner in order to extract the surface information. This is most easily done in RAS where just a fast switching of the polarisation vector is necessary (Fig. 2a). In SPA, working near or at the Brewster angle, the low reflected intensity in p-polarisation is utilised for differentially monitoring by changing the surface conditions through pulsed operation of the fluxes (Fig. 2b). This differential mode, however, makes determining a certain surface status difficult because the difference between two surface states is measured. The method turns out to be quite sensitive and is the choice for pulsed growth methods like atomic layer epitaxy [4]. SE finally, (Fig.2c) measures an effective dielectric function within the penetration depth of the light and determines in general essentially bulk (isotropic) dielectric properties. However, when applied in a differential manner

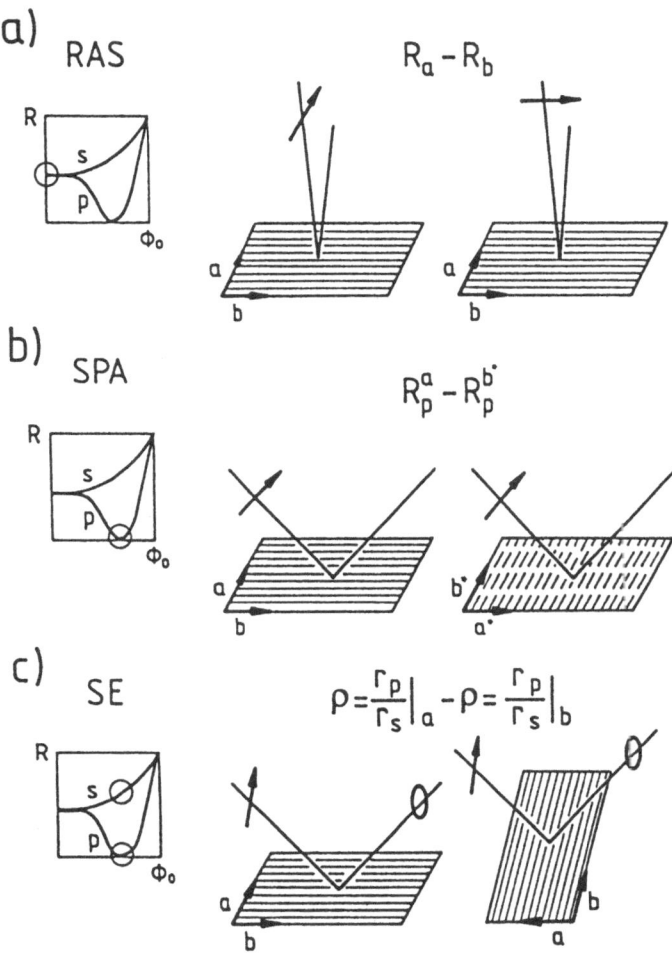

Figure 2 . Schematic representation of the different linear optical reflectance techniques in use for growth monitoring. Left side diagrams show operation point with respect to Fresnel`s formulas. Right side indicates how these methods can be used in a differential manner for obtaining an anisotropic surface sensitive signal.

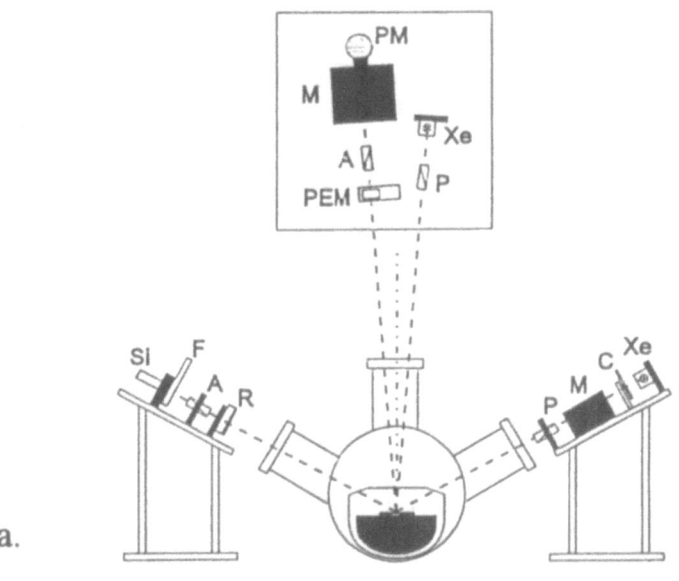

a.

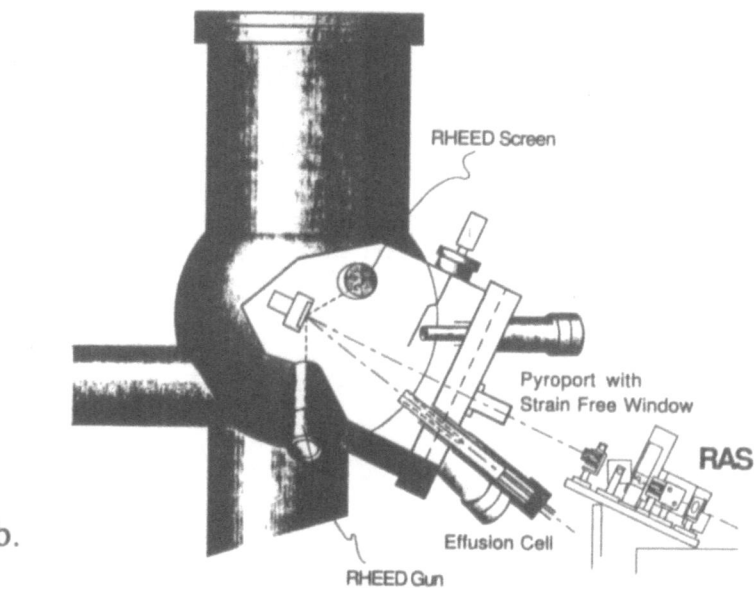

b.

Figure 3. Experimental arrangements for optical monitoring **a.** on a horizontal MOVPE reactor (SE and RAS) and **b.** on a MBE (RAS).

between two identical samples oriented differently, the surface dielectric function as well as the bulk contribution can be extracted [5]. Nevertheless, the sensitivity is less than in the other two methods. Experimentally, SPA and SE are more difficult to perform since two windows instead of one are needed and the oblique angle alignment is more difficult. RAS is therefore the preferred method of choice as far as surface sensitivity and experimental simplicity is concerned. However, in order to evaluate the surface dielectric function $\varepsilon_0 \cdot d$ from the measured differential reflectivity

$$\Delta r / r = 2 (r_{-110} - r_{110}) / (r_{-110} + r_{110}) \tag{1}$$

the substrate dielectric function ε_b is needed

$$\Delta \varepsilon_0 \cdot d = (\varepsilon_b - 1) \cdot \lambda \cdot (1/4\pi i) \; \Delta r/r \tag{2}$$

which in most cases will not be available in the literature for the specific temperature and stochiometry under consideration. At present it is therefore advantageous to measure simultaneously to RAS the dielectric function by ellipsometry. For MOVPE this can be done in a set-up as shown in Fig. 3. Similar configurations can be used for MBE. In both methods low strain ("strain free") windows should be used since otherwise the polarisation might be strongly affected by the window and less by the sample thus decreasing the sensitivity of measurement. Further details concerning the principle and realisation of an RAS apparatus can be found in Ref. 6.

3. RAS spectra from surfaces in equilibrium before growth

RAS spectra on (001)-surfaces depend essentially on the number, direction and kind of dimers present on the surface. For this reason they are sensitive to surface reconstructions. Correlations between surface reconstructions and RAS spectra have been established under ultrahigh vacuum conditions (MBE) with the help of simultaneously performed electron diffraction measurements (RHEED, LEED) [7]. A relatively complete database exists already by now for GaAs(001) surfaces (Fig. 4) but data for other III-V-semiconductors become more and more available [8]. For binary compounds these data turn out to be already quite useful, especially in MOVPE, for the purpose of adjusting the group V element stabilisation of the surface before growth.

The more interesting case for growth monitoring arises when different dimers are present. This occurs for example in ternary or quaternary compounds and allows for a determination of stochiometry. Fig. 4 gives an example. The shifts observed turn out to be in the same direction as the bulk energy gaps. They are less in those cases where only the group III element changes, because the top dimers are made up by the same group V atoms and are only influenced by the underlying III-atoms (AlGaAs, InGaAs). When the top dimers are made up of different V-atoms (InGaAsP) energy shifts and signal changes are considerably stronger (Fig. 5).

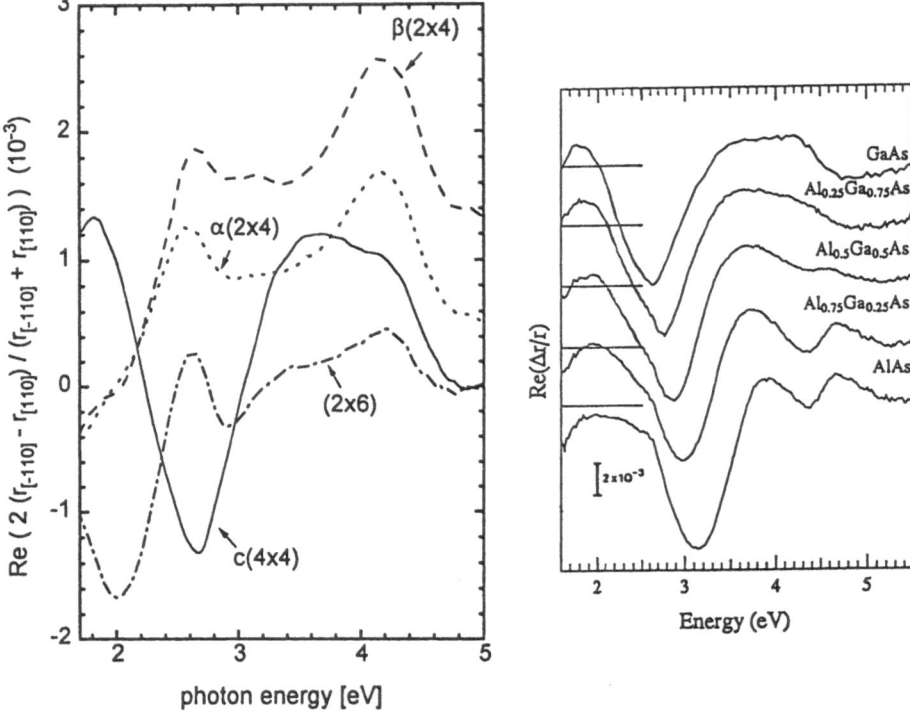

Figure 4. Dependence of RAS spectra on surface reconstruction and stochiometry: left side: GaAs(001): surfaces with different reconstructions , right side: $Al_x Ga_{1-x} As$ surfaces all with c(4x4) reconstruction but different Al content.

Shifts of energetic positions and signal changes are also observed with changing temperature. After calibration they can be used for temperature measurements [8]. They turn out partly to be similar to those of the bulk gaps but can be quite large and different in cases where the surface reconstruction changes with temperature.

Further changes are of course observed when the surface is passivated or oxidised. The latter surface condition shows besides the disappearance of the spectral line shape that is given by the specific surface reconstruction in general also the appearance of an oxide related structure around photonenergies of 5 eV. Both the disappearance of the surface reconstruction as well as the oxide related structure can be used to monitor the substrate deoxidation process and the correct conditioning of the surface before growth [10].

4. RAS spectra from surfaces during growth

The RAS spectra during growth differ from those before growth depending on the flux of the group III - element i.e. on the growth rate. In MBE as well as in MOMBE the picture looks relatively simple. With increasing gallium flux (growth rate) the surface changes from a c(4x4) to a (2x4) reconstructed one as can be seen for MBE in Fig. 6a.

This leads to the plausible conclusions that just the As coverage decreases with increasing Ga flux. In contrast in MOVPE the RAS spectrum changes from a c(4x4) to one dominated by a (nx6) response. This has to be interpreted in terms of Ga-dimers occurring during growth. Thus surface processes for the UHV based methods MBE and MOMBE turn out to be significantly different as compared to MOVPE. In both cases however, it seems that the regions of reduced As coverage occur near step edges [10].

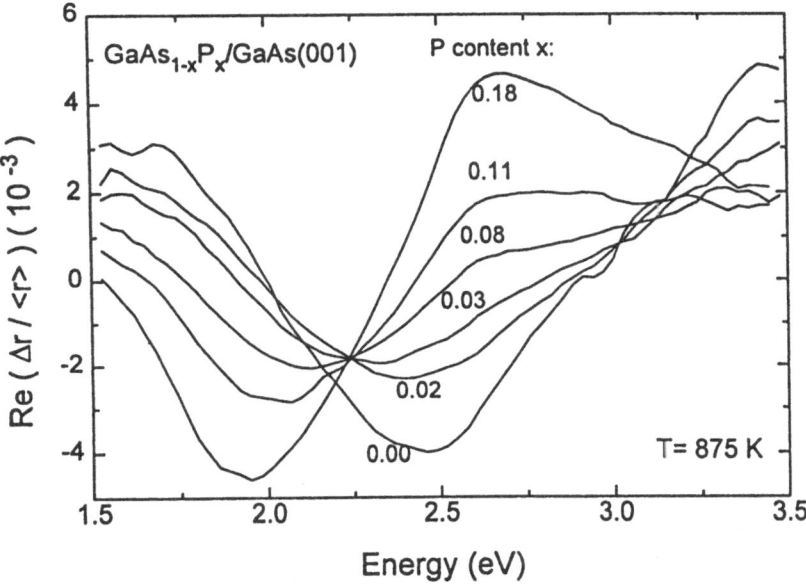

Figure5. RAS spectra from GaAs$_{1-x}$P$_x$ layer grown by As/P exchange.

A further difference between MOVPE and the other two growth methods can be deduced from the RAS signal oscillations with monolayer periodicity. These monolayer oscillations are connected with the island growth mode and disappear at the transition temperature to step flow growth. At that temperature the surface diffusion length equals the average terrace width between surface steps and allows an estimate of the diffusion length of the growth determining species. It turns out from the much lower oscillation disappearance temperature in MOVPE that diffusion lengths are considerably larger in MOVPE than in MBE or MOMBE [10]. This should have consequences especially for the growth of low dimensional structures.

5. Aspects for epitaxial growth monitoring

We have already discussed the aspects of stoichiometry and temperature determination as well as the monitoring of the deoxidation process. During growth the changes in spectra with Ga- flux or temperature can be used to establish phase diagrams which in turn allow to define surface conditions independent of specific readings of

18

thermocouples or mass flow controllers. Such phase diagrams have been published for MBE [11] some time ago but also recently for MOVPE [12].

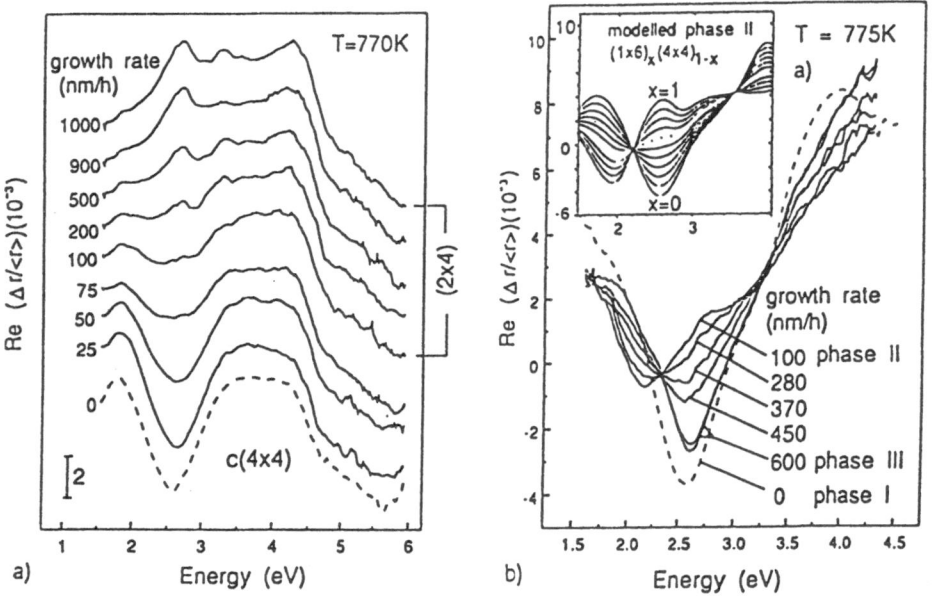

Figure 6. RAS spectra during growth for different growth rate (Ga supply to the surface). Left: MBE, right: MOVPE.

A very important aspect during growth is of course the layer thickness. There are two possibilities and length scales RAS can help in determining the thickness. The first and classical optical procedure is the utilisation of Fabry-Perot interferences. This technique is also called reflectometry or dynamical optical reflectivity and gives an accuracy in the order of a few percent of the wavelength of the light i.e. in the range of a few nanometer. More accurately of course is the second possibility of using the monolayer oscillations if one works in parameter regions were these oscillations occur. In this case submonolayer resolution can be obtained and applications such as phase locked epitaxy or the exact completion of monolayers become possible. Both the Fabry-Perot and the monolayer oscillations can be seen in Fig. 7 [13]. The former as the modulating envelope function in the top (with a period of a few hundred seconds) and the latter in the magnification of one superlattice period (with a period of 2s).

The last aspect we would like to emphasise are morphology related signal changes. After first observations in MOVPE and MBE it has become clear recently that these variations can be utilised to monitor in real time for example the creation of quantum dots or possibly other surface structures appearing in growth processes aiming at low dimensional structures.

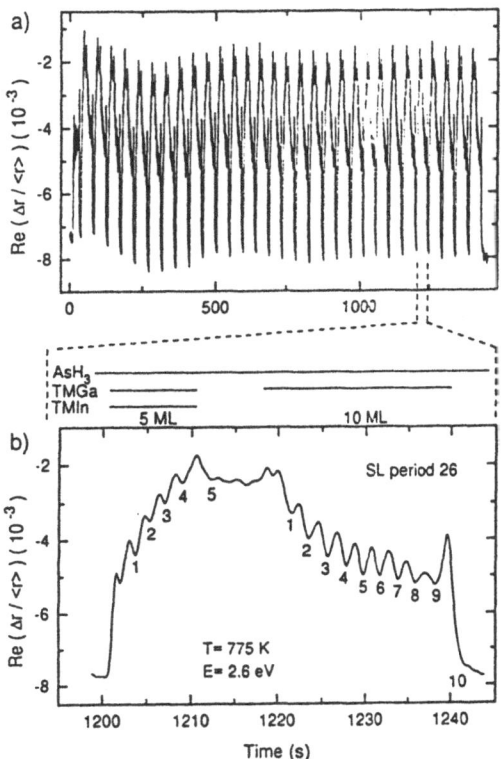

Figure 7. Time resolved RAS signal at a fixed photon energy (2.6 eV) for a superlattice consisting of 30 periods each containing 5ML InGaAs and 10ML GaAs. Top: total response for 30 periods showing also the Fabry-Perot oscillations, bottom: magnification of one period with monolayer oscillations (indicated by numbers).

6. Conclusions

With linear optical techniques especially with RAS it is possible to characterise a surface under growth conditions. It turns out that the variation of many growth relevant parameters causes significant and reproducible changes in the RAS signal. They can be measured with a sufficient signal to noise ratio to allow for real time monitoring and perhaps at some later stage also for real time control.

On the other hand a basic understanding of the growth processes can be obtained and for the first time this can be done in all three growth methods MOVPE, MBE and MOMBE. Thus comparisons become now possible and reveal already principal differences in the growth mechanisms.

For practical growth purposes it seems today already feasible to use RAS (possibly in connection with ellipsometry) for continuous growth monitoring. Such a diagnostic

20

tool will be helpful in signalising at any time the deviation of any growth influencing parameter from ist preset value and can be the basis for correcting the growth parameter. The application of RAS in the stage of developing a new growth process or procedure should also be emphasised. This application seems to be already quite useful today since a large part of the characterisation can be performed easily in situ without the long time constants occurring with ex-situ characterisation.

Acknowledgments

The work reported here was funded by the EU (ESPRIT Basic Research Action 7688 "EASI" and the German Ministry of Education and Research (contract BMFT 01 BT 310/835). We also very gratefully acknowledge the cooperations we had with the PDI (Berlin), UWCC (Cardiff) and the U of Liverpool concerning RAS measurements on their respective epitaxial growth equipment. These cooperations were partly made possible through support for travel from the DAAD.

References

1. Herman, M.A. and Sitter, H. (1989) H., *Molecular beam epitaxy,* Springer Series in Materials Science, Berlin
2. Kisker, D., Stephenson, G., Fuoss, P.H. Lamelas, F.J., Brennan, S., and Imperatori, P., (1992), J. Crstal Growth 124, 1
3. McGilp, J.F., Weaire, D., and Patterson, C. Eds. (1995), EPIOPTICS, Esprit Basic Research Research Series, Springer, Berlin
4. Kobayashi, N., and Horikoshi, Y., (1991) Jap. J. Appl. Phys. 30, L319
5. Wassermeier, M., Behrend, J., Zettler, T., Stahrenberg, K., Richter, W., and Ploog, K. (1995), to be published
6. Aspnes, D.E., Harbison, J.P., Studna, A.A., and Florez, L.T. (1988), J. Vac. Sci. Technol. A6, 1327
7. Kamiya, I., Aspnes, D.E., Tanaka, H., Florez, L.T., Harbison, J.P., and Bhat, R. (1992), Phys. Rev. Lett, 68, 627
8. Richter, W., and Zettler, T., Appl. Surface Science accepted
9. Ploska, K., Richter, W., Reinhardt, F., Jönsson, J., Rumberg, J., Zorn, M. (1993), MRS Proceedings 334, 155
10 Knorr, K., Jönsson, J., Pristovsek, M., Kamiya, I., Zettler, T., and Richter, W. (1995), Phys. Stat. sol. (a), to be published
11. Däweritz, L., Hey, R., Surf. Sci. 236, 15 (1990)
12. Reinhardt, F., Jönsson, J., Zorn, M., Richter, W., Ploska, K., Rumberg, J., Kurpas, P., (1994), J. Vac. Sci. Technol. B12, 2541
13. Zorn, M., Jönsson, J., Krost, A., Richter, W., Zettler, T., Ploska, K., Reinhardt, F. (1994), J. Crystal Growth 145, 53

INFLUENCE OF CARRIER GAS ON AlAs, GaAs AND InP MOCVD GROWTH

S. HASENÖHRL[1)], HILDE HARDTDEGEN[2)], CH. UNGERMANNS[2)]

1) Institute of Electrical Engineering
Slovak Academy of Sciences
Dúbravská cesta 9
SK 842 39 Bratislava Slovak Republic

2) Institut für Schicht und Ionentechnik
Forschungszentrum Jülich GmbH
Postfach 1913
D-52425 Jülich Germany

1. Introduction

In semiconductor technology H_2 is nearly exclusively used as an ambient or as the carrier gas. In some cases the replacement of H_2 by an inert gas or the combination of H_2 with an inert gas brings advantages.

The first results in [1] and [2] showed the suitability of N_2 as a carrier gas in the MOCVD process for GaAs, AlGaAs and for C doped GaAs growth. Previous reports [1], [4] show that the carrier gas does not play an active role in the growth process. On the other side, in [3] was reported, that for InP growth from TEIn and PH_3, the best results were obtained by using H_2:N_2 (1:1) mixture as a carrier gas. In that case, the presence of H_2 was necessary to avoid the deposition of C and N_2 avoided the parasitic reactions of TEIn with PH_3.

The purpose of this study was to find out, if change of the carrier gas from H_2 to N_2 has an influence on the chemism of the reactor reactions. If not, then the growth rate r_g should change only due to the change of the diffusion coefficient of the source molecule in a carrier gas mixture of various composition. For the case, when the growth rate limiting factor is the transport of the source molecule to the growth interface, the growth rate can be expressed as a function of the carrier gas mixture and source molecule properties.

2. Experimental

The growth of AlAs, GaAs and InP was performed in a horizontal LP MOVPE reactor. The substrates were semiinsulating (100) 2° off towards <110> GaAs wafers cleaned in concentrated H_2SO_4 and semiinsulating (100) InP of exact orientation cleaned in 10% HF. AsH_3 and TEGa (Triethylgallium) or TMGa (Trimethylgallium) were the sources for GaAs and AsH_3 and DMEAA (Dimethylethylaminalane) or TMAl (Trimethylaluminum) for AlAs. InP was deposited using PH_3 and TMIn (Trimethylindium). The process pressure was 20 hPa. The H_2/N_2 carrier gas mixture

J. Novák and A. Schlachetzki (eds.), Heterostructure Epitaxy and Devices, 21–24.

varied from pure H_2 to pure N_2. The gases were purified by a palladium membrane purifier and by the getter column for H_2 and N_2 respectively.

The layer morphology was determined by a Nomarski interference microscope, the thickness by weighing.

2.1. CHOICE OF THE GROWTH PARAMETERS

First, growth studies were performed with the different source molecules to identify the diffusion controlled growth regime. The Arrhenius plots for all sources present a wide regime (above 500 °C), in which growth rate is nearly independent on temperature. The temperature for the further investigation was chosen so, that it was in the middle of this regime: 600 °C for the alane, TMIn and TEGa and 650 °C for the TMGa and TMAl.

2.2. INFLUENCE OF CARRIER GAS COMPOSITION ON GROWTH RATE

A series of experiments was done to investigate the dependence of r_g on the composition of the carrier gas. The growth parameters for each growth series were the same, to ensure the comparability of the results. Table 1. presents an overview of all growth parameters used for the successive investigations of samples deposited using gas mixtures.

TABLE 1. Partial vapour pressures of the sources in the reactor. Reactor gas velocity was 0.9 ms^{-1}, reactor pressure 20 hPa, growth temperature 600°C resp. 650°C.

	p_{AsH_3}	p_{PH_3}	p_{TMGa}	p_{TEGa}	p_{TMAl}	p_{DMEAAl}	p_{TMIn}	V/III
	(Pa)	(Pa)	(Pa)	(Pa)	(Pa)	(Pa)	(Pa)	
GaAs : AsH$_3$+ TMGa	34,4		7,2 10^{-1}					47,7
GaAs : AsH$_3$ + TEGa	36,0			6,9 10^{-1}				52,0
AlAs : AsH$_3$+ TMAl	103,4				1,05 10^{-1}			984,7
AlAs : AsH3+ DMEAAl	105.5					5,3 10^{-1}		200,0
InP : PH3 + TMIn		266					5,36 10^{-1}	496.2

3. Theorethical

In the diffusion limited growth regime, the growth rate is controlled by the transport of the respective source molecule through the boundary layer. The driving force for this diffusion through the boundary layer toward the substrate is the existent gradient of the source molecule concentration c within the boundary layer. The ratio of the source molecules number to the number of the carrier gas molecules in the reactor is only 10^{-2} for hydride source and 10^{-4} for MO source. It can be supposed that the boundary layer consists exclusively of carrier gas and only the properties of the H_2/N_2 carrier gas mixture will determine the rate of source molecule diffusion. To be able to calculate the diffusion coefficients the following suppositions need to be made: the complete undissociated source molecule diffuses through the boundary layer without any interactions with other molecules in the boundary layer. The source molecule concentration outside the boundary layer, c^0, is to be constant. At last thermodiffusion and the source molecule stream towards the substrate under the precondition of laminar flow are to be neglected. In MOCVD process the source molecule concentration is expressed by the partial pressure of the source vapour, so the substitution of the concentration c by the partial pressure p is convenient. In the typical cases, when the group V source inlet partial pressure p_V^0 is much higher than the group III source inlet partial pressure p_{III}^0, the growth rate does not depend on p_V^0.

For investigation, how r_g should change with composition of the carrier gas, the expression for r_g as a function of the properties of the carrier gas was used. The expression was derived by using the 1-st Fick's law, the expression for the thickness of the diffusion boundary layer d_D [5] and the Blanc's law [6] expressing the diffusion coefficient of the group III source molecule in a mixture of H_2+N_2.

The final expression for the growth rate used in this work is a function of:

- v velocity of the gas stream in the reactor
- P_{proc} process pressure
- p_{III} partial pressure of the group III source in the reactor
- x distance from the front of the susceptor in the direction parallel to the gas stream
- T growth temperature
- M_{III}, M_{H2}, M_{N2} molecular weights of group III source, H_2 and N_2 molecules
- V_{III}, V_{H2}, V_{N2} molar volumes of group III source, H_2 and N_2 in the fluid phase
- y_{H2}, y_{N2} molar fractions of the components in a gas mixture

$$r_g = K \left[\frac{\left(M_{III}^{-1}+M_{H_2}^{-1}\right)^{0.5}\left(M_{III}^{-1}+M_{N_2}^{-1}\right)^{0.5}\left(\sqrt[3]{V_{III}}+\sqrt[3]{V_{H_2}}\right)^{-2}\left(\sqrt[3]{V_{III}}+\sqrt[3]{V_{N_2}}\right)^{-2}}{\left(1-y_{N_2}\right)\dfrac{\left(M_{III}^{-1}+M_{N_2}^{-1}\right)^{0.5}}{\left(\sqrt[3]{V_{III}}+\sqrt[3]{V_{N_2}}\right)^2}+y_{N_2}\dfrac{\left(M_{III}^{-1}+M_{H_2}^{-1}\right)^{0.5}}{\left(\sqrt[3]{V_{III}}+\sqrt[3]{V_{H_2}}\right)^2}} \right]^{-0.5} \tag{1}$$

$K = \text{const.}\ f\,(v^{1/2}, p_{proc}^{-1/2}, p_{III}, x^{-1/2}, T^{-0.175})$

4. Results and discussion

For the set of experiments, where the gas velocity in the reactor v, the process pressure p_{proc}, the partial pressure of the group III source in the reactor p_{III} and the growth temperature T are the same, providing that the distance of the substrate from the inlet edge of the susceptor is long enough or the substrate rotation is used, K becomes to be a constant and expression (1) determines, how the growth rate should change due to

24

the change of the composition of the carrier gas. In our experiments we determined the value of K experimentally from the run with the pure H_2 as a carrier gas.

Fig.1 shows the comparison of the theoretical growth rate calculated from (1) with experimental values.

In the case of AlAs, GaAs, and InP growth from Trimethyl- and Triethyl- sources the experimental and theoretical values are corresponding.

Another situation occurs during the AlAs growth from Dimethylethylaminealane. In this case the presence of N_2 causes that DMEAAl molecule dissociates outside the boundary layer and the group III molecule flux consists of several types of molecules with higher diffusion coefficient. The composition of this flux depends on the composition of the carrier gas mixture. In this case the influence of the carrier gas on the growth process has to be taken into account.

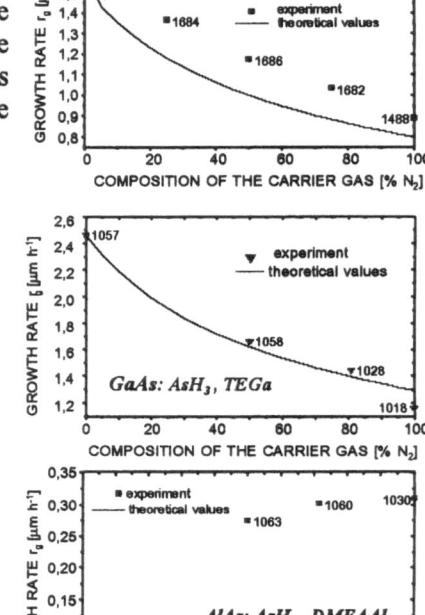

Figure 1. Comparison of the theorethical growth rates calculated from (1) with experimental values.

References

1. Hardtdegen, H., Hollfelder, M., Meyer, R., Carius, R., Münder, H., Frohnhoff, S., Szynka, D. and Lüth, H. (1992) J. Crystal Growth **124,** 420
2. Hardtdegen, H., Hollfelder, M., Ungermanns Chr., Wirtz, K., Carius, R., Guggi, D., and Lüth, H. (1994) *GaAs and related compounds 1993*, Inst. of Physics Conference Series **136,** 625
3. Razeghi, M. (1989) *MOCVD Challenge. Vol.1.*, Adam Hilger
4. Arens, G., Heinecke, H., Pütz, N., Lüth, H. and Balk, P. (1986) J. Crystal Growth **76,** 305
5. Ghandhi, S. K. and Field R. J. (1984) J. Crystal Growth **69,** 619
6. Reid, R. C., Prausnitz, J. M. and Sherwood, T. K (1977) *The Properties of Gases and Liquids*, McGraw-Hill Book Company

24

LP-MOVPE OF III-V SEMICONDUCTORS USING HIGHLY PURE N₂ AS THE CARRIER GAS

H. HARDTDEGEN, M. HOLLFELDER, S. HON, R. CARIUS and
CHR. UNGERMANNS
Institute of Thin Film and Ion Technology, Research Center Jülich
D-52425 Jülich, Germany

Abstract. The advantages of highly pure nitrogen as the carrier gas in LP-MOVPE to hydrogen are demonstrated for different III-V materials. It will be shown, that the use of N_2 improves all three weakpoints in MOVPE: decomposition of stable source compounds, layer homogeneity and process safety and therefore leads to a higher cost efficiency of the growth process.

1. Introduction

Metalorganic vapor phase epitaxy (MOVPE) is one of the most important deposition methods for III-V semiconductor device structures from the industrial point of view above all due to its high throughput. Weakpoints of the process that need attention, however, are insufficient purity of Al-containing compounds in terms of carbon and oxygen contamination, insufficient homogeneity on large wafer areas and problems due to the gases used in the process: the pyrophoric metalorganic compounds, the toxic hydrides and the highly explosive carrier gas hydrogen. Solving these problems leads to lower production costs. Either the (cost) efficiency of the growth process or that of the devices is increased.

In this report we will show that the replacement of the commonly used carrier gas hydrogen by nitrogen addresses all three problematical subjects due to the physical and chemical properties of nitrogen. The reasons why hydrogen should be replaced by nitrogen are its lower thermal conductivity and heat capacity which should lead to more homogeneous layers, its inertness which leads to higher process safety, its higher impact number as an impact partner in the gas phase which should lead to a better decomposition of the source compounds and to purer materials and last but not least because it is cheap and can be purified easily [1].

J. Novák and A. Schlachetzki (eds.), Heterostructure Epitaxy and Devices, 25–28.
© 1996 *Kluwer Academic Publishers.*

2. Experimental

Growth was carried out in a conventional horizontal cold wall double reactor system (AIX 200) equiped with gas foil rotation. The gas velocity was kept at 0.9 and 2.1 m/s for nitrogen and hydrogen as the carrier gas, respectively and the reactor pressure at 20 mbar. The following source chemicals were employed: TMAl, TMGa, TMIn (Heraeus), 100% AsH_3 and PH_3 (Praxair). The carrier gases H_2 and N_2 as well as the AsH_3 were purified with a Pd-cell, a getter column (SAES Getters) and a resin purifier (Millipore), respectively. Typical growth rates were 1 µm/h. The layers were characterized morphologically by Nomarski interference microscopy, electrically by Hall effect studies at 300 and 77 K, structurally by x-ray diffraction (XRD) and optically by photoluminescence (PL) at 2 K.

3. Results and discussion

3.1. Source decomposition

The enhancement of source decomposition is demonstrated best when layers are deposited using relatively stable source compounds. First of all this is the case for InP, where PH_3 is the stable source. The morphology of InP layers is strongly dependent on sufficient phosphorous overpressures i.e. on sufficient PH_3 decomposition. Therefore relatively high deposition temperatures or extremely high V/III ratios are needed. Using nitrogen as the carrier gas the growth temperature

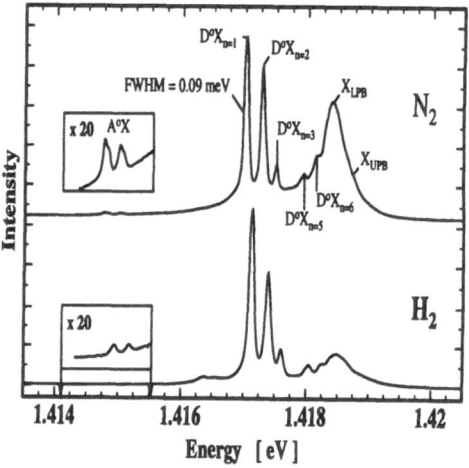

Figure 1. 2 K PL spectra of InP deposited with N_2 (upper curve) and H_2 (lower curve) as the carrier

can be lowered by at least 20°C or the V/III-ratios can be greatly reduced by at least a factor of 2 compared to the employment of hydrogen at otherwise the same deposition conditions. The optical quality of the layers is comparable as can be seen from figure 1. The background carrier concentration of the layers mentioned is so low, that undoped 2 µm thick layers could not be evaluated electrically.

A second example for the enhanced decomposition by the carrier gas nitrogen is the growth of AlGaAs using the relatively stable source TMAl. The high Al-C bond strength leads to carbon levels in AlGaAs in the mid 10^{16} cm^{-3} range. Additionally, oxygen incorporation is observed due to the affinity of Al towards oxygen and to the reactivity of Al-organic compounds towards oxygen and water. The exothermic reaction of Al with oxygen to Al_2O_3 can be reversed either by using high deposition temperatures or by large amounts of hydrogen, which reduces the oxide to the metal and which is supplied by the AsH_3 in the process. These two impurities have a big influence on the electrical and optical data of the AlGaAs

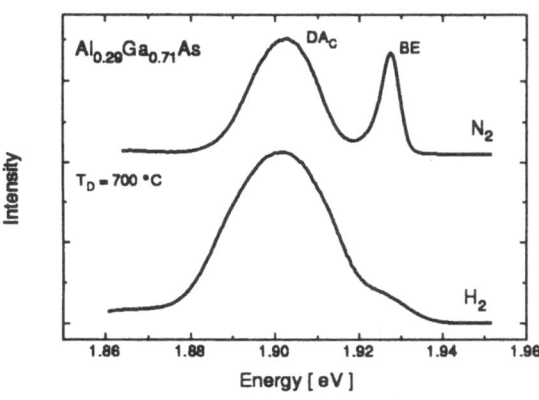

Figure 2. 2 K PL spectra of $Ga_{0.29}Al_{0.71}As$ deposited with N2 (upper) and H2 (lower curve) as the carrier

investigated, carbon being an acceptor and oxygen a deep level impurity. Photoluminescence spectra recorded at 2 K are shown in figure 2 for layers deposited at the same growth conditions such as V/III ratio, reactor pressure and growth rate but with different carrier gases. The higher intensity ratio from the bound exciton to donor acceptor transition documents the distinctly lower carbon and deep level impurity (oxygen) incorporation into the layers, when nitrogen is used as the carrier gas.

3.2. Layer homogeneity

For industrial needs, layer homogeneity in terms of layer thickness, composition and conductivity should be within a standard deviation range of ± 1 %. This is difficult for materials grown by conventional MOVPE. The use of nitrogen as the carrier gas indeed improves homogeneity as comparative growth studies of GaInAs using both carrier gases show. The thickness homogenity increases from a standard deviation of ± 1.5 % for H_2 to ±1.1 % for N_2. Figure 3 shows the lattice mismatch over a full 2" wafer. Larger areas of the same composition are at one's disposal when nitrogen is used.

Homogeneous temperature distri-

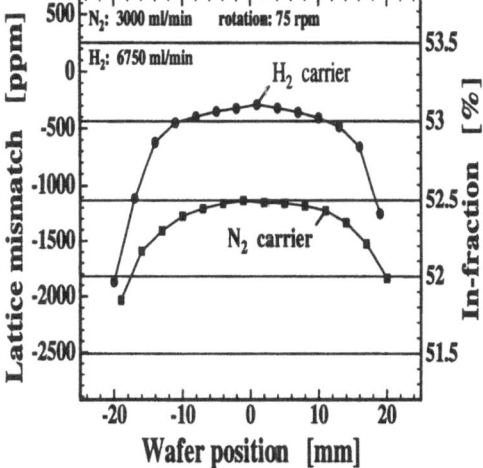

Figure 3. Homogeneity studies: lattice mismatch as a function of wafer position for GaInAs layers deposited with H_2 (upper) and N_2 (lower curve)

bution on and above the substrate during deposition seems to be the key point, why layers grow more homogenously in a nitrogen atmosphere. The temperature distribution improves. Two materials that are extremely sensitive to temperature differences are chosen to prove this supposition. The height of donor concentration in Si-doped materials such as GaAs using SiH_4 is extremely temperature dependent. Therefore the uniformity in conductance on a full wafer is usually well above ± 1 %. With nitrogen as the carrier gas, we achieved ± 0.6 %. Last but not least: the band gap of GaInP is extremely dependent on deposition temperature. The growth temperature influences the amount of ordering on the group III sublattice. The difference in band gap between completely ordered and completely disordered material has been observed to be about 200 meV. Room temperature PL measurements on our samples deposited using nitrogen as the carrier gas show that the energy of the peak maximum is within a standard deviation of ± 0.26 nm and an absolute deviation of ± 1.2 nm. This high uniformity has up to now not yet been obtained in our reactor type with hydrogen as the carrier gas.

3.3 Economical aspects

Due to the better uniformity of the layers a larger area on the wafer can be used for devices. Additionally, the layer purity of Al-containing materials is enhanced. These features already lead to higher (device) efficiency. The costs for the process are, in addition, reduced with respect to hydride consumption due to their better decomposition: for As-containing compounds, 30 % less AsH_3 and for phosphorus containing materials 60 % less PH_3 is needed to obtain equal quality layers. The reduction in gas scrubbing costs corresponds to the lower hydride consumption. The costs for the carrier nitrogen, when evaporated from the liquid, are 96 % lower than for hydrogen. The costs for the metalorganic compounds, however, is increased by 25 % due to the slower diffusion of the molecules through the more viscous nitrogen. All in all the cost reduction is predominate. Due to the colder gas phase, less side wall deposition and deposits in the exhaust system are observed, leading to longer uptimes and lower maintenance costs. Last but not least hydrogen detection equipment and its maintenance becomes superfluous.

4. Conclusion

It has been demonstrated, that the use of highly pure nitrogen as the carrier gas is a great improvement of MOVPE. Higher layer purity (i.e. lower C and O contamination), increased layer uniformity and higher process safety lead all in all to a more efficient growth process.

5. Reference

1. Briesacher, J.L., Nakamura, M. and Ohmi, T. (1991) Gas purification and measurement at the PPT level, J. Electrochem. Soc. 138, 3717 - 3723

DEPENDENCE OF PROPERTIES OF LP MOCVD InGaP LAYERS ON GROWTH CONDITIONS

R. KÚDELA, I. VÁVRA, J. NOVÁK and M. KUČERA
Institute of electrical engineering, SAS, Dúbravská 9
842 39 BratislavaSlovak Republic

1. Introduction

$In_xGa_{1-x}P$ is a very promising material for optoelectronic and micro-electronic applications. This ternary compound can be grown lattice-matched to GaAs substrates with $In_{0.485}Ga_{0.515}P$ composition and the band-gap energy near 1.9 eV at 300 K. It is prepared by different epitaxial methods. $In_xGa_{1-x}P$ layers grown grown by MOCVD are usually prepared in the temperature range from 600 to 700 ºC and they exhibit an ordering [1-4]. Over 700 ºC ordering was lower usually, but interdiffusion during the epitaxial growth of multilayer structures plays a negative role [5-7] at these temperatures. Besides this fact, very high partial pressures of DMZn are required for obtaining a higher doping levels if p-type layers are necessary. At low temperatures below 600 ºC these effects are not critical and high concentration of Zn can be achieved, but there is necessary to modify the growth parameters and adjust them to the growth temperature . We have studied the preparation and properties of $In_xGa_{1-x}P$ zinc doped epitaxial layers grown by low-pressure MOCVD technique at temperatures below 600 ºC in this paper. These growth temperatures were chosen to avoid the negative consequences mentioned above and some samples were grown at 720 ºC to compare their properties. We have measured the dependence of Hall concentrations and mobilities on the growth temperature and on the DMZn partial pressure at different temperatures. Low temperature photoluminescence and transmission electron diffraction (TED) were used to measure the layer quality and ordering in the layers.

2.Experimental

$In_xGa_{1-x}P$ epitaxial layers were grown in a low-pressure MOCVD equipment AIX 200 on (100) oriented semiinsulating GaAs substrates with nonintentional misorientation up to 0.2 °. The substrates were cleaned and etched in the mixture NH_4OH + H_2O_2 + H_2O before loading into the reactor. We have used trimethylgallium, trimethylindium and dimethylzinc as precursors and hydrogen as a carrier gas. The growth procedure consisted of heating the GaAs substrate in the overpressure of AsH_3, epitaxial growth and cooling phase with the overpressure of PH_3. Total pressure inside the reactor was 50 mbar typically. This pressure was chosen as an optimal value for all the growth temperatures (720 ºC, 560 ºC and 520ºC) with respect to the surface morfology and uniformity of the grown layers. At

J. Novák and A. Schlachetzki (eds.), Heterostructure Epitaxy and Devices, 29–32.
© 1996 *Kluwer Academic Publishers.*

these preliminary experiments the total pressure was changed from 20 to 80 mbar. The corresponding optimal flow velocity inside the reactor was near 0.8 m/s. Typical mole fractions of TMGa, TMIn were of the order of 10^{-5} and the corresponding growth rates varied from 0.6 to 1.2 um/hour. The optimal TMIn/TMGa ratio was found for each growth temperature and its value varied in the range from 1.19 at 520 ℃ to 1.42 at 720 ℃. The V/III ratio was in the range from 216 to 555 at our experiments. The higher values from the given range were necessary to reach good surface morphology at the growth temperatures 720 ℃ and 520 ℃. All the growth parameters were adjusted according to the growth temperature and a very low growth rate of 0.6 um/h was optimal at 520 ℃.

When the growth conditions for a given a temperature were found, the experiments with doping by adding the DMZn into the gas phase were made. As the activation coefficient of Zn was measured in previous works [8,9,14] and its value was near unity (in the range of accuracy of measurements), the hole concentration has been taken to directly represent the Zn concentration in the epitaxial layer. Electrical parameters were measured by Van der Pauw method.

It is known from the published experimental results that besides the mole fraction of DMZn the concentration of Zn depends on the V/III ratio and on the growth rate. The dependence on the V/III ratio can be derived from thermodynamic equations [10] but there exists also the dependence on the growth rate, which can be explained only by kinetic factors. The kinetic effects were considered in the theoretical model suggested in [9]. According to this model the concentration of Zn in the $In_{1-x}Ga_xP$ layer can be written as:

$$c_{Zn} = K_{Zn} \cdot \Theta_P \cdot p_{Zn} \cdot \frac{110GR}{110GR + 110GR_0 / 3} \qquad (1)$$

where c_{Zn} is the Zn concentration, K_{Zn} is the equilibrium constant, p_{Zn} is the partial pressure of Zn (DMZn), Θ_P is the coverage of surface by phosphorus, and $110GR$ is the growth rate in the [110] direction and $110GR_0$ the growth rate in perpendicular direction. The values in (1) were found by fitting the experimental data. A very important factor is also the growth temperature, which was changed significantly in our experiments. The dependence of the hole concentration on the growth temperature can be seen in fig. 1.

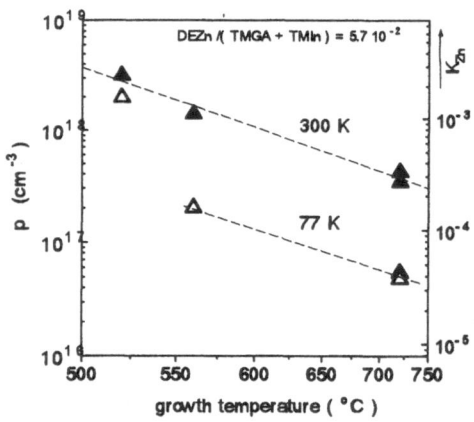

The hole concentration increases by about one order when the growth temperature decreases from 720 to 520 ℃ and practically no influence of the low growth rate at 520 ℃ was observed at 300 K . Because the V/III ratio was higher than 200 at our experiments,

Fig.1 The dependence of hole concentrations in the epitaxial layers on the growth temperature and the distribution coefficient K_{Zn}.

distribution coefficient K_{Zn} was supposed to be saturated with respect to this parameter [9]. Neglecting the small influence of the V/III ratio and the growth temperature we can write the dependence of K_{Zn} for equilibrium conditions in a general form as:

$$K_{Zn} = c \cdot \exp\left\{\frac{\Delta G}{kT}\right\} \qquad (2)$$

where ΔG is the difference of Gibbs energy of Zn atoms between the gas phase and solid, k is the Boltzmann constant, T is the temperature and c a normalization constant. According to this equation we can find from the temperature dependence of K_{Zn} the value of ΔG or desorption energy of Zn atoms E^{Zn}_d respectively. A value of 0.55 eV was calculated from our dependence in fig.1. The difference of the hole concentratios measured at 300 and 77K was high in samples grown at 560 and 720 °C as we ca see from this figure, and smaller for 520 °C. This effect can possibly be explained by the high activation energy of the zinc in the epitaxial layer. This

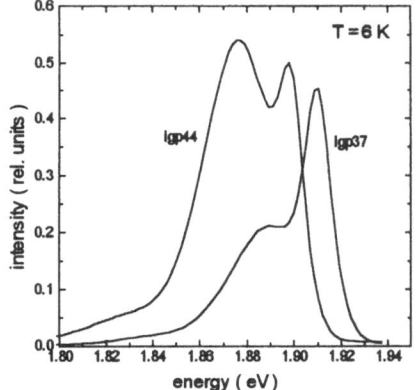

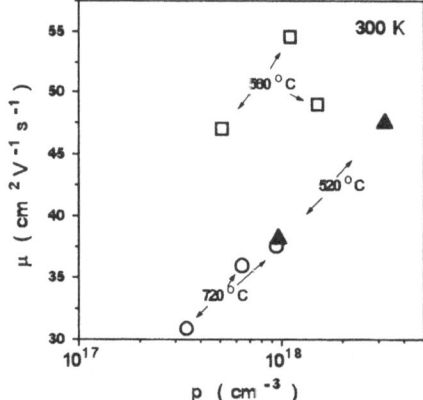

Fig. 2 The low temperature photoluminescence of Zn doped layers. From the peak positios estimated activation energy is 21 meV.

Fig. 3 Dependence of hole mobilities on the concentrations for different growth temperatures.

activation energy for Zn in the ternary $In_{.49}Ga_{.51}P$ calculated in [11] is E_A (mV) $=45.75 - 8.2\times10^{-6}\ p^{1/3}$, where p is the hole concentration. We have measured the photoluminescence spectra to confirm this relation and the measured PL spectra are shown in fig. 2. It can be seen from the PL measurements, that the difference between acceptor and valence band is about 21 meV, a smaller value than predicted. Hole mobilities in our samples ranged from 30 to 55 cm^2/Vs. As it follows from fig. 3, the highest values of the hole mobility were measured in the samples grown at 560 °C.

Ordering was observed in most of epitaxial layers grovn at different temperatures. It was detect by X-ray measurements, where some asymetric reflections were much less intensive ((115) reflections) then symetric ones. Such a behaviour can be caused by the CuPt- type ordering, which is most common for this ternary. More precisely ordering was studied by transmision electron diffraction (TED). Observations were performed on a JEOL 1200 EX microscope operating at 120 keV. The most intesive ordering was observed in the layers grown at 640 °C. In the layers grown at 560 and 720 °C ordering was not so intensive, but evident and the layers grown at 520 °C were completly disordered. Disordering caused by Zn incorporation was not observed. Zn doped epitaxial layers were ordered, if they were grown at the temperatures where undoped layers were ordered.

Extra diffractions caused by the ordering were observed in several rojections. Unusual diffraction patterns were observed near the [001] pole where only "diagonal " $\left\langle \frac{1}{2}, \frac{1}{2}, 0 \right\rangle$ extra spots ere observed. It was concluded that this can be caused by combining CuPt and Y2 type ordering with the APBs paralell with the surface.

3. Conclusions

The dependence of epitaxial layer properties on growth conditions was investigated. Experimental data are compared with the data from the literature. The desorption energy of Zn atoms of 0.55 eV is estimated from the temperature dependence of hole concentration . An unusual type of ordering was observed in our samples .

References

1. Gomyo,A., Suzuki,T.,Kobayashi,K., Kawata,S.,Hino,I, and Yuasa,T. (1987) Appl. Phys. Lett. 48 p. 1603.
2. Su,L.C., Ho,I.H., and Stringfellow,G.B. (1994) Effect of substrate misorientation and growth rate on ordering in InGaP, J. Appl. Phys. 75 p. 5135-5141.
3. Kurtz,S.R.,Olson,J.M., and Kibbler,A. (1990) Effect of growth rate on the band gap of GaInP, Appl. Phys. Lett. 57p.1922-1924.
4. Kurtz,S.R.,Olson,J.M.,Arent,D.J., Bode, M.H.,and Bertness, K.A. (1994) Low-band-gap InGaP grown on (511)B GaAs substrates, J. Appl. Phys. Incorporation of zinc in MOCVD growth of InGaP 75 p. 5110-5113.
5. Guimaraes, F.E.G., Elsner,B., Westphalen, R., Spangenberg,B., Geelen,H.J., and Balk, P.(1992)LP-MOVPEgrowth and optical characterization of GaInP/GaAs heterostructures, quantum wells and quantum wires, J. Crystal Growth 124 p.199-206.
6. Francis,C.,Bradley,M.A,Boucaud,P.,Joulien,F.H., and Razeghi,M. (1993)Intermixing of GaInP/GaAs multiple quantum wells, Appl. Phys. Lett. 62 P.178-180.
7. R.Bhat, M.A.Koza, M.J.S.P.Brasil, R.E.Nahory, C.J.Palmstrom, and B.J.Wilkens(1992) Interface control in GaAs/GaInP superlattices grown by OMCVD, J.Crystal Growth 124 p. 576-582.
8. Ikeda,M., and Kaneko, K. (1989)Selenium and zinc doping in GaInP and AlGaInP grown by metal-organic chemical vapor deposition, J. Appl. Phys. 66 p.5285-5289.
9. Kurtz,S.R., Olson,J.M, Kibbler,E.A., and Bertness ,K.A.(1992)Incorporation of zinc in MOCVD growth of InGaP J. Crystal Growth 124 p. 463-469.
10. G.B.Stringfellow (1989) Organometallic Vapor-Phase Epitaxy:Theory and Practice, (Academic Press, Inc.) p.129.
11. Chang,C.Y., Wu,M.C., Su,Y.K., Nee, C.Y.,and Cheng,K.Y. (1985)The doping concentration dependence of zinc acceptor ionization energy in InGaP, J. Appl. Phys. 58 p.3907-3908.

GROWTH OF GAN MOCVD LAYERS ON GAN SINGLE CRYSTALS

K. PAKUŁA[a], A. WYSMOŁEK[a], K.P. KORONA[a],

J.M.BARANOWSKI[ab], R.STĘPNIEWSKI[a],

I. GRZEGORY[b], M. BOĆKOWSKI[b], J.JUN[b],S.KRUKOWSKI[b],

M.WRÓBLEWSKI[b], S. POROWSKI[b]

a. *Institute of Experimental Physics, Warsaw University,*
Hoża 69, 00-681 Warszawa, Poland

b. *High Pressure Research Center, Polish Academy of Sciences,*
Sokołowska 29/37, 01-142 Warszawa, Poland

Abstract: We report photoluminescence (PL) and reflectance of GaN homoepitaxial layers grown by metalorganic chemical vapour deposition on GaN substrates. Very narrow (FWHM = 1.0meV) PL lines at 3.4661eV and 3.4714eV were observed. Energies of free excitons were found: E_A= 3.4785eV, E_B= 3.483eV and E_C= 3.501eV. The energy gap temperature dependence was obtained: $E(T) = E(0)- 0.00181*T^2/(T+2300)$.

We would like to report growth of the high quality lattice matched GaN layers showing very sharp excitonic structure in luminescence and reflection. Bulk GaN crystals which have been used as substrates were grown by the high pressure method described previously[1]. They were in a form of platelets with the hexagonal c axis perpendicular to the surface. Electron concentration was above 10^{19} cm^{-3} most likely resulting from a high concentration of nitrogen vacancies [2]. The GaN platelets used as substrates had area of a few mm^2 and relatively flat surfaces which have been used for the MOCVD growth without any mechanical or chemical treatment. The GaN epitaxial layers have been grown in a horizontal atmospheric pressure Epigress system

J. Novák and A. Schlachetzki (eds.), Heterostructure Epitaxy and Devices, 33–36.
© 1996 *Kluwer Academic Publishers.*

adopted for growth of nitrides. The trimethylgallium (TMG) and NH_3 have been used as sources of Ga and N respectively, in addition to H_2 as a carrier gas. The growth took place at temperatures close to 1000°C and the growth parameters have been adjusted in such way that a layer of about 2 μm thickness was expected. The growth of GaN layers was realised directly on single crystals substrates without deposition of a low temperature nucleation layer.

Scanning electron microscopy (SEM) was performed using Zeiss DSM-942 electron microscope. The surface of the GaN layer is flat and randomly decorated with pinholes. These hexagonal pinholes are possibly caused by incomplete coalescence of the GaN during layer by layer growth. The density of the pinholes ranges between 10^3-$10^5 cm^{-2}$.

The first luminescence of crystalline GaN layers grown on GaN single crystals substrate has been recently reported by us [3]. The significant progress in the quality of obtained layers allow us to perform the detail study of the excitonic spectrum visible in luminescence and reflectance.

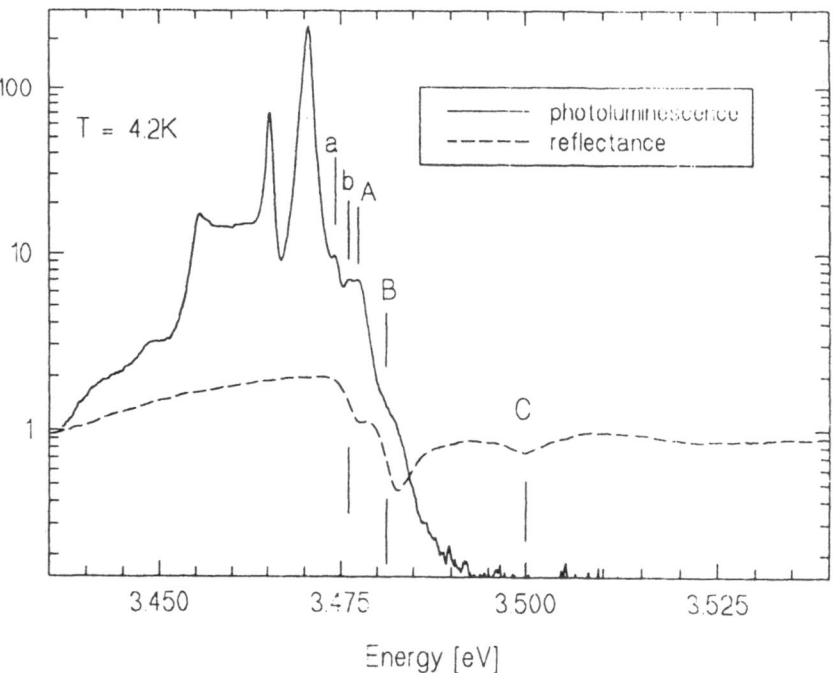

Figure 1. The luminescence spectrum and reflectance measured on the same surface of the sample.

The luminescence spectrum (Fig.1.) in the exciton region is shown in Fig. 1. The low temperature spectrum is dominated by two strong and narrow lines at energies 3.4661eV and 3.4714eV. Since their FWHM are 1.0meV and 1.8 meV respectively, these lines are the most narrow ones up to now observed in GaN. Moreover four weak lines: a, b, A, B with energies $E_a= 3.4749(3)eV$, $E_b= 3.4769(3)eV$, $E_A= 3.4785(5)eV$, and $E_B= 3.483(1)eV$ are observed. The two luminescence lines A and B correspond exactly with the reflectance dispersion lines (see Fig.1.). In addition, in reflectance spectrum, there is the third one at $E_C= 3.501(1)eV$. These are the free excitons lines formed with participation of holes originating from the crystal-field and spin-orbit splitted valence band. Comparison of our results with positions of free excitons in MOCVD GaN layers grown on saphire [4] shows that the center of gravity of excitons lines is shifted to higher energies: $E_A= 3.482eV$, $E_B= 3.4885eV$ and $E_C= 3.52eV$[4]. This effect would be an indication of strain existing in GaN layers grown on sapphire.

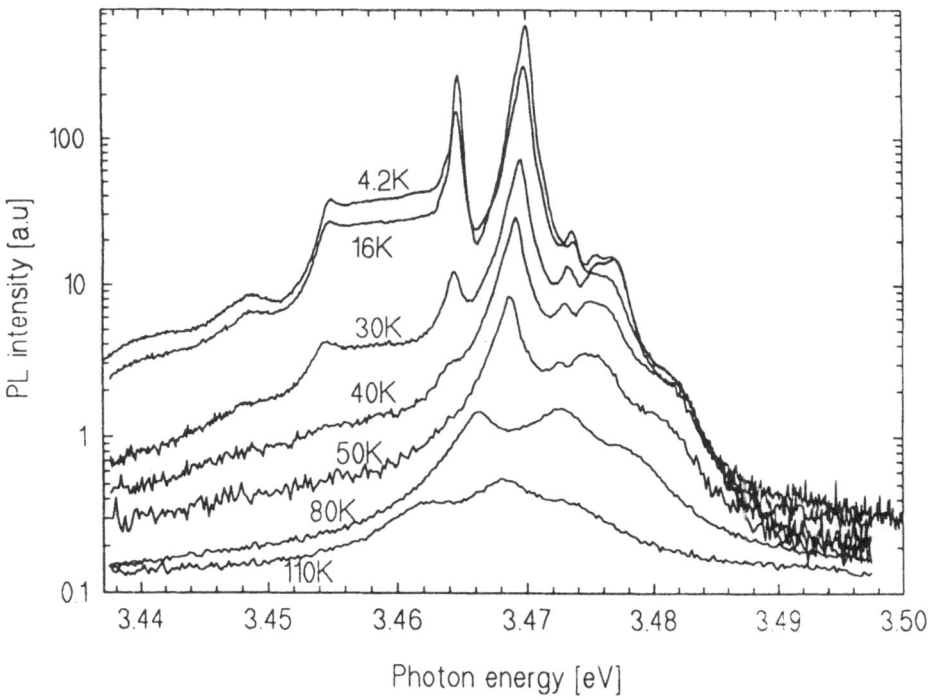

Figure 2. The temperature dependence of the luminescence spectrum.

In temperature range from 4.2K to 390K the free excitons energy determined from the reflectance fits to the following curve: $E(T)=E(0)-0.00181*T^2/_{(T+2300)}$.

The temperature dependence of the luminescence spectrum (Fig.2.) shows that the free exciton lines grow up with increse of temperature. In the temperature range between 70K-100K the line A together with line b become even the dominant ones. It seems that the line b is due to donor bound exciton formed with participation of a hole from the deeper valence band.

The remaining part of the luminescence spectrum which consists of the narrow line at 3.4661eV and a band of a strange rectangular shape disappears above 40K. This part of the spectrum seems to be connected with acceptors.

In summary GaN (0001) homoepitaxial layers have been grown via MOCVD process on single crystal GaN substrates. The low temperature photoluminescence spectrum shows intense emission in the exciton region with a narrow line of a FWHM of 1meV. These indicates that homoepitaxial layer of GaN has a good structural and optical properties.

The authors would like to express their gratitude to SEZAM program organised by the Foundation for Polish Science for the purchase of the scrubbing system for the MOCVD system. This research was partially supported by KBN grant PBZ-101.12.

References

1. S. Porowski, J.Jun, M.Boćkowski, M.Leszczyński, S.Krukowski, M.Wróblewski, B.Łucznik, I. Grzegory, Proc. 8th Conf. on Semi-Insulating III-V Materials, Warsaw, Poland 1994, ed.M.Godlewski, Word Scientific, p.61

2. D.Jenkins, J.D.Dow, Ming-hsiung Tsai, J.Appl.Phys. 72, p.4130 (1992)

3. K. Pakuła, J.M. Baranowski, R.Stępniewski, A. Wysmołek, I. Grzegory, J.Jun, M.Sawicki, S. Porowski, K.Starowiejski, to be published in Acta Phys.Pol.

4. W.Shan, T.J.Smith, X.H.Yang, S.J.Hwang, J.J.Song, B.Goldenberg, Appl.Phys.Lett. 66, p.985 (1995)

ELECTRICAL AND OPTICAL PROPERTIES OF TELLURIUM-DOPED GALLIUM ANTIMONIDE GROWN BY MOVPE

KARI HJELT, TURKKA TUOMI
Optoelectronics Laboratory
Helsinki University of Technology, 02150 Espoo, Finland

1. Introduction

GaSb and related ternary and quaternary compounds have received much attention due to their utilization in optical devices operating in the 2-4 μm region. The growth of n-type GaSb layers is essential in device fabrication. The growth of GaSb by MOVPE has been studied by several groups, but there are only few papers about n-type GaSb [1, 2]. In this paper we investigate electrical and optical properties of tellurium-doped GaSb grown on GaAs with atmospheric MOVPE using diethyltellurium as a dopant source.

2. Experimental

The layers were grown at atmospheric pressure in a horizontal MOVPE reactor made by Thomas Swan. The precursors used were trimethylgallium (TMGa) and trimethylantimony (TMSb) cooled to -10°C. The dopant source was diethyltellurium (DETe) diluted in hydrogen (100 ppm). The carrier gas flow velocity was 7.5 cm/s. Undoped, semi-insulating GaAs (100) wafers were used as substrates. Before growing the GaSb-layer a 70 nm thick undoped GaSb buffer layer was grown on top of the GaAs substrate at the temperature of 560 °C to stop the dislocations caused by the large lattice mismatch between GaAs and GaSb. The DETe mole fraction (MF) was varied between $6 \cdot 10^{-10}$ and $3 \cdot 10^{-6}$. The doped GaSb layers were grown at 600 °C and they were typically 2 μm thick.

The carrier concentration and mobility were determined with van der Pauw and Hall technique. The photoluminescence (PL) spectra were measured at the temperature range from 12 K to 350K and with different excitation intensities at 12K. Argon-ion laser at the wavelength of 488 nm was used for excitation. The PL spectra were detected with a liquid-nitrogen-cooled germanium and a room-temperature PbS detector using standard lock-in technique.

37

J. Novák and A. Schlachetzki (eds.), Heterostructure Epitaxy and Devices, 37–40.
© 1996 *Kluwer Academic Publishers.*

3. Results and discussion

Figures 1 and 2 show the electron concentration and mobility at room temperature of the n-type samples. Three different series with V/III ratios of $R_{V/III}$=1.05, 1.3 and 1.5 were investigated. Figure 1 shows the effect of TMGa concentration, i.e. the growth velocity on the electron concentration and mobility. The growth velocity depends nearly linearly on the TMGa mole fraction and is about 6 µm/h at the TMGa MF of $8 \cdot 10^{-6}$.

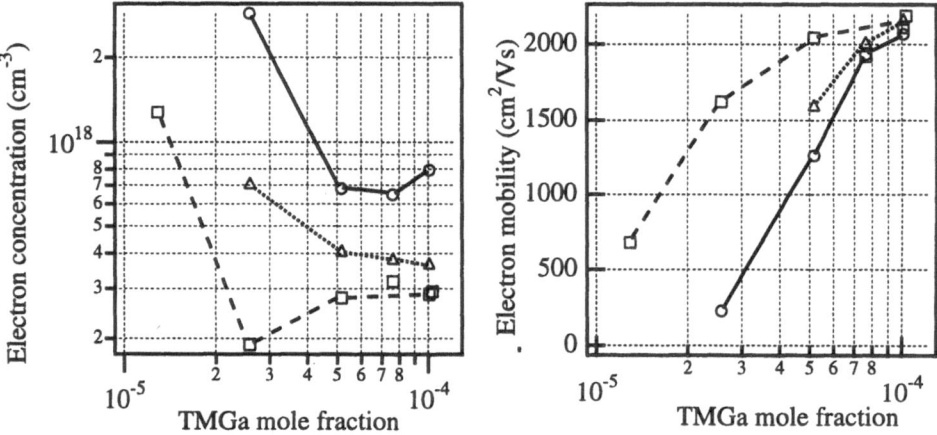

Figure 1. Electron concentration and mobility as a function of TMGa mole fraction in growth gas. Samples with different V/III ratio ($R_{V/III}$) are marked with circles, $R_{V/III}$=1.05; triangles, $R_{V/III}$=1.3 and squares, $R_{V/III}$=1.5.

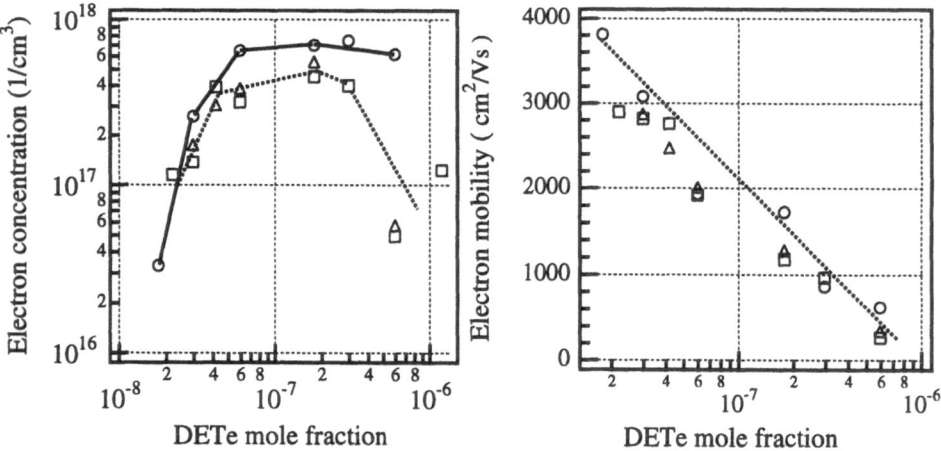

Figure 2. Electron concentration and mobility as a function of DETe mole fraction in growth gas. Samples are marked as in *Fig. 1*.

The optimal TMGa MF was found to be about $7.6 \cdot 10^{-5}$. Figure 2 shows the electron concentration and mobility as a function of the DETe concentration in the growth gas. The TMGa mole fraction was $7.6 \cdot 10^{-5}$ for all the samples in Fig. 2. At the DETe mole fraction of $2 \cdot 10^{-8}$ a rapid change of the layers from p-type to n-type occurs. The electron concentration increases at least linearly up to 10^{18} cm^{-3} with increasing DETe concentration, after which it saturates and starts to decrease. The electron mobility decreases logarithmically with increasing DETe mole fraction. The morphology of the layers grown with DETe mole fraction less than $6 \cdot 10^{-8}$ was good. At higher DETe concentrations the morphology of the layers deteriorated rapidly. We were not able to conduct the electrical measurements with the samples grown with DETe MF greater than $6 \cdot 10^{-6}$. The largest mobility achieved at room temperature was 3800 cm^2/Vs with electron concentration of $3 \cdot 10^{16}$ cm^{-3} and the largest electron concentration achieved was $3 \cdot 10^{18}$ cm^{-3}, as can be seen from Figures 1 and 2.

Figure 3a shows the PL spectra of three samples grown with different DETe mole fractions in the growth gas and with $R_{V/III}=1.05$. Samples i and ii are n-type and sample iii is p-type. The change of the PL spectra with increasing DETe concentration is clearly seen. The very broad transition band at low energies is characteristic of Te-doped n-type GaSb-layers. The p-type sample shows two peaks, which are a band-to-native acceptor transition at 775 meV and an acceptor-bound exciton transition at 800 meV [3]. Fig. 3b shows the relative integrated intensities of the samples grown with different V/III ratios and DETe concentrations. The integrated PL intensity increases by a factor of 100 when the layer changes from p-type to n-type. The TMGa mole fraction is $7.6 \cdot 10^{-5}$ for all the samples in Fig. 3.

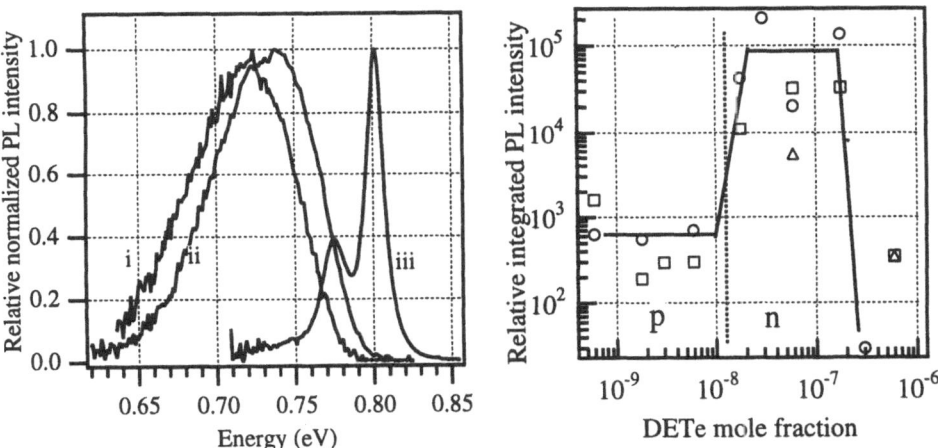

Figure 3. a) PL spectra of three GaSb-layers with $R_{V/III}=1.05$. The DETe mole fractions in growth gas were: i) $6 \cdot 10^{-8}$; ii) $3 \cdot 10^{-8}$, iii) $6 \cdot 10^{-10}$. b) Relative integrated intensities of samples grown with different V/III ratios and DETe concentrations. Solid line is drawn to clarify the behavior of the integrated intensity. Dotted vertical line separates p-type and n-type samples.

The integrated intensity decreases rapidly when the DETe concentration exceeds $2 \cdot 10^{-7}$ MF. This coincides with the decrease of the electron concentration in Fig. 2. The luminescence quenching and the decrease of the carrier concentration can be attached to the formation of defect complexes. This explains also the bad morphology at higher doping concentrations. The optimal DETe concentration for n-type doping determined with the aid of Figures 1-3 is about $6 \cdot 10^{-8}$ MF in the growth gas.

The broad PL peak of Fig. 3a has its maximum at energies from 0.68eV to 0.74 eV. Usually this Te-related luminescence band moves to higher energies when Te-concentration is increased due to the rising of Fermi level [4]. In our samples, however, the peak energy varied greatly from sample to sample making it difficult to find a clear trend. The peak energy increases first with increasing doping and then decreases rapidly towards 0.70 eV. In the highly doped samples the maximum occurs at lower energies than in moderately doped ones. This can be explained by the formation of different complexes that create different deep levels. The low-energy PL band is probably a superposition of several competing radiative transitions. We also measured the PL spectra as a function of temperature to determine the thermal quenching of the transitions. The activation energies calculated ranged from 60 meV to 140 meV and could not be explained by the simple model of a band to deep acceptor transition alone. We also observed that the native acceptor-related transition band disappeared with increasing Te-doping. We thus believe, that the Te-related deep level PL is caused by the tellurium-native acceptor complex ($V_{Ga}Ga_{Sb}Te_{Sb}$) together with band-to-native acceptor transitions, as reported in [5]. With increasing Te-concentration in the complex there can be one, two or three tellurium atoms together with the native acceptor changing the energy levels of the complex. The behavior of carrier concentration as a function of tellurium concentration in Fig. 2 can also be explained with this model.

4. References

1. Pascal, F., Delannoy, F., Bougnot, J.,Gouskov, L., Bougnot, G., Grosse, P., and Kaoukab, J., (1990) Growth and Characterization of Undoped and N-Type (Te) Doped MOVPE Grown Gallium Antimonide, *J. Electron. Mater.* **19** (2), 187-195.
2. Nakamura, F., Taira, K., Funato, K., and Kawai, H., (1991) Se and Te doping in LP-MOCVD-grown GaSb using H_2Se and DETe, *J. Cryst. Growth* **115**, 474-478.
3. Chidley, E. T. R., Haywood, S. K., Henriques, A. B., Mason, N. J., Nicholas, R. J., Walker, P. J., (1991) Photoluminescence of GaSb Grown by Metal-Organic Vapour Phase Epitaxy, *Semicond. Sci. Technol* **6**, 45-53.
4. Lazareva, I. K. and Stuchebnikov, (1970) Photoluminescence of Gallium Antimonide Doped with Tellurium, *Sov. Phys. Semicond.* **4** (4), 550-554.
5. Lebedev, A.I., Strelnikova, A., (1979) About the Nature of Effective Recombination Centre in GaSb and Solid Solutions, *Sov. Phys. Semicond.* **13**, 229-231.

COMPUTER SIMULATIONS OF EPITAXIAL GROWTH, SURFACE KINETIC PROCESSES AND RHEED INTENSITY OSCILLATIONS

D. PAPAJOVÁ, Š. NÉMETH AND M. VESELÝ

Department of Microelectronics,
Faculty of Electrical Engineering and Information Technology, STU
Ilkovičova 3, SK-812 19 Bratislava, SLOVAKIA

1. Computer simulation of epitaxial growth

It is well known, that under suitably-chosen operating conditions the MBE (Molecular Beam Epitaxy) growth proceeds in a layer-by-layer mode [1], that is, a given layer is almost completed before the growth of next layer is initiated. Such a growth mode allows to grow atomically-abrupt interfaces required in the growth of low-dimensional semiconductor structures with novel electronic and optical properties. While the basic physical principles of the process are known, the details of the manner in which the interface is formed vary from one system to another and thus do not allow to create a single universal theory. This fact as well as the inability to probe small length- and time-scale kinetic phenomena have led to a development of computer simulations enabling a better understanding of the formation of microscopic structures. In the description of growth kinetics several models are currently employed. Two principal techniques that have been utilized are Monte Carlo (MC) simulations and molecular dynamics (MD).

1.1. SIMULATIONS USING KINETIC RATE EQUATIONS - THE RE MODEL

The MBE growth in RE model is described by a set of differential equations [2,3]. It calculates at each time step the change in the N_{kj} number of adatoms and two-dimensional islands of varying k sizes in every j-th growing layer and sequentially the whole coverage of each layer, Θ_j. This change is due to surface kinetic processes occurring on a surface during the growth. Their substrate temperature dependence is described by Arrhenius form [4]. In Fig. 1a the results of a typical two-dimensional (2D) growth are obtained for substrate temperature 775K. Every curve of Θ_j coverage describes one growing layer. As it can be seen every next layer starts to grow after previous layer is nearly completed. Both, surface roughness (IW) and RHEED intensity exhibit the time dependence having a period which is coincident with that of the growing layer. In Fig. 1b results of a three-dimensional (3D) growth using low 'substrate' temperature (T<600K) are presented. In this growth mode several layers are grown at the same time. The RHEED oscillations are damped and their amplitude decreases with each subsequent growing layer. This is a consequence of increasing surface roughness.

J. Novák and A. Schlachetzki (eds.), Heterostructure Epitaxy and Devices, 41–44.

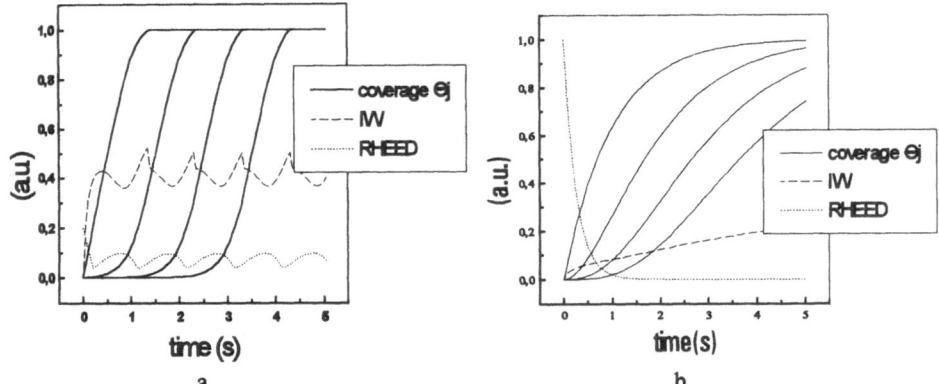

Figure 1. a, Output profiles of θ coverage, IW and the RHEED oscillations for 2D growth
b, Output profiles of θ coverage, IW and the RHEED oscillations for 3D growth

1.2. MONTE CARLO SIMULATIONS

Atomistic model using Monte Carlo (MC) method has been applied for simulations of GaAs epitaxial growth. This model [5] describes the GaAs growth on ideal (100) monocrystalline substrate in several independent processes: impingement of Ga atoms and As_2 molecules on the substrate, their surface migration, the formation and the growth of islands, their mutual connections and atom desorption from the surface. The model is extended with evolution of reflected beam intensity from the surface of grown crystal which makes it possible to compare the simulation results with experiments. The simulations were done at two different substrate temperatures (723K and 858K). The growth rate was 1ML/s. The simulation results are in good agreement with our experiments (see Fig. 2) [6]. The obtained correlation may be used for better adjustment of experimental growth conditions mainly for a growth rate of 1ML/s.

The morphological evolution of vapour deposited aluminium thin films with impurity atoms (carbon) is explored using the enhanced atomistic model based on MC method [7]. The ratio of the number of impinging C and Al atoms was in both cases the same, namely 0.15. The growth rate was 0.85ML/s. As we can see (Fig. 3), through varios substrate temperatures we have obtained 2D (673K) and 3D (373K) growth modes and also the impurity atom segregation which is larger at higher temperatures.

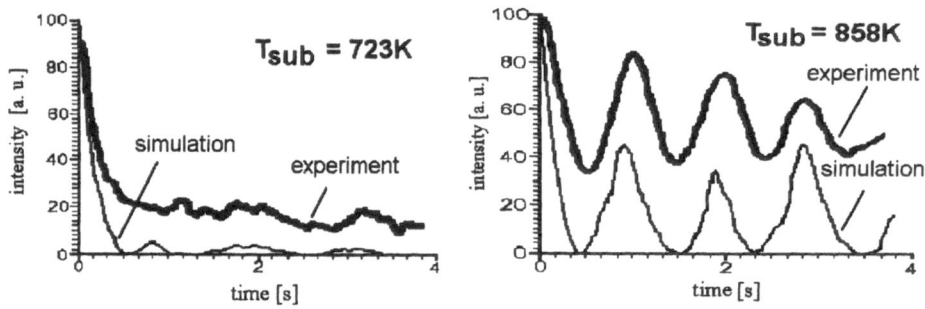

Figure 2. A comparison between RHEED intensity simulations and experiments

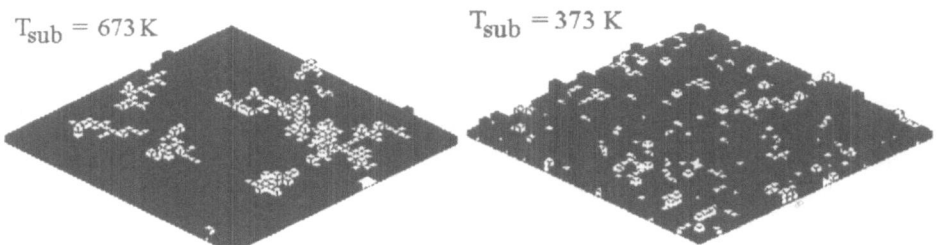

$T_{sub} = 673\,K$ $T_{sub} = 373\,K$

Figure 3. The surface configuration at different substrate temperatures at 1,3 sec after onset of the simulation, dark cubes represent Al atoms, white ones represent C atoms

Since each of the models is based on a different method there is a great interest to know if the models give the same results using the same input parameters for simulations (namely substrate temperature, T). In this, a comparison of results of the RE model with MC simulations has been done and gives agreement in the coverage output profiles (see Fig. 4).

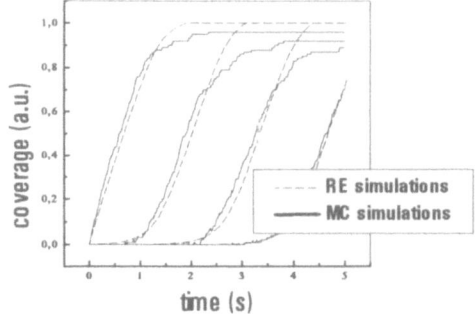

Figure 4. The comparison of output coverage profiles from MC and RE simulations at 'substrate' temperature T=775K

2. Model and simulation of RHEED intensity oscillations

We present a new simple model of RHEED intensity oscillations based on a reflectivity of the growing surface. The model calculates the change of electron beam intensity reflected from unscreened surface atoms in each *j-th* layer, also called "exposed coverage" (EC_j) (see Fig. 5a). Fig. 5b shows schematically the principle of the model. As it can be seen exposed coverage in each *j-th* layer is decreased by atoms which are not touched by analyzing electron beam or which are in "shadow" and so do not contribute to reflected intensity. The amplitude of reflected intensity represented by the real fraction of exposed atoms from which the primary electron beam is reflected can be expressed as

$$I_1(t) = I_o \cdot \sum_{j=1}^{M} \left[EC_j - shadow \right] \qquad (1)$$

where I_o is the intensity of primary electron beam, M is the number of growing layers, *shadow* is an area which is shielded from primary electron beam

$$shadow = \sum_{k=1}^{S} N[k, j+1] \cdot \sqrt{k} \cdot cotg\phi \cdot a \qquad (2)$$

44

where ϕ is an incident angle of primary electron, a is the lattice constant, S is the substrate size, $N_{kj}\cdot\sqrt{k}$ is the length of islands edges.

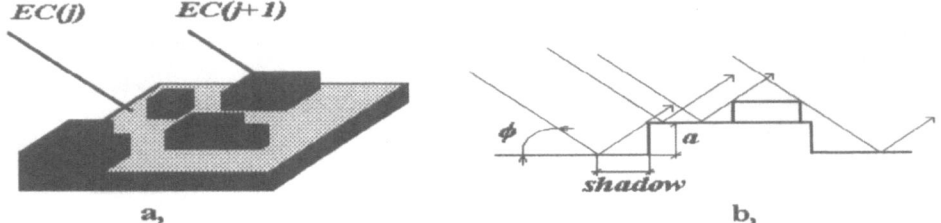

Figure 5. a, exposed coverage of each *j-th* layer
b, schematical view of shadow area around islands

For our simulations the incident angle of primary electron beam was taken to be 5°. Results of this model can be seen in Fig. 1.

3. Conclusion

We have presented the simulations of epitaxial growth using two models. For both, MC and RE simulations, we have obtained 2D and 3D growth choosing appropriate 'substrate' temperature (input parameter for simulations). Using simulating model of RHEED intensity oscillations we were able to determine the growth modes and compare results with experiments. A comparison gives a good agreement. Using new extended atomistic model we have studied the temperature dependence of impurity segregation process in growing aluminium layers. We have also presented a new simple model of RHEED intensity oscillations based on surface reflectivity.

References

[1]   Neave, J.H., Joyce, B.A., Dobson, P.J., and Norton, N. (1983) *Appl. Phys. A* **31**, 1.
[2]   Kariotis, R., Lagally, H.G. (1989) Rate equation modeling of epitaxial growth, *Surf. Sci.* **216**, 557-578.
[3]   Papajová, D., Hagston, W.E., Harrison, P. (1994) The simulation of single-crystal growth by molecular beam epitaxy using a kinetic rate-equation model, *Appl. Phys. A* **59**, 215-222.
[4]   Papajová, D., Németh, Š., Hagston, W.E., Sitter, H., and Veselý, M. (1995) A study of kinetic rate equation model for simulations of molecular beam epitaxy crystal growth: The temperature dependence of surface kinetic processes, paper has been sent to *J. of Appl. Phys. A.*
[5]   Kawamura, T., Kobayashi, A., and Das Sarma, S. (1989) Stochastic simulation of molecular-beam epitaxial growth of a model compound semiconductor: Effects of kinetics, *Physical Review B* **39**, 12723-12734.
[6]   Németh, Š., Harman, R., Veselý, M. (1993) Correlation between the stochastic simulation of molecular beam epitaxy growth and experiments, *9-th International Conference on Thin Films*, September 6-10, Vienna, Austria.
[7]   Németh, Š., and Veselý, M. (1994) Monte Carlo simulation of an enhanced solid-on-solid model for epitaxial growth: Effects of impurities, *Workshop on Computer Simulation of the Growth of Semiconductor Materials*, May 30 - June 2, Lyon, France, pp. 11.

GROWTH AND CHARACTERIZATION OF INP/GA$_{0.47}$IN$_{0.53}$AS DEPOSITED BY MOMBE

M. v. d. AHE, CHR. UNGERMANNS, H. HARDTDEGEN,
A. FÖRSTER, B. SETZER, H. LÜTH
Institute of Thin Film and Ion Technology, Research Center Jülich
D-52425 Jülich, Germany

Abstract. MOMBE growth of InP and Ga$_{0.47}$In$_{0.53}$As was studied with respect to heterostructural device application. A set of deposition parameters was developed with which good structural, optical and electrical quality InP as well as GaInAs can be grown. At these growth conditions interface abruptness was studied.

1. Introduction

The most consequent realization of LP-MOVPE (Low Pressure Metal Organic Vapor Phase Epitaxy) leads to MOMBE (Metal Organic Molecular Beam Epitaxy). MOMBE combines the advantages of MOVPE and MBE in III-V semiconductor growth. On the one hand the deposition of phosphorous containing materials as well as selective area growth is possible, on the other hand *in-situ* characterization methods like RHEED (Reflection High Energy Electron Diffraction) and mass spectrometry can be applied. We will use these advantages for the growth optimization of InP/GaInAs heterostructures for HEMT (High Electron Mobility Transistor) application - a device, which depends on an abrupt heterostructural interface and good bulk material quality. To this end InP and GaInAs growth was investigated with respect to deposition temperature, to find a common growth parameter window for heterostructures, which fulfills the needs for HEMT application.

2. Experimental

Growth was carried out in a Varian Gen II CBE machine using TEGa, TMIn and precracked AsH$_3$ and PH$_3$. For bulk material characterization, 1 μm thick GaInAs

45

J. Novák and A. Schlachetzki (eds.), Heterostructure Epitaxy and Devices, 45–48.

46

as well as InP layers were deposited on semi-insulating exactly oriented epi-ready (100) InP substrates; a 200 nm thick InP buffer layer was additionally deposited for the ternary. V/III-ratios of 5.6 and 2.3 were used for InP and GaInAs, respectively. When InP was switched to GaInAs, growth was interrupted for 30 s under phosphorous stabilization. The switching sequence for GaInAs to InP was as follows: growth interruption for 30 s under arsenic stabilization and an additional 30 s under phosphorous stabilization.

The layers were characterized *in situ* by RHEED for growth rate and structural investigations and *ex situ* electrically by Hall effect measurements at 300 and 77 K, optically by photoluminescence spectroscopy (PL) at 2 K, structurally by X-ray diffraction (XRD) using the (400) reflection and morphologically by Nomarski interference microscopy.

3. Results

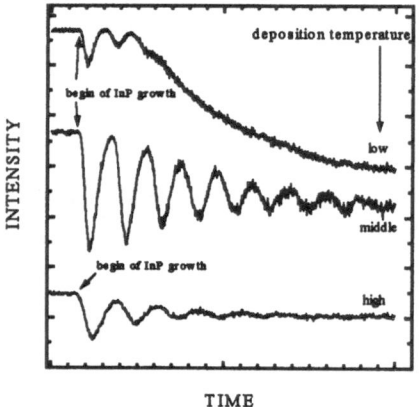

Figure 1. RHEED intensity oscillations as a criterion for determination of good starting conditions for the epitaxy of InP

3.1. InP and GaInAs growth

First, the influence of deposition temperature on InP growth was studied with RHEED. Fig. 1 shows an example of RHEED intensity oscillations of the specular beam for a (100) InP surface along the [001] azimuth and for different temperatures at a fixed III/V-ratio. The oscillations are strongly dampened at the extreme deposition temperatures, which can be tracked to a three dimensional growth mode. The RHEED oscillations in the middle temperature range, however, indicate a two dimensional layer by layer growth [1], which is a prerequisite for obtaining optimal material properties with respect to device application. The influence of deposition temperature on optical properties was also studied. As can be seen in figure 2, extremely intensive excitonic PL at 2 K, also indicative of good crystal quality, is observed in the same temperature range between 460 and 480°C. In this range layers also had excellent surface morphology. Obviously, RHEED is a powerful time saving method to establish a set of good starting conditions for the epitaxy of binary materials such as InP, which makes the use of *ex situ* characterization superfluous. RHEED oscillations were also used to obtain lattice matched GaInAs. It was possible to achieve lattice matching by this method within 1000 ppm.

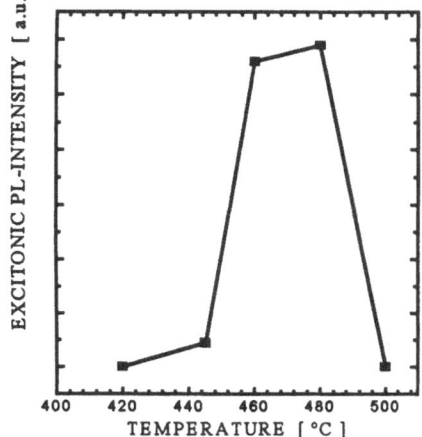

Figure 2. Excitonic PL intensity as a function of deposition temperature

In contrast to InP, the optimal GaInAs growth temperature was judged by *ex situ* characterization methods, since RHEED intensity oscillations are much weaker for ternary compounds. Best morphology was observed for layers grown at 480 and 500°C. In this temperature range we also found very small FWHM (Full Width at Half Maximum) of only a little over 20 arcseconds for the layer peaks in HRXRD. This result indicates, that our material is of excellent structural quality and of homogeneous composition- Figure 3 also shows the dependence of FWHM of the bound exciton transition

peak as a function of deposition temperature. Here we found the smallest FWHMs for layers deposited between 480 and 520°C. Hall measurements showed that all layers were n-type. The carrier concentration decreases from 3.2 *10^{16}cm^{-3} to 5.3 *10^{15}cm^{-3} as the growth temperature increases from 460 to 540°C. This indicates that incomplete decomposition of the metalorganic sources and therefore carbon must be responsible for the background carrier concentration. Room temperature electron mobilities of around 9,000 cm^2/Vs for growth temperatures from 480 - 520°C document the high quality and low

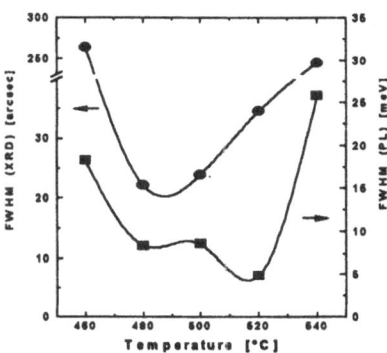

Figure 3. FWHM of 2 K PL bound exciton peaks and of the GaInAs (004) XRD reflection at various deposition temperatures

compensation ratio of the material for the respective carrier concentration. The best liquid N$_2$ mobilities of 40,000 cm^2/Vs were achieved for layers deposited at 480 and 500°C. These mobilties indicate, that heterostructures with good interface quality should be deposited in this temperature range.

48

3.2 heterostructural growth

Abrupt interfaces, as needed for HEMT-type structures, can best be obtained at relatively low deposition temperatures, due to diffusion and exchange of atoms at the heterointerface. With the additional prerequisite of excellent bulk InP and GaInAs material quality, 480°C was chosen for heterostructural deposition. At this temperature a tenfold superlattice consisting of 10 nm thick $Ga_{0.47}In_{0.53}As$ and 40 nm InP layers was deposited and evaluated by XRD [2]. The aim of this investigation is not only to see whether the deposition temperature is suitable, but also whether the growth interruptions at the interfaces are sufficient and whether GaInAs and InP layers can be stacked reproducibly. Figure 4 shows the X-ray pattern measured in comparison to the simulated rocking curve. Best agreement

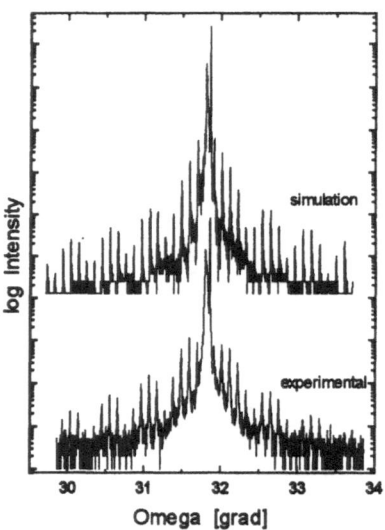

Figure 4. Simulated and measured HRXRD of a GaInAs/InP superlattice deposited at 480°C

between simulation and measurement was obtained, when the intended layer thicknesses were used for the calculation and when a thin layer of only one monolayer was inserted between the layers in which carry-over of phosphorous in GaInAs and of arsenic in InP was assumed. All in all the growth conditions and growth interruptions employed lead to very sharp heterointerfaces.

4. Summary and Conclusions

The growth temperature for InP and GaInAs was optimized. 480°C was determined as best compromise for heterostructural growth with respect to HEMT application. At this growth temperature interface abruptness was studied. The transition from one material to the other and vice versa takes place within one monolayer.

5. References

1. J.H. Neave, B.A. Joyce, P.J. Dobson, N. Norton (1983) *Appl. Phys.* **A31**, 1

2. S. Takagi (1969) *J. Phys. Soc. Japan* **26**, 1239 - 1253.

INVESTIGATION OF THE EFFECT OF GaAs BUFFER LAYERS GROWN BY MBE AT DIFFERENT TEMPERATURES ON THE PERFORMANCE OF GaAs MESFET's

M. LAGADAS, M. ANDROULIDAKI, G. CONSTANTINIDIS, N. KORNILIOS and Z. HATZOPOULOS
Foundation for Research and Technology-Hellas, Institute of Electronic Structure & Laser, P.O. Box 1527, 711 10 Heraklion Crete, Greece

ABSTRACT

We have investigated the degradation of the mobility in MBE nGaAs epilayers, grown on top of GaAs buffer layers. The growth temperature of the buffer layer varied between 200°C and 580°C. For T_{gr} lower than 500°C a decrease in the mobility and the concentration of the free carriers in nGaAs epilayers have been observed. PL measurements reveal the presence of V_{Ga} in epilayers grown on top of Low Temperature (LT) buffer (T_{gr} =200°C-300°C for the LT growth). Electrochemical C-V profiling have shown a decrease in the channel depth . The degradation of the electrical properties of epilayers reduces the extrinsic transconductance g_m of MESFET's. In order to improve the performance of these devices we use intermediate layers (GaAs and GaAs/Al$_{0.5}$ Ga$_{0.5}$ As superlattices grown at 580°C-600°C) between the LT buffer (grown at 250°C) and the nGaAs active layers. The structures with superlattices gives higher g_m than the devices with GaAs intermediate layers.

INTRODUCTION

Much attention has been given lately in the use of LTGaAs as buffer layers in FET's devices[1]. The high resistivity[2] of low temperature grown layers provide elimination of parasitic effects on the device performance. However, point defects[3] and precipitates outdiffusion[4] reduce the extrinsic transconductance g_m .

In this work we investigate the effect of the growth temperature of the buffer layer on the electrical properties of the nGaAs epilayers grown by MBE and the use of GaAs and AlGaAs/GaAs superlattices layers in order to compress defect outdiffusion.

SAMPLE GROWTH AND CHARACTERIZATION

All samples were In bonded and grown in a MBE system, using As$_4$ beam. (001) LEC S.I. GaAs substrates were degreased in organic solvents and etched in a 60°C-70°C

J. Novák and A. Schlachetzki (eds.), Heterostructure Epitaxy and Devices, 49–52.
© 1996 Kluwer Academic Publishers.

solution of 5:1:1 (H_2SO_4:H_2O_2:H_2O) for 1 min. After the oxide desorption 0.25µm GaAs was grown at 580°C in order to minimize possible surface asperities and contaminations due to substrate preparation. The growth was then interrupted, the substrate temperature was ramped down to the desired Low Temperature and the growth of the LTGaAs was commenced with 1µm/hr growth rate and V/III beam equivalent pressure (BEP) in the range of 10 to 30.

In order to investigate the effect of the LTGaAs buffer layer on the electronic quality of the doped active layer, a 0.3µm (0.5µm) nGaAs layer was grown on the top of the LT buffer at T_{gr}=580°C, BEP=15 and n=2x10^{17}cm^{-3} (1.5x10^{18}cm^{-3}). Also FET structures [70nm n$^+$GaAs(1.5x10^{18}cm^{-3}) / 180nm nGaAs(4x10^{17}cm^{-3})], with GaAs and AlGaAs/GaAs intermediate layers between the 0.7µm LT buffer (BEP=30 during LT growth) and the active layer, were examined. AlGaAs layers were grown at 600°C and GaAs at 580°C. The structures of these intermediate layers are shown in the Table.

Electrolytic C-V profiling and Hall mobility at 300K with In contacts annealed at 400°C for 2mins were measured. PL measurements were performed at 18K with a power density 0.5W/cm^2. Ohmic contacts on MESFET devices fabricated with Au/Ge/Ni annealed at 410 °C for 20 secs and gate contacts with Ti/Pt/Au.

RESULTS AND DISCUSSION

Figure 1 illustrates the Hall mobility and the Electrolytic C-V profiling dependence of the nGaAs (2x10^{17}cm^{-3}) epilayers on the growth temperature of the

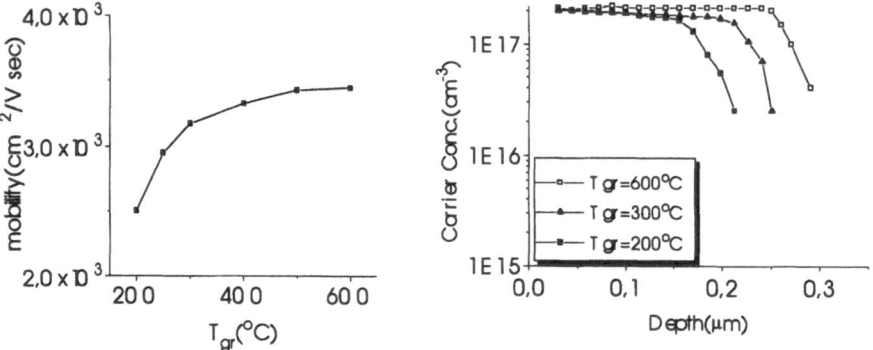

Fig. 1. Hall mobility and C-V profilign of 0.3µm nGaAs (n=2x10^{17}cm^{-3}) epilayers as a function of the growth temperature of the buffer.

buffer layer. For T_{gr} lower than 500°C there is a decrease in the mobility which becomes more pronounced for T_{gr}<300°C. The doping profile also depends strongly on growth temperature. As T_{gr} decreases we observed a decrease in the channel depth and a sloped profile on the carrier concentration. The effect is stronger for lower T_{gr}. Previous investigations on nGaAs/LTGaAs structures by TEM [4] have revealed As-precipitates outdiffusion to a depth of 200nm into the epilayer, after annealing during the growth at 580°C for 1.5hrs. Due to their mettalic nature, As precipitates forms Schottky contacts with the surrounding GaAs crystal and as a result they trap free carriers. We attribute the channel depth shorting to the presence of As-precipitates in the nGaAs epilayer whose density and diameter are higher in the region close to the doped layer-LTGaAs interface. Annealing at 580°C for a maximum of 4 hrs, eliminates the sloped doping profiles and decreases the carrier concentration to ~1.3x10^{17}cm^{-3}. The channel depth,as determined from C-V profiling, decreases from ~0.19μm (after 20 mins annealing) to ~0.15μm (after 4 hrs annealing).

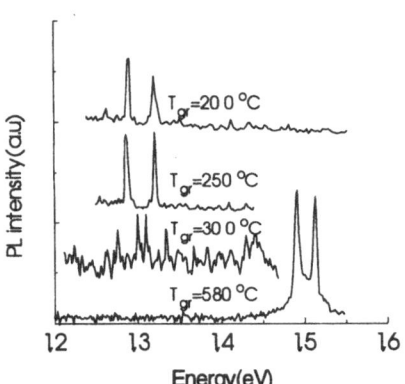

Fig.2 PL spectra from 0.5μm n⁺GaAs with buffer layers grown at different T_{gr}

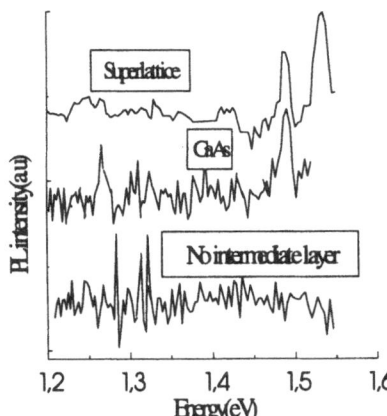

Fig.3 PL from FET structures.

PL signals (fig.2) from 0.5μm n⁺GaAs with the buffer layer grown at T_{gr}=580°C includes only two peaks due to (e, A⁰) transition (1.495eV) and (D⁰X/D⁺h) transition (1.514eV). Decreasing the growth temperature of the buffer layers the PL spectrum becomes noisy with small peaks in 1.45eV-1.44eV and 1.275eV-1.335eV regions. P.W. Yu et. al. [5] have observed several peaks in the PL spectrum of LTGaAs in the range of 1.42eV-1.47eV which have been attributed to the presence of As$_i$ and V_{Ga} point defects. In accordance to their observations these peaks dissapears for LT growth lower than 300°C. The peaks at 1.31eV-1.32eV are caused by ionised states of gallium vacancies [6]. At T_{gr}=200°C and 250°C there are two strong peaks at 1.32eV and 1.29eV. The second one is caused by electronic transitions from the conduction band to acceptor states located ~0.23eV above the valence band. By DLTS measurements an acceptor state at 0.25eV[7] has been found, very close to our measurements. The higher peak intensities, for the LT layers grown at T_{gr} lower than 300°C, are attributed to the higher concentration of these acceptor states in the epilayer.

Due to the decreased mobility in the channel, the FET structures with LT layers have lower g_m than conventional samples with GaAs buffer grown at 580°C. The use of intermediate layers results in the recovery of the mobility and the g_m values. Althouth the channel depth and carrier concentration, as determined by C-V profiling, for structures with inermediate layer is the same as the conventional ones, the mobility is lower in FET with 0.1μm GaAs between LT buffer and nGaAs. This can be explained by the PL signal (Fig.3) which reveals the presence of small peaks at

1.28eV and 1.32eV,due to the remaining outdiffusion of point defects. On the other hand the spectrum from the sample with the 10nm AlGaAs superlattice layer reveals only peaks due to nGaAs layer and quantum well structure (peak at 1.52eV). The absence of point defects in that structure has been attributed to the strong Al-As bond, which hinders V_{Ga} and excess As outdiffusion, as already have been observed[8,9].

Table: Intermediate layer, Hall mobility and g_m for FET structures

Samples	Intermediate layer	μ $(cm^2/V\ sec)$	g_m (mS/mm)
Conventional	-	2730-2760	180-190
L.T. buffered	-	2400-2420	140-160
L.T. buffered	0.1 μm GaAs	2650	165-175
with	0.6 μm GaAs	2760	175-185
additional	0.4μm GaAs/(30nm Al$_0$ $_5$Ga$_0$ $_5$As/10nm GaAs)x5	2750	200-220
layers	20nm GaAs/(10nm Al$_0$ $_5$Ga$_0$ $_5$As/10nm GaAs) x5	2700	-

CONCLUSION

Mobility, carrier concentration and channel depth of nGaAs/LTGaAs structures deteriorate as the growth temperature of LT layer decreases. PL measurements reveal peaks in the region 1.27eV-1.44eV due to the presence of point defects in the doped epilayer. The lower the T_{gr} of the buffer layers, the higher density of the point defects.The degradation of the electrical properties is also caused by the As precipitates outdiffusion.The use of 0.1μm intermediate layers, especially AlGaAs/GaAs superlattice, acts as a barrier in the point defects diffusion in nGaAs epilayer and results in an increase in the mobility and the transconductance of the device.

REFERENCES

1. Smith,F.W., Calawa,A.R., Chen,C.L., Manfra,M.J. and Mahoney,L.J. (1988) New MBE Buffer Used To Eliminate Backgating in GaAs MESFET's, IEEE Electr. Dev. Lett. **9**, 77-80.
2. Chu,T.Y., Dodabalapur,A., Srinivasan,A., Neikirk,D.P. and Streetman,B.G. (1991) Properties and applications of Al$_x$Ga$_{1-x}$As(0≤x≤1) grown at low temperatures, J. Cryst. Growth **111**, 26-29.
3. Streit,D.C., Hoppe,M.M., Chen,C.H., Liu,J.K. and Yen,K.H (1992) Are low temperature buffer MESFET's good for microwave applications?, J.Vac.Sci.Technol. **B10**, 819-821.
4. Ginoudi,A., Paloura,E.C., Theys,B., Chevallier,J., Kalomiros,I., Lagadas,M. and Hatzopoulos,Z. (1995) Hydrogen-Indused Passivation of Deep Traps in n-GaAs:Si Grown on LTGaAs,Mat. Res. Soc. **378**, in press.
5. Yu,P.W., Reynolds,P.C. and Stutz,C.E. (1992) Sharp-line photoluminescence of GaAs grown by low-temperature molecular beam epitaxy, Appl. Phys. Lett. **61** , 1434-1437 .
6. L. Pavessi and M. Gussi, (1994) Photoluminescence of Al$_x$Ga$_{1-x}$As alloys, J. Appl. Phys. **75** , 4779-4842.
7. Darmo,J., Dubecky,F., Kordos,P., Forster,A. and Luth,H. (1993) Deep Traps in LT-GaAs, Mater. Sci. Engin. B28, 393-396.
8. Ohbu,I., Takahama,M. and Imamura,Y. (1992) Diffusion of Gallium Vacancies from Low-Temperature-Grown GaAs, Jpn. J. Appl. Phys. **31**, L1647-L1649.
9. Melloch,M.R., Otsuka,N., Mahalingam,K., Warren,A.C.,Woodall,J.M. and Kirchner,P.D. (1992) Incorporation of excess As in GaAs and AlGaAs epilayers grown at low temperatures by MBE, Mat. Res. Soc. **241**, 113-124.

MULTILAYERED GaAs VPE STRUCTURES FOR MICRO MACHINING

K. SOMOGYI, SZ. VARGA, CH. GRATTEPAIN*, and L. DOBOS
Research Institute for Technical Physics of the Hungarian Academy of
Sciences, H-1325 Budapest, P.O.B.76., Hungary
*Laboratoire de Physique des Solides of CNRS, Meudon-Bellevue, 92125
Meudon Cedex, 1. Pl. A. Briand, France

1. Introduction

Recently the development of different micro-mechanical devices based on semiconductors (micro-sensors, -actuators, -vacuum devices) is widely studied. Since these
devices require microscopically micro structured construction, one can not usually apply
mechanical methods of the preparation. Mainly chemical methods are used exploiting
the properties of selective etching. The reasons of the differences in the etching rates
can be differences in crystallographic orientations, doping levels, etc. Such wet etching
techniques are well established and widely used in the case of the silicon [1, 2].

The physical properties of the III-V semiconductors, including also GaAs are much
more appropriate in many respects for such purposes, e. g. piezoelectric properties,
higher working temperatures, wider range of electric properties, etc. Though some
advantages are evident, much less work was devoted to the research and development of
micro-machined structures in the practice of compound semiconductors. Only a few
works report on the application of III-Vs and GaAs [3].

In these cases the selectivity is based on the composition differences. Binary and
ternary compound layers have to be grown in one structure and an etchant has to be
used, which exhibits different etch rates relative to the given binary and ternary
compounds [3, 4]. However the presence of the ternary compounds usually requires
expensive, sophisticated (or dangerous) epitaxial growth techniques (MBE, MOCVD).

In this work results are presented using much simpler epitaxial technique and a very
highly selective electrochemical method to obtain thin GaAs sheets, lamellae, lamellar
structures and needles. The aim is to demonstrate the feasibility of such structures,
which can serve further for micro-machined devices of several kinds.

2. Experimental Background

If using solely GaAs, high selectivity can be achieved by wet chemical method applying
electrochemical method. As it was shown earlier [5], an electrochemical etching can be
performed due to holes. For this n-GaAs is illuminated by visible light of a wavelength

J. Novák and A. Schlachetzki (eds.), Heterostructure Epitaxy and Devices, 53–56.
© 1996 *Kluwer Academic Publishers.*

54

corresponding to the band gap in order to generate holes. In p-type material holes are given directly and no illumination is necessary. Therefore having n- and p-type regions together in the semiconductor, then only p-type regions will be etched off applying this electrochemical attack under dark conditions. The selectivity ratio, by principle could be as high as the ratio of the hole concentrations in the p- and n-type materials (e. g. 10^{10}:1). In practice it is much less, of course, but it will be much higher, than e. g. 10:1, typical for the selective chemical etching of ternary to binary compound combinations [4]. Such electrochemical method based on 0.1 M solution of tiron was chosen in this work [5].

For the development of this versatile and flexible method multilayered VPE structures have been prepared first with subsequent growth of p- and n-type layers. Variable doping has to be solved first. In these experiments classical $AsCl_3$/Ga/H_2 hot wall setup was used [6]. The n-type doping was performed from gas phase using SF_6 diluted in He. For the p-type doping GaAs:Zn was used as auxiliary solid source [7]. Only the S doping level was variable, while the Zn doping remained constant during a growth process. S doping was switched on and off in a "yes-no" way to obtain n- ("yes") and p-type ("no") layers.

These wafers were then cut into parallelepiped bricks. On one end ohmic contacts were made to each layer in order to avoid influence of the electric properties of the p-n junctions on the lateral current flow. The opposite end was immersed into the tiron solution containing a graphite electrode besides the saturated calomel electrode. The removal of the p-type layers from in-between the n-type ones was performed applying appropriate current between the graphite electrode and the ohmic contacts. The etch time depends on the given conditions and varied between 0.5 and 10 hours for different cases.

3. Results and Discussion

Figures 1 and 2 show scanning electron microscopic (SEM) pictures of the cut end of a structure, where p-type layers were etched out. Views are of different magnifications. Both types of layers have a thickness of 3 μm. The lamellar structure contains n-type layer sheets with a length and a width of several mm. The latter parameters can be varied in a wide range. Similar SEM views are shown in figures 3 and 4, but the thickness of the layers was about 0.27 μm (and the number of the layers was 3 and 4). Notches on the edges of the n-type lamellae are caused probably by the conditions, which can be improved.

The figures demonstrate not only the feasibility of such thin GaAs sheets and the perfect selectivity of the etching, but also one more advantage of the method. There are no problems with the removal of the GaAs and the electrolyte from the narrow grooves. Proper electric contacts are necessary to the subsequent layers and no additional tricks are needed, like e. g. mixing of the liquid, bending of the layers, etc.

Figure 5 shows SEM picture of a tip made from an n-type and 3 μm thick lamella. Figure 6 shows the point of the tip closer having a radius of curvature smaller than 15 nm. This curvature is quite enough for cold electron emission applied in micro-vacuum devices.

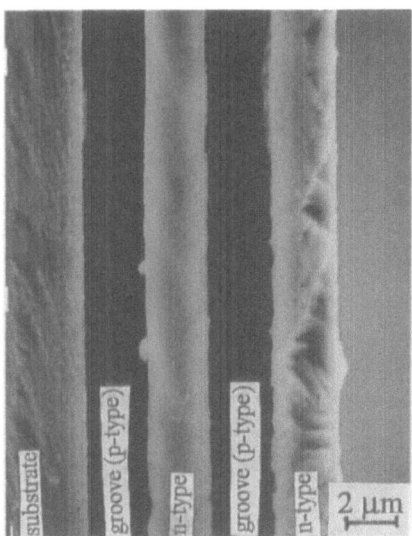

Figure 1. SEM picture of the etched out edge of a GaAs p-n multilayered structure. Black stripes are the grooves in the place of the p-layers

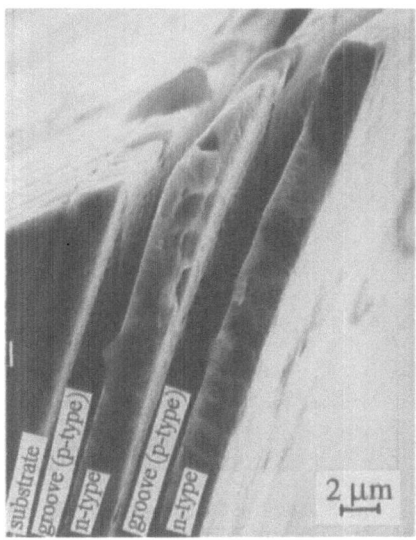

Figure 2. SEM picture of a corner of the etched edge of the former structure. The thickness of the n- and p-layers is about 3 μm

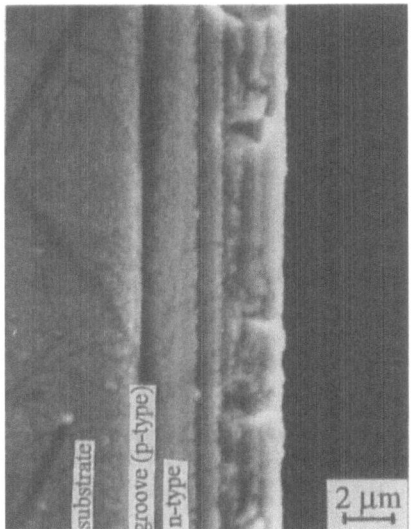

Figure 3. SEM picture of an etched structure as in figs. 1 and 2, but with 3 p-type layer and with thicknesses of 0.27 μm.

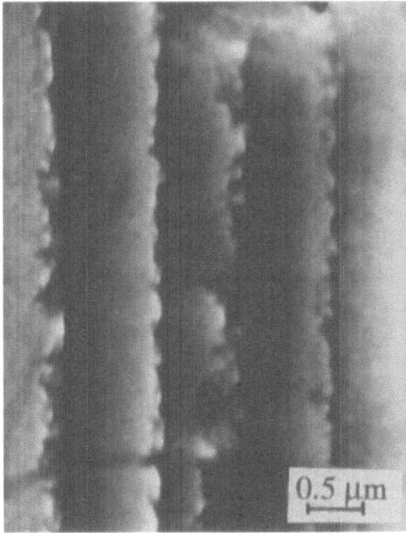

Figure 4. SEM picture of the edge of the same sample as in fig. 3, but with higher resolution

56

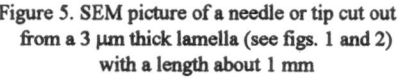

Figure 5. SEM picture of a needle or tip cut out from a 3 μm thick lamella (see figs. 1 and 2) with a length about 1 mm

Figure 6. SEM picture of the point of the tip shown on fig. 5. The radius of the curvature of the point is less than 15 nm

4. Conclusions

It was demonstrated that GaAs multilayered structures can be applied for preparation of selectively etched systems used in different micro-machined devices. Also removal and implantation onto another substrates/holders of chips and layers can be solved in this way for e.g. integrated (opto-)electronic purposes, etc. All this is based on the possibility of codoping of Zn and S, because no interference between the dopants was observed in the VPE growth process.

Results are preliminary, but promising for the development of micro-machined GaAs devices.

5. Acknowledgements

The authors are indebted to Gy. Kiss, E. Jakocska and K. Szedlacsek for their assistance. This work was supported a part by the Hungarian National Research Foundation (OTKA, grants No. T4178, E12012, and T15619).

6. References

1. Seidel, H.G. and Voss, R. (1991) in *Micro Systems Technologies*, VDE Verlag, Berlin, 291.
2. Hjort, K., Schweitz, J.A., Andersson, S., Kordina, O., and Janzén, E. (1992) in *Proc. Conf. Micro Electro Mechanical Systems*, Travemünde, Germany, 83.
3. Yamaguchi, K., Okamoto, K., and Yugo, S. (1995) *J. Applied Physics* 77, 6061.
4. Novák, J. (1995) HEAD `95, (invited) same book
5. Ambridge, T. and Faktor, T. (1975) *J. Appl. Electrochem.* 5, 319.
6. Görög, T., Gyúró, I., and Somogyi, K. (1985) *Acta Physica Hungarica* 57, 223.
7. Somogyi, K., Varga, Sz., Grattepain, Ch., and Dobos, L. (1995) *Fizika A* 4, in press.

GROWTH OF InP AND GaInAsP LAYERS BY LIQUID–PHASE EPITAXY USING HOLMIUM GETTERING AND DOPING

O. PROCHÁZKOVÁ, F. ŠROBÁR, J. NOVOTNÝ, and J. ZAVADIL
Institute of Radio Engineering and Electronics, Czech. Acad. Sci.
Chaberská 57, CZ–182 51 Praha 8, Czech Republic

1. Introduction

Epitaxial layers and heterostructures of the $A^{III}B^{V}$ semiconductor compounds form the basis of the most of the contemporary micro– and optoelectronic devices (radiation sources, detectors, modulators etc.). A special area of progress in this field that lies on the sidelines of the mainstream development but which nevertheless holds a potential for considerable advances in some specific branches of micro– and optoelectronics is the study of the influence of rare earth (*RE*) elements on the physical and chemical properties of the $A^{III}B^{V}$ semiconductor materials.

In this paper we report results of the study of liquid–phase epitaxial (*LPE*) growth of InP and GaInAsP layers on (100)–oriented substrates from melts containing, in addition to the constitutive elements, also Ho admixture. The observed data indicate that Ho atoms operate as efficient gettering agents. No hints of Ho incorporation were present in the experimental results.

2. Behaviour of *RE* Elements in $A^{III}B^{V}$ Crystals

RE elements can be used in the role of gettering agents and dopants. Various of elements (e.g. S, Si, C, Te) which act as donors in $A^{III}B^{V}$ semiconductors, form stable compounds with *RE* atoms, insoluble in growth solution. The *RE* elements can therefore be used to remove these impurities from the melt and to reduce the background electron concentration of the grown layers. In this way, high–purity InP layers with electron concentration inferior to $7 \times 10^{13} \mathrm{cm}^{-3}$ at 300K were prepared. Such layers possess high saturation values of the drift velocity, a property valuable for GHz microelectronics. Similarly, GaInAsP layers with electron concentrations one to two orders of magnitude lower than in the case of conventionally prepared undoped layers were obtained employing this approach.

Another aspect is the incorporation of *RE* ions into the crystal lattice of the $A^{III}B^{V}$ compounds which may result in the rise of radiative efficiency of light emitting devices. *RE* atoms introduce electron transitions with good radiative yield. A common outer electronic configuration of the *RE* atoms is $5s^2 5p^6 6s^2$. Their inner 4f shell is being filled as we progress from lanthanum ($4f^0$) to lutetium ($4f^{14}$). *RE* atoms usually

57

J. Novák and A. Schlachetzki (eds.), Heterostructure Epitaxy and Devices, 57–60.

58

form trivalent cations whereby two of the 6s electrons and one of the 4f electrons are given off while the outer 5s and 5p shells are left intact. The remaining 4f electrons are therefore to a considerable degree (but not completely) shielded from external influences. The crystal field of the host lattice, for instance, can lift degeneracy of these levels.

Some intra–centre transitions between 4f levels of such RE ions are of interest for optical pumping and stimulated emission. Although 4f–4f transitions are forbidden by the Laporte's rule, if the RE ion is located in a non–centrosymmetric site, odd–order terms in the expansion of the static (or dynamic) crystal field admix states of higher, opposite–parity configurations, such as $4f^{n-1}5d$, into $4f^n$, and the transitions become allowed. Trivalent lanthanide ions have been used extensively as ingredient of the active medium in the optically pumped solid–state lasers because they possess suitable absorption bands and numerous fluorescent lines of high quantum efficiency in the visible and near–infrared region. The second most extensively exploited RE ion has been holmium. Stimulated emission has been observed in various crystalline hosts for nine different transitions ranging from 0.55 to 2.9μm [1]. The most common laser transition involving Ho^{3+} is $^5I_7 \rightarrow {}^5I_8$.

The first results relative to observation of the 4f–4f emission from III–V semiconductors were gained on InP, GaP, and GaAs layers doped with Yb and Er . Tsang and Logan [2] reported the preparation of injection laser with Er–doped active layer. Single–mode lasing occurred, the authors claimed, via the 4f–4f transitions. Recent results [3] suggest that excitation transfer involving RE ions may be accompanied by nonradiative transitions which tend to decrease the overall luminescence efficiency. A possible way for suppressing these de–excitation processes has been indicated. Apparently, these questions require further detailed study, including the extension of the set of RE elements under investigation.

3. Experimental

For the preparation of InP and GaInAsP semiconductor layers the LPE growth method has been chosen for its flexibility and the near–equilibrium operating conditions. InP and GaInAsP epitaxial layers were grown on (100)–oriented InP substrates from the melt containing Ho. Composition of the growth melts was fixed, the Ho concentrations were varied in the range from 0 to 0.4 wt per cent.

The epitaxy was carried out in a conventional graphite sliding boat under a hydrogen ambient by the supercooling process. The chemical activity of Ho addition with respect to hydrogen and oxygen was reduced by embedding the Ho pieces in the In melt. Ho being denser than In, the pieces remained beneath the melt surface and were screened from the ambient at the stage prior to dissolution. The growth was commenced at 640°C by supercooling the melt by 10°C, with a cooling rate of 0.7°C/min. The thickness of the InP and GaInAsP epitaxial layers has been dependent on the growth period and was typically 2–3μm. The details of the LPE process were given elsewhere [4]. This growth technique produced epitaxial layers with sufficiently good surface morphology. Metallographic quality, thickness and overall morphology of the layers have been monitored employing the scanning electron microscopy (SEM).

Chemical composition including impurity concentrations has been determined via Rutherford backscattering spectrometry (*RBS*), proton–induced X–ray emission (*PIXE*), and electron probe X–ray microanalysis (*EPMA*).

The results obtained with the help of these methods have been compared with those produced by measurement of the free–carrier concentration and of the photoluminescence (*PL*) spectra taken at various temperatures. Free–carrier concentration was monitored as a function of Ho content by the C–V method using mercury probe. The *PL* spectra were measured by exciting the wafers with the 632.8–nm line of the He–Ne laser. The beam intensity was about $1.5 \, \text{W/cm}^2$.

4. Results and Discussion

Two sets of InP and GaInAsP ($\lambda_g \approx 1.3 \mu m$) samples grown from the melt containing Ho in concentrations ranging from 0 to 0.4 wt% were investigated. The layers showed the desired smooth surface when chemical activity of Ho was reduced in the way indicated above. The density of microparticles (consisting perhaps of Ho oxides, hydrides, arsenides, and phosphides) was low. The interface between epilayer and substrate was always flat and free of inclusions. The typical thickness of InP and GaInAsP epilayers was $2-3 \mu m$. The composition of GaInAsP layers varied slightly with the Ho content. It was suggested that formation of insoluble Ho compounds, probably HoAs and HoP, could change the composition of solid phase. Presence of Ho atoms in the crystal lattice of InP and GaInAsP was not confirmed by the *RBE* or *PIXE* method. Free–electron concentrations were determined as a function of the Ho amount and correlated with *PL* spectra. InP and GaInAsP layers prepared from the melts both with and without Ho were n–type. The background concentration decreased with the rising Ho content from $1.64 \times 10^{17} \text{cm}^{-3}$ to $1.60 \times 10^{16} \text{cm}^{-3}$ at room temperature [5]. Table 1 summarises some properties of GaInAsP layers in dependence on the Ho content in the growth melt.

TABLE 1. Summary of the parameters of $Ga_x In_{1-x} As_y P_{1-y}$:Ho layers

Sample	Ho content	Solid composition		Wavelength	n	Conduction
	(wt %)	x	y	λ_g (nm)	10^{17}cm^{-3}	type
1	0.00	0.268	0.566	1256	1.68	n
2	0.12	0.268	0.560	1254	1.64	n
3	0.21	0.265	0.557	1250	1.17	n
4	0.30	0.260	0.550	1247	1.06	n
5	0.35	0.260	0.542	1240	0.16	n

In addition, low temperature *PL* spectra were measured for a set of GaInAsP samples grown from melts with Ho content values of 0.0, 0.1, and 0.2 wt%, respectively. At room temperature all sampes exhibit only one emission band due to free electron–to–free hole (band–to–band) recombination. *PL* half–width was decreased from 57 meV to 45 meV and the *PL* peak was slightly shifted from 1254 to 1251 nm with increasing the Ho amount. It is suggested that wavelength of the *PL* peak shifts due to gettering of the As from the melt. Similarly, the half–width of the *PL* curve will

60

decrease due to donor gettering during the epitaxial growth, manifesting the decrease of free–carrier concentration. The photoluminescence spectra of two samples taken at 300 and 10 K are shown in Fig.1. In this figure the 300–K *PL* peak intensity is two orders of magnitude weaker than that measured at 10 K and all curves are normalized to the same peak intensity. The gettering effect described above at room temperature is more pronounced at low temperature. The half–width of band–to–band transition line (peak A) is reduced from 22 to 8 meV and also the bound–state related luminescence (peak B) is resolved on Ho–doped samples. Thus both the Ho gettering of the As and the background donors from the melt takes place. We conclude therefore that Ho atoms can act as efficient gettering centres for shallow donors present in the growth solution. The second manifestation of the *RE* ions' presence mentioned in section 2 – radiation originating from the 4f–4f transitions – did not materialize in the present study.

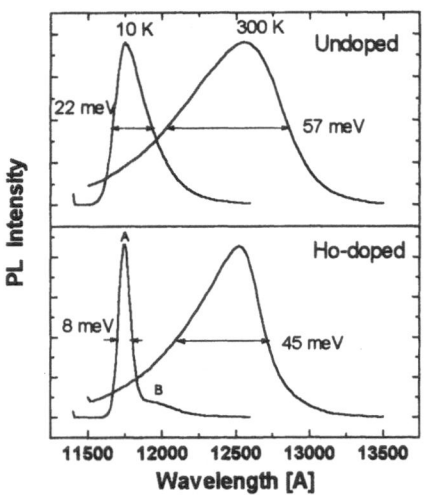

Fig.1 The 300 and 10 K photoluminescence spectra of undoped and Ho–doped (0.2 wt%) InGaAsP samples. All the *PL* peak intensities are normalized to the same value.

This work has been supported by the Grant Agency of the Czech Republic, project No. 102/93/0642.

5. References

1. Gschneider Jr., K.A. and Eyring, Le Roy (eds.) (1978) *Handbook on the Physics and Chemistry of Rare Earths*, North Holland, Amsterodam, New York, Oxford.
2. Tsang, W.T., and Logan, R.A. (1986) Observation of enhanced single longitudinal mode operation in 1.5μm GaInAsP erbium–doped semiconductor injection lasers, *Appl. Phys. Lett.* **49**, 1686–1688.
3. Takahei, K. and Tagushi, A. (1995) Photoluminescence–excitation analysis of Er–doped GaAs grown by metalorganic vapor phase deposition, *J. Appl. Phys.* **77**, 1735–1740.
4. Novotný, J. and Procházková, O. (1992) LPE growth of GaInAsP/InP heterostructures for 1.3μm planar buried mesa lasers, *Cryst. Res. Technol.* **27**, 481–489.
5. Procházková, O., Šrobár, F., Zavadil, J., and Žďánský, K. (1995) Characterisation of GaInAsP layers prepared by LPE using holmium doping, *Photonics 95*, EOS Annual Meeting Digest Series: **2B**, 405– 408.

MEANDER TYPE LPE AND HIGH TEMPERATURE STABILITY OF ELASTICALLY STRAINED GaInAsP/InP LAYERS

D. NOHAVICA*, K.P. HOMEWOOD, W.P. GILLIN,
M.A. LOURENCO, Z YANG, J. OSWALD** and
D.EHRENTRAUT***
*Institute of Radio Engineering and Electronics, Czech Academy of
Sciences,
Chaberska 57, 182 51 Prague 8, Czech Republic
Department of Electronic and Electrical Engineering, University
of Surrey,
Guilford, Surrey GU2 5XH United Kingdom
**Institute of Physics, Czech Academy of Sciences,
Cukrovarnicka 10, 162 53 Prague 6, Czech Republic
***Institut für Kristallzüchtung
Rudower Chaussee 6, D-12489 Berlin, Germany

The meander type technique of liquid phase epitaxy (LPE) is used for preparation of quaternary solid solution in the GaInAsP/InP material system. Surface morphology of the layers prepared by meander type LPE has been compared with ones prepared by conventional LPE. Quaternary strained layers with composition near to $Ga_{.21}In_{.79}As_{.75}P_{.25}$ were grown with perpendicular lattice mismatch up to 1.24 % in compression. High temperature processing the samples was investigated in 7 annealing steps. The results indicate high temperature stability of the prepared structures.

1. Introduction

Layer growth from the narrow slot of the melt onto the moving substrate is first of all very suitable to shorten the growth time [1]. However, additional flexibility is provided by the melt slot shaping [2] when the system of slots filled with the melt from the only melt bin is created. So, it is possible to grow a lot of thin layers from one initial melt.
This paper describes a development of this growth method to prepare strained layers in the GaInAsP/InP material system.

2. Growth system

Basic types of meanders under which the growth substrate is being moved during the growth have been shown in [2]. Variant labelled DM LPE is suitable in the case when we match two meanders containing the melts of different composition to create periodic

61

J. Novák and A. Schlachetzki (eds.), Heterostructure Epitaxy and Devices, 61–64.
© 1996 Kluwer Academic Publishers.

structure. The meander filled with one melt (SM LPE) can be used for growth of thicker layers composed from many individual segments. So, SM LPE enables the preparation of thin layers with properties different from those obtained by diffusion limited growth in traditional LPE technology. The general arrangement of the growth cassette offers the possibility of a combination of growth onto the moving substrate with „thick" layer growth by traditional LPE technology [3].

Two sets of InP samples have been prepared for the comparative study. The first set has been prepared by conventional LPE in supercooling regime with thicknesses 1 - 1.5 μm and the second one by SM LPE composed from 31 segments with growth time of each individual segment 50 ms. The total thicknesses were the same in both cases. SM LPE layers have been grown in well defined graphite boat system (see Figure 1), where substrate of InP, after each movement (Figure 1,a) was hold in position with phosphorus protective atmosphere (Figure 1,b,c,). Phosphorus overpressure has been realised by InP-Sn melt situated above the grown layer. The pause duration when the grown layer was in the protective atmosphere was changing from 1 to 3 sec.

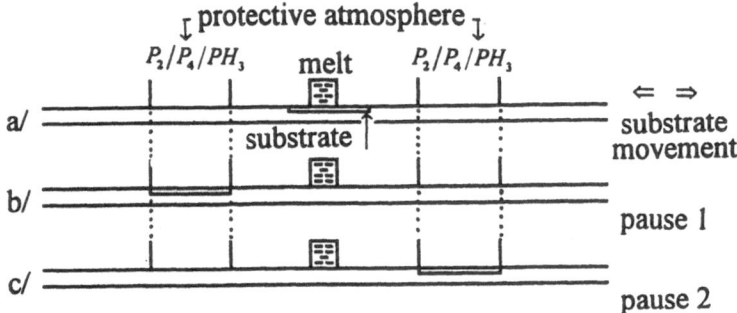

Figure 1 . Grown layer protection in the phosphorus atmosphere

Surface morphology of the layers grown by conventional LPE was microscopically smooth with very weak surface waves with wavelength 10 - 40 μm. Theoretical prediction of the minimal wavelength (λ_o), according to the Nishinaga's model [4] gives value λ_o = 16 μm, which is in reasonable agreement with the experiment. Surface morphology of the SM LPE layers, observed by atomic force microscopy (AFM), is shown in Figure 2. The surface morphology wavelength is about 200 nm. Theoretically calculated value of λ_o for 40 nm thick layer segment is 1.8μm. The observed considerable difference suggests significant deviation from the equilibrium LPE growth mechanisms. This morphology is more similar to MO VPE than to conventional LPE case.

In Figure 3 there is a design of melt slot configuration combining SM (quaternary GaInAsP layer with emission wavelength 1.55 μm, designed as Q, 1.55) , DM LPE (Q,1.55 and Q, 1.3 μm) and „traditional" (last InP layer). Thicknesses are influenced by different melt widths in first InP ;Q,1.1μm ; Q,1.3μm ; Q,1.55μm ;Q,1.3μm and Q,1.1μm , when the substrate is moved (from the left to the right) with constant velocity to the last InP melt. The melt slot configuration provides growth in three symmetrical steps waveguiding structures comprising a core layer composed from 1 , 2 or 3 individual layers, depending on quantity of the melt introduced to SM.

Figure 2. Surface morphology of the InP layer grown by SM LPE

The layer growth occurred in the supercooling regime from the single-phase melt with the initial supersaturation with $\Delta T = 4$ to $12°$ C and a cooling rate of $0.6°C/min$ at a growth temperature $641°C$. The epitaxial layers were grown on Sn-doped InP substrates with misorientation less than $0.3°$ from the exact (100) orientation. Melt slot width is in the range of 2 - 11 mm. Growth substrate movement velocity is from 1m/sec to 1m/15sec.

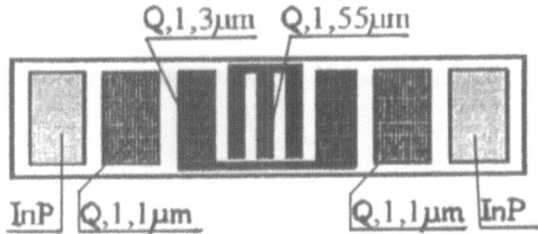

Figure 3. Melt slots configuration for growth in three steps of symmetrical waveguiding structure with a core layer composed from 1, 2 or 3 individual layers

3. Results and discussion

Strained layers were prepared with a similar melt configuration. Buffer InP, symmetrical unstrained GaInAsP layers and capping InP were grown on InP (100)

substrate to surround strained core layers grown from the one melt segment. Composition of the barrier / well layer near to $Ga_{.34}In_{.66}As_{.75}P_{.25}$ / $Ga_{.21}In_{.79}As_{.75}P_{.25}$ (with constant group V sublattice composition) was grown . Values of the perpendicular lattice mismatch up to 1.24% in compression has been measured by double crystal x-ray diffraction. To our knowledge this is the first published successful preparation of the strongly strained layers from the liquid phase in the GaInAsP/InP material system .

Increasing the GaAs contained in the melt from 0.58 mol % to 1.1 mol % shifted the 80K PL maxima from 1347 nm to 1411 nm with small decrease of the perpendicular lattice mismatch from (1.2-1.3) % to 0.8 %.

We have attempted the growth of strained ternary GaInAs/InP structures but so far this has been not successful. Detailed study of the tolerance of thin layers deposited from the liquid phase to the introduced strain is intended to be investigated in the future. Quantity of components of the solid solution and their covalent radii seems to be very important.

The thermal stability of these samples has also been investigated. Prepared strained samples were encapsulated with Si_3N_4 and treated by rapid thermal annealing (RTA) in 7 steps. The observed dependence of the 80 K PL maxima and PL intensity (affected by different opaqueness of RTA Si_3N_4 / InP interface) of the sample compressively strained with perpendicular lattice mismatch 1.12 % after the individual RTA steps are in Table I . Core layer thickness of the sample is approx. 19 nm.

TABLE 1

RTA step [°C]		750	750	800	800	800	800	900
[min]		2.5	5	1	3	5	2.5	1
PL max. at 80 K [nm]	1347	1343	1343	1341	1340	1340	1337	1337
PL int. at 80 K [rel. u.]		1166	1097	1077	1091	820	704 Si_3N_4 destr.	

The high temperature stability of the samples may be positively influenced by low point defect concentration of LPE growth materials.

This work was supported by the Grant Agency of the Czech Rep., Contract N.106/94/1058

References

1. Garbuzov, D.Z., Zhuravkevich, E.V, Zhmakin, A.I., Makarov, Yu.N., Ovchinnikov, A.V. (1991) On the peculiarities of the short-time solid solution LPE growth onto the moving substrate *J. Crystal Growth* **110**, 955-959.

2. Nohavica, D., Oswald,J (1995) Preparation of periodic structures by meander type liquid phase epitaxy *J.Crystal Growth* **146**, 287-292

3. Nohavica, D., Teminova, J., (1991) Controlled epitaxial growth of GaInAsP/InP from the liquid phase *Crystal Properties and Preparation* **32-34**, 630-634

4. Nishinaga, T., Pak, K., Uchiyama, S., (1978) Studies of LPE Ripple Based on Morphological Stability Theory *J. Crystal Growth* **43**, 85-92

SCANNING TUNNELING MICROSCOPY CHARACTERIZATION OF HETEROSTRUCTURES

EICKE R. WEBER, JUN-FEI ZHENG,* AND XIAO LIU**

Materials Science Division, Lawrence Berkeley Laboratory, and Department of Materials Science, University of California, Berkeley, CA 94720, USA.

1. Introduction

The study of heteroepitaxially-grown semiconductor interfaces supports advances in both fundamental scientific understanding and technological applications of heterostructure devices. Better understanding of the fundamental problems of interface formation can significantly impact thin film growth processes by controlling the quality of interfaces. With state-of-the-art techniques such as molecular beam epitaxy (MBE), epitaxial films with high structural quality can be grown. Besides analyzing the structure of interfaces, it is important to determine the chemical interactions taking place at interfaces between two heteroepitaxial layers.

It is obviously desirable to have an experimental technique that is capable of performing both chemical and structural analysis on an atomic scale. The fundamental interface phenomena, in terms of chemical fluctuations and gradients in heterostructures, may be classified as (1) interface segregation, (2) interface intermixing, and (3) interface clustering. These phenomena determine the structure or the roughness of III-V heterostructure interfaces on the atomic scale, a subject of intensive studies in the last 20 years [1,2]. Until now, all interface problems have been studied with techniques that obtain averaged information about the interface. For example, segregation has been studied using surface-sensitive techniques such as photoelectron spectroscopy applied in-situ during epitaxy growth [3], and intermixing has been analyzed by secondary ion mass spectroscopy [4]. Clustering in III/V heterostructures has been studied by indirect methods such as Raman spectroscopy [5,6]. Although transmission electron microscopy (TEM) provides structure analysis with atomic resolution, TEM is a technique that projects the interfaces averaged over the many layers of the sample thickness. TEM has chemical sensitivity with atomic plane resolution [7], but it does not have the chemical sensitivity to differentiate individual atoms in random arrangements at the intermixed interfaces. Scanning tunneling microscopy applied in cross-section (X-STM) is found to be a very powerful technique in imaging III-V heterostructure interfaces [8]. In this paper, we discuss chemical and structural analysis of various III-V heterostructures using X-STM.

* Now with INTEL, San Jose
** Now with American Crystal Technology (AXT), Dublin

J. Novák and A. Schlachetzki (eds.), Heterostructure Epitaxy and Devices, 65–74.
© 1996 *Kluwer Academic Publishers.*

2. Experimental

Cross-sectional STM studies were performed on (110)-type surfaces after cleavage in an ultrahigh-vacuum (UHV) chamber with a base pressure of 6×10^{-11} Torr. The STM head was designed similarly to that developed by Frohn et al. [9], with specific modifications to allow for easy in-situ cleavage of the heterostructures. The STM tip can be translated on the sample surface over a distance of millimeters. Once interfaces are located, the tip can be positioned along the interface with nanometer stepping resolution. Constant current images were taken with a typical tunnel current of about 1 nA, using electrochemically etched Pt-Rh tips. Samples were grown by molecular beam epitaxy (MBE) on n^+ GaAs (001) substrates. InGaAs/GaAs layers were grown at 540°C, AlAs/GaAs layers at 620°C. The GaAsP sample was grown by the liquid encapsulated Czochralski method [10].

3. Results and Dicussion

3.1. INTERFACE SEGREGATION AND CLUSTERING IN InGaAs/GaAs

InGaAs/GaAs strained-layer heterostructures have received increased attention in recent years because of their potential application in both optoelectronics and field effect transistors. These applications make use of the narrow band gap of InGaAs for infrared optoelectronic devices and the small electron effective mass for high frequency transistors. However, interface quality has been a crucial issue in the use of this heterostructure system. Indium atoms have a larger atomic size than gallium atoms, so that the lattice parameter of InGaAs is larger than that of GaAs, causing lattice strain. Indium atoms are known to segregate from the InGaAs alloy at the InGaAs/GaAs interface. Dislocation-free InGaAs layers can only be prepared by the controlled pseudomorphic growth of a layer with the InGaAs thickness below the critical thickness so that a strained-layer structure is formed. The interface quality is influenced by the indium segregation during the growth and has been actively studied [11-16]. Figure 1 presents a X-STM image of a vertical-cavity surface emitting laser structure consisting of an active region of three pseudomorphically grown InGaAs quantum wells of 8nm width with an In concentration of 20%, separated by 10-nm-thick GaAs barrier layers. The small insert shows a part of one quantum well in high magnification. The expanded image is taken from the first two quantum wells with the first GaAs barrier layer.

In the low-magnification image, depicting 100nm x 100nm sample area, the three InGaAs wells are imaged as bright bands separated by two darker GaAs layers. The black band from the bottom right to the top left is caused by a single atomic step created by the cleavage. In the constant current mode, the InGaAs region is imaged 0.5 Å higher than the GaAs. This can be qualitatively explained by the smaller energy bandgap of the InGaAs compared with the GaAs substrate and barrier layers. The smaller bandgap results in a lower tunnel barrier and thus enhanced tunnel current that is compensated by a retrieval of the tip above the quantum wells. In the higher resolution images, bright, atom-size spots appear inside the quantum wells that correspond in density to the In fraction of 20%.

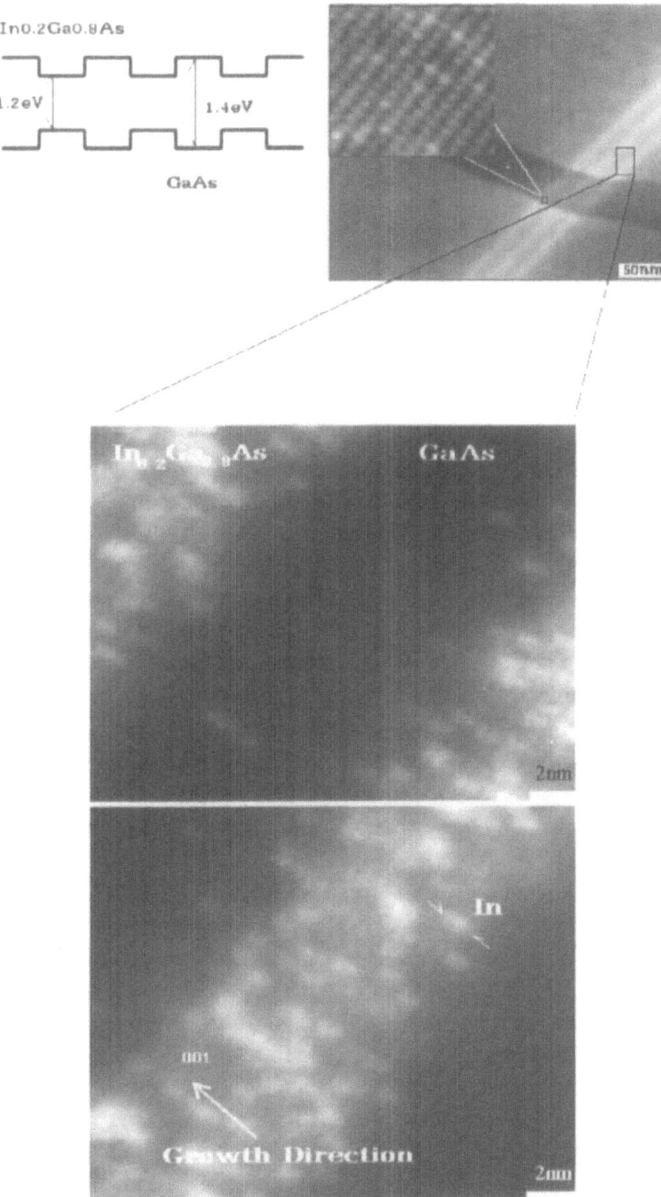

Figure 1. Cross-sectional STM image of three InGaAs quantum wells grown by MBE on GaAs (100). The low-magnification image shows the three quantum wells and a monoatomic surface step created by the cleavage. The insert and the expanded image show the arrangement of individual In atoms as bright spots within the quantum wells.

The image is taken at positive sample bias, corresponding to electrons tunneling from the tip to the empty states of the sample surface. Since empty states are located preferentially at the group III (cation) sites, the atoms imaged in the GaAs region are gallium and those in the InGaAs region are gallium or indium Thus it is possible to identify the bright atoms with locally enhanced tunnel current (resulting in a corresponding tip retraction) with the individual In atoms, allowing for a direct resolution of the group-III sublattice in this ternary alloy [17]. This identification is based on the following arguments. First of all, the fraction of bright spots is constant in all STM images. The brightest spots correspond to 20±5% of all the group III sublattice sites in one monolayer, which agrees well with the nominal concentration of 20% In. Secondly, since indium is associated with empty states with lower energy than gallium, the tunneling probability is larger when the tip is located at the indium site rather than at the gallium site. The difference between the atomic radii of indium and gallium is too small to account for the observed corrugation difference, indicating that this must be the result of an electronic effect. In addition, different apparent brightness in the constant-current image can be ascribed to In atoms within the first few monolayers beneath the surface. These subsurface In atoms obviously enhance the local tunneling of the neighboring surface gallium atoms.

Close inspection of many images reveals that the ternary InGaAs alloy region tends to have InAs clusters containing 2-3 indium atoms each, see Fig. 1. Statistically, 90 percent of the indium clusters contain 2-3 atoms. The clusters are aligned in strings preferentially along the [001] growth direction. The possibility of two-dimensional, plate-like clustering on the (1-10) planes, perpendicular to the cleavage surface, that are imaged edge-on, is discarded. This plate-like clustering should also be observable on the (110) cleavage plane in the form of extended islands, but such islands were never seen. Thus the clustering is concluded to be chain-like along the [001] growth direction.

It is worthwhile to consider the most likely mechanism of the formation of the 2-3 atom strings along the growth direction. It is proposed that this is the result of local strain rather than random clustering or chemical effects. In substitutional ternary III/V alloys with two group III elements, the clustering probability as a result of random arrangement is high when the percentage of the substitutional group III element is large enough. However, random clustering should lead to a wide distribution of cluster sizes [18]. Figure 2 shows a comparison of experimentally observed cluster sizes in ternary alloys compared with a random distribution. In the case of MBE-grown InGaAs, clustering far exceeding the random values is clearly observed. Chemical effects may enhance the possibility of clustering, especially since the larger bonding energy of indium to indium relative to that of indium to gallium would tend to produce indium adatom aggregation on a growing surface [15]. If such aggregates are the reason for clustering, it should be possible to observe clustering both in the growth direction and in the transverse direction. The fact that preferential clustering is only observed along the growth direction indicates that other effects are operative. The following explanation is based on a local strain model. During the MBE growth of InGaAs on GaAs, incorporation of indium causes strain because its size is larger than that of gallium. Indium atoms tend to segregate on the growing surface and lead to a high concentration of surface indium adatoms [15]. Near equilibrium, indium atoms can be incorporated into the lattice by kinetic freezing [16]. Once an indium atom is incorporated at a group III site in the GaAs, strain is built up locally [19]. This strain will locally distort the lattice and can thus, during overgrowth of the next layer, induce nucleation of another In atom on top of the previous one. However, this process is not likely to incorporate more than a few indium atoms in a string because the increasing local strain along the growth direction will limit the cluster size to a few atoms, making it more favorable for indium atoms to be incorporated at a new site rather then in a long string.

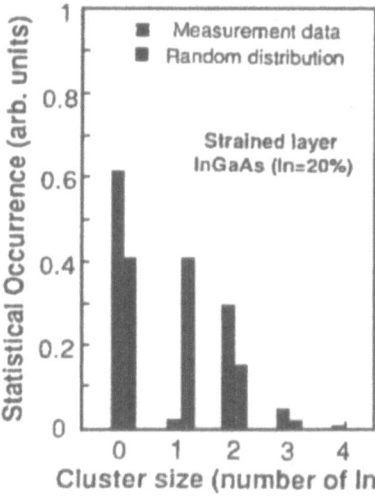 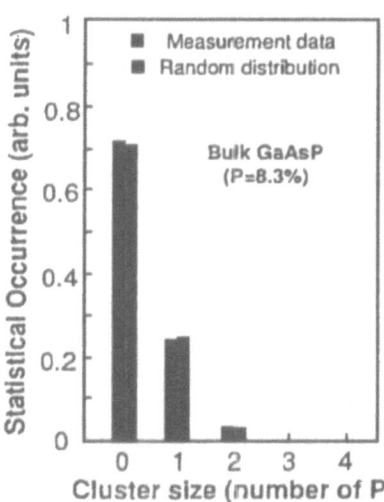

Figure 2. Measured size distribution of In clusters in MBE-grown InGaAs (20% In) and of P clusters in bulk-grown GaAsP (8.3% P), compared with the expected random distribution of cluster sizes.

In addition to observing InAs clustering, the ability to differentiate indium from gallium allows the study of atomic-scale roughness at the GaAs/InGaAs/GaAs interfaces [17]. The interface of InGaAs grown on GaAs is about 2-4 layers wide, whereas the interface of the GaAs grown on InGaAs is more diffuse, extending over 5-10 layers. The interfaces are clearly asymmetrically broadened. This result can be easily understand by the segregation of indium atoms on the growing surface [15]. At the upper interface, after closing the In shutter, excess indium on the surface will result in its incorporation into the growing GaAs, resulting in a very diffuse interface. Quite a few indium atoms can still be found deep in the GaAs layer, close to the next quantum well, see Fig. 1. It is interesting to notice that even these indium atoms in the GaAs barrier layer also form InAs clusters containing ~2 indiums oriented along the growth direction, similar to those inside the InGaAs quantum well.

These results allow us to conclude that clustering and segregation are the main reason for interface roughness in the InGaAs/GaAs system. The asymmetrical interface broadening can be observed here on a truly atomic scale. It is consistent with previous studies by photoelectron spectroscopy [3] and by TEM chemical lattice image studies of InGaAs/AlGaAs heterostructures [20]. The interface roughness has a length scale of about ~2nm, as directly seen here in the X-STM image. The interface observed by TEM, however, is generally averaged over the ~30 layers of a high-resolution TEM cross-sectional specimen and therefore cannot reveal such details of the interface. While photoluminescence (PL) indicates apparently smooth interfaces of GaAs/InGaAs/GaAs heterostructures [14], PL does not probe fluctuations smaller than the confined exciton diameter (~10nm) [21].

3.2. AlAs/GaAs SUPERLATTICES

Using the same experimental approach as in the previous section for InGaAs/GaAs structures, superlattices of AlAs/GaAs were investigated in order to study dopant-dependent superlattice intermixing [22]. Figure 3 shows as an example a structure consisting of a repetition of six GaAs layers alternating with two layers of AlAs.

Figure 3. X-STM image of a n-type AlAs/GaAs superlattice taken at -2V sample bias.

Due to the larger bandgap of AlAs, the surface As atoms visible in this filled state image appear dark, i.e., with reduced tunnel current, as compared to GaAs. The image contrast of the dark and bright As atoms provides a direct atomic-scale mapping of the Al concentration, although a more direct way of mapping Al concentration in AlGaAs alloys is possible by imaging at positive sample bias (empty state image) [23]. In layers grown with different shallow doping levels, dopant-dependent intermixing could be directly observed by X-STM of n^+-doped structures [22].

3.3. GaAsP TERNARY ALLOY

The examples discussed up to now included only alloys with group-III sublattice replacements. In the case of InGaAs, In atoms revealed a larger tunnel current than that of Ga atoms, in the case of AlAs/GaAs layers, the AlAs layers appeared with smaller tunnel current corresponding to a dark contrast in the constant current image. Both these observations were attributable to the smaller and larger bandgap of InGaAs and AlAs as compared to GaAs, respectively.

Therefore it is interesting to compare these results with a ternary alloy with group-V substitution. Figure 4 shows an example of a X-STM image of GaAsP with 8.3%P. It is obvious that in this filled state image roughly 10% of the sites have a reduced tunnel current, so that an identification with the P atoms in the GaAsP alloy is straightforward. In another specimen with smaller P content, a correspondingly smaller amount of atoms with the typical, reduced tunnel current was found. The analysis shown in Figure 2 demonstrates that the P atoms in this bulk-grown alloy were quite randomly distributed, however, in specific crystal regions clustering of P in clusters to 10-20P atoms could be observed, demonstrating the unique power of X- STM in the study of atomic clusters in crystalline solids.

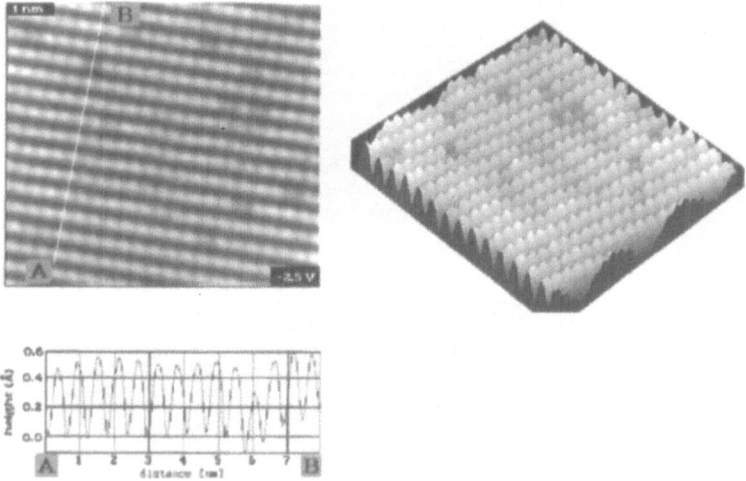

Figure 4. Filled-state X-STM image of $GaAs_{0.9}P_{0.1}$ grown by the LEC technique. The atoms with smaller apparent corrugation, e.g., in the line scan along AB, can be identified with P atoms in this alloy system.

3.4. As ANTISITE DEFECTS IN N-TYPE AlGaAs

An unexpected result was obtained from the cross-sectional STM image of $n^+ - Al_{0.67}Ga_{0.33}As$: within the layer, a high concentration of defects very similar to the characteristic image of As_{Ga} antisite defects [24] was found, see Fig. 5. This result is very surprising, as generally, in MBE growth of GaAs and related compounds at temperatures above 400°C, no excess As is incorporated, so that the As_{Ga} -related EL2 defect in GaAs grown by MBE is typically not detected.

The explanation for this phenomenon might be found in recent work by Newman et al. [25]: Based on local-vibrational-mode studies of silicon δ-doped GaAs these authors propose that, as a result of Si diffusion, pairs of Si_{Ga} donor atoms might convert to Si_{Ga}-Si_{As}-As_{Ga} complexes. This reaction would require a DX-like transition state of a Si_{Ga} donor as proposed in the Chadi and Chang model [26]:

$$Si_{Ga} \Rightarrow Si_i + V_{Ga}.$$

Subsequently, a displacement of a substitutional As might create an antiste defect via:

$$Si_i + V_{Ga} + As_{As} \Rightarrow Si_{As} + As_{Ga}.$$

This step is speculated to be more probable in the presence of an additional Si donor atom nearby [25]. Si donor atoms in AlGaAs with high Al content do indeed form DX-centers. Although a further systematic study of the effect shown in Figure 5 as a function of shallow doping and Al content is necessary before final conclusions can be drawn, it is worth noting that the present results seem to support Newman's model.

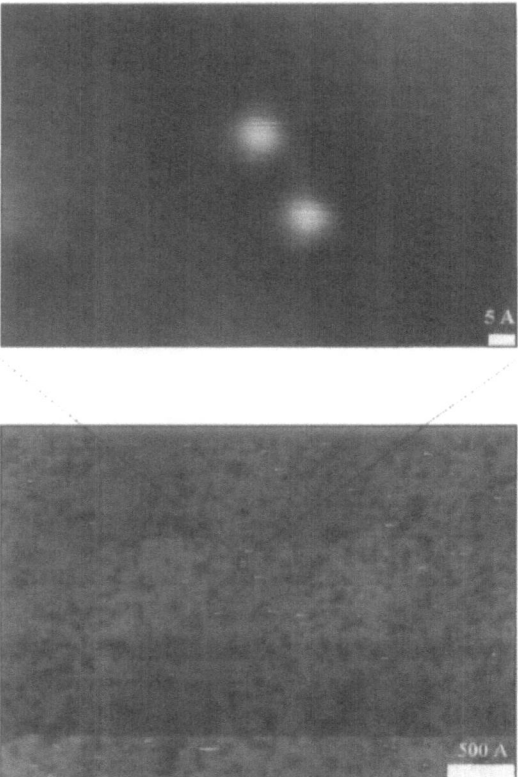

Figure 5. X-STM image of n^+ - $Al_{0.67}Ga_{0.33}As$. The lower image shows the presence of a high concentration of point-like defects; the upper image shows an enlargement of two of these defects revealing details typical of STM images of As_{Ga} antisite defects [24].

4. Summary and Conclusion

Cross-sectional scanning tunneling microscopy has been shown to be a powerful new method for the study of defects and interfaces in heterostructures. Its capability to distinguish gallium and indium in InGaAs permit the study of segregation and clustering at the atomic scale across interfaces. In strained-layer InGaAs/GaAs heterostructures, InAs clusters are found for the first time; they are preferentially aligned along the growth direction in the form of chains containing 2-3 indium atoms. The segregation-induced asymmetrical broadening of the two interfaces of GaAs/InGaAs/GaAs is directly revealed in real space. In cross-sectional STM studies of intermixing of Al and Ga in a short period AlAs/GaAs superlattice, the STM chemical sensitivity can be applied to analyze superlattice intermixing on an atomic scale. The study of GaAsP confirms the possibility of distinguishing atoms on the group-V sublattice as well. In highly doped n-AlGaAs grown by MBE, evidence for the existence of a high concentration of As antisites has been found for the first time. These studies show clearly that X-STM can be used to directly study the structure of point and interfacial defects on a truly atomic scale. In future, these structural studies may be complemented by electrical and optical spectroscopy of individual atom sitese.

5. Acknowledgements

The authors acknowledge Drs. Miquel Salmeron and Frank Ogletree of Lawrence Berkeley Laboratory for many useful discussions in the course of this project, and Nikos Jaeger for his technical assistance. This research was supported by the Director, Office of Energy Research, Office of Basic Energy Sciences, Materials Science Division of the US Department of Energy under contract No. DE-AC03-76SF00098.

6. References

1. Duke, C.B. (1993) Twenty years of semiconductor surface and interface structure determination and prediction, *J. Vac. Sci. Technol.* **B11**, 1336-1346.
2. Bode, M.H., and Ourmazd, A. (1992) Interfaces in GaAs/AlAs: perfection and applications, *J. Vac. Sci. Technol.* **B10**, 1787-1792.
3. Larive, M., Nagle, J., Landesman, J.P., Marcadet, X., Mottet, C., and Bois, P. (1993) In situ core-level photoelectron spectroscopy study of indium segregation at GaInAs/GaAs heterojunctions grown by molecular-beam epitaxy, *J. Vac. Sci. Technol.* **B11**, 1413-1417.
4. Kawabe, M., Shimizu, N.; Hasegawa, F. and Nannichi, Y. (1985) Effects of Be and Si on disordering of the AlAs/GaAs superlattice, *Appl. Phys. Lett.* **46**, 849-850.
5. Parayanthal, P., and Pollak, F.H. (1980) Raman scattering in alloy semiconductors: 'spatial correlation' model, *Phys. Rev. Lett.*, **52**, 1822-1825.
6. Hull, R., and Bean, J.C. (1990), in T.P. Pearsall (Ed.) Strained-Layer Superlattices, Semiconductor and Semimetals, Vol. 32 and 33, (Academic, San Diego).
7. Ourmazd, A., Taylor, D.W., Cunningham, J., Tu, C.W. (1989) Chemical mapping of semiconductor interfaces at near-atomic resolution, *Phys. Rev. Lett.* **62**, 933-6.
8. Salemink, H. and Albrektsen, O. (1991) Tunneling microscopy and spectroscopy on cross-sections of MBE-grown (Al)GaAs multilayers, *J. Vac. Sci. Technol.* **B9**, 779-82.
9. Frohn, J., Wolf, J.F., Besocke, K. and Teske, M. (1989) Coarse tip distance adjustment and positioner for a scanning tunneling microscope, *Rev. Sci. Instr.* **60**, 1200-1201.

74

10. Slupinski, T., Przybytek, J., Wysmolek, A., Lesczynski, M., Babinski, A., Borysiuk, J., Kurpiewski, A., Barcz, A., and Stepniewski, R. (1994) in M. Godlewski (Ed.): Semi-Insulating III-V Materials, World Scientific, Singapoure, pp.39-42.

11. Jogai, B., and Yu, P.W. (1990) Energy levels of strained $In_xGa_{1-x}As$-GaAs superlattices, *Phys. Rev.* **B41**, 12650-12658.

12. Devine, R.L.S., and Moore, W.T. (1987) Effect of interface structure on photoluminescence of InGaAs/GaAs pseudomorphic single quantum wells, *J. Appl. Phys.* **62**, 3999-4001.

13. Kirby, P.B., Constable, J.A., and Smith, R.S. (1989) Photoluminescence study of undoped and modulation-doped pseudomorphic $Al_yGa_{1-y}As/In_xGa_{1-x}As/Al_yGa_{1-y}As$ single quantum wells, *Phys. Rev.* **B40**, 3013-20.

14. Muraki, K., Fukatsu, S., Shiraki, Y., and Ito, R. (1993) Surface segregation of In atoms and its influence on the quantized levels in InGaAs/GaAs quantum wells, *J. Crystal Growth*, **127**, 546-549.

15. Moison, J.M., Houzay, F., Barthe, F., Gerard, J.M., Jusserand, B., Massies, J., and Turco-Sandroff, F.S. (1991) Surface segregation in III-V alloys, *J. of Crystal Growth*, **111**, 141-150.

16. Gerard, J.-M. and Marzin, J.-Y. (1992) Monolayer-scale optical investigation of segregation effects in semiconductor heterostructures, *Phys. Rev.* **B45**, 6313-6.

17. Zheng, J.F., Walker, J.D., Salmeron, M.B. and Weber, E.R. (1994) Interface segregation and clustering in strained-layer InGaAs/GaAs heterostructures studied by cross-sectional scanning tunneling microscopy, *Phys. Rev. Lett.* **72**, 2414-2417.

18. Holonyak, N. Jr., Laidig, W.D., Camras, M.D., Morkoc, H., Drummond, T.J., Hess, K., and Burroughs, M.S. (1981) Clustering in molecular-beam epitaxial $Al_xGa_{1-x}As$-GaAs quantumwell heterostructure lasers, *J. Appl. Phys.* **52**, 7201-7.

19. Proietti, M.G., Martelli, F., Turchini, S., Alagna, L, Bruni, M.R., Prosperi, T., Simeone, M.G., and Garcia, J. (1993) Microscopic investigation of the strain distribution in InGaAs/GaAs quantum well structures grown by molecular beam epitaxy, *J. Cryst. Growth* **127**, 592-595.

20. Kim, J., Alwan, J.J., Forbes, D.V., Coleman, J.J., Robertson,I.M., Wayman, C.M., Baumann, F.H., Bode, M., Kim, Y., and Ourmazd, A. (1992) Chemical characterization of (In,Ga)As/(Al,Ga)As strained interfaces grown by metalorganic chemical vapor deposition, *Appl. Phys. Lett.* **61**, 28-30.

21. Warwick, C.A., Jan, W.Y., Ourmazd, A., and Harris, T.D. (1990) Does luminescence show semiconductor interfaces to be atomically smooth?, *Appl. Phys. Lett.*, **56**, 2666-2668.

22. Zheng, J.F., Salmeron, M., and Weber, E.R. (1995) The effect of shallow donors and acceptors on AlAs/GaAs superlattice intermixing studied on the atomic scale, *Solid State Commun.*, **93**, 419-423.

23. Salemink, H.W.M. and Albrektsen, O. (1992) Atomic scale survey of III-V epitaxial interfaces, *J. Vac. Sci. Technol.* **B10**, 1799-1802.

24. Feenstra, R.M., Woodall, J.M., and Pettit, G.D. (1993) Observation of bulk defects by scanning tunneling microscopy and spectroscopy: arsenic antisite defects in GaAs, *Phys. Rev. Lett.* **71**, 1176-1179.

25. Newman, R.C., Ashwin, M.J., Wagner, J., Fahey, M.R., Hart, L., Holmes, S.N., and Roberts, C. (1995) Silicon delta doping in GaAs: An ongoing enigma, *MRS Symp. Proc.* **vol. 378**, 567-578.

26. Chadi, D.J., and Chang, K.J. (1988) Theory of the atomic and electronic structure of DX centers in GaAs and $Al_xGa_{1-x}As$ alloys, *Phys. Rev. Lett.* **61**, 873-876.

MICROSCOPIC ORIGIN OF FEMTOSECOND SPECTRAL HOLE BURNING IN QUANTUM WELLS

A. MOŠKOVÁ AND M. MOŠKO

Institute of Electrical Engineering, Slovak Academy of Sciences
Dúbravská cesta 9, SK-842 39 Bratislava, Slovakia

Abstract. The 200–fs plasma thermalization in GaAs quantum wells d-educed from previous spectral–hole burning measurements cannot be explained in terms of two–dimensional carrier–carrier scattering due to its weak thermalizing efficiency. This conclusion is drawn from the Monte Carlo simulation which simulates the two–dimensional carrier–carrier scattering beyond the Born approximation. Some other mechanisms which might be responsible for the observed thermalization time are proposed.

When a two–dimensional (2D) electron–hole (e–h) plasma is excited into the intrinsic GaAs quantum well by a short ($\sim 100 fs$) quasi–monoenergetic laser pulse, energy distributions of electrons and holes resemble the Gaussian shape of the excitation pulse spectrum. When the absorption spectrum of the quantum well is probed by a 50–fs broad–band pulse immediately after the excitation, a spectral hole is burned into the spectrum because the absorption of probe photons is suppressed for interband transitions into the states occupied by the nonthermal carriers. The spectral hole reflects the shape of the nonthermal carrier distribution and its evolution probed at various delays reflects the carrier thermalization towards a Maxwell distribution [1]. In Ref. 1 the 2D plasma density of $2 \times 10^{10} cm^{-2}$ was found to thermalize within 200 fs. Assuming the carriers excited below the energy threshold for optic phonon emission, the observed 200–fs thermalization was ascribed to 2D carrier–carrier (c–c) scattering and the excite–probe experiments have been thought to measure the 2D c–c scattering rate [1, 2]. Our Monte Carlo / molecular dynamics analysis has shown [3], that the 2D c–c scattering gives a much slower plasma thermalization than the observed one. In this work some other thermalization mechanisms are discussed.

J. Novák and A. Schlachetzki (eds.), Heterostructure Epitaxy and Devices, 75–78.
© *1996 Kluwer Academic Publishers.*

We present the Monte Carlo (MC) technique, which simulates the 2D c–c scattering beyond the Born approximation. To avoid a difficult implementation of the exact quantum cross section for 2D c–c scattering we implement the classical 2D c–c scattering cross section. (In the experimental conditions of Ref.1 the classical cross section is quite close to the exact quantum cross section, while the Born approximation strongly overestimates the exact result [3].) The scattering angle between the initial (**g**) and final (**g′**) relative wave vectors of two colliding classical carriers is [4]

$$\Theta = \pi - 2 \int_{r_{min}}^{\infty} dr \frac{b}{r^2 \{1 - (\frac{b}{r})^2 - \frac{4V(r)}{\mu g}\}^{1/2}} , \qquad (1)$$

where b is the impact distance, r the distance between the carriers, r_{min} the minimum distance, $\mu = 2m_e m_h / (m_e + m_h)$, m_e (m_h) is the electron (hole) effective mass, and the intercarrier Coulomb interaction

$$V(r) = \frac{1}{(2\pi)^2} \int d\mathbf{Q} \ e^{i\mathbf{Q}\cdot\mathbf{r}} \frac{q_e \cdot q_h}{2KQ\epsilon(Q)} . \qquad (2)$$

Here q_e (q_h) is the electron (hole) charge, K the material permittivity, and $\epsilon(Q)$ the static 2D screening function [3] in the limit of strictly 2D plasma. The 2D c–c scattering rate of the carrier of wave vector **k** reads

$$\lambda^{c-c}(k) = \frac{1}{2} \sum_{\mathbf{k}_0, s_0} f(k_0)(\frac{\hbar g}{\mu}) 2b_{max} , \qquad \mathbf{g} = \mu(\frac{\mathbf{k}_0}{m_h} - \frac{\mathbf{k}}{m_e}) , \qquad (3)$$

where $\mathbf{k}_0$ (s_0) is the wave vector (the spin) of the scattering partner, $f(\mathbf{k}_0)$ the occupation number at $\mathbf{k}_0$, b_{max} the maximum impact distance, and the extra factor of 1/2 is introduced to have a correct MC algorithm [5].

The classical carrier is scattered to zero scattering angle when $V(r) = 0$, i.e., when $r = \infty$. Therefore (1) gives $b_{max} = b(\Theta = 0, g) \to \infty$. To avoid the divergency of (3) we take the total scattering cross section $2b_{max}$ as the mean intercarrier distance $d = (2N_S)^{-1/2}$, where N_S is the e–h pair density. As in previous MC simulations [5], we interrupt the carrier free flights according to a constant c–c scattering rate Γ^{c-c}, which is greater than λ^{c-c}. When the free flight is interrupted, the scattering partner is selected at random from simulated ensemble and g is evaluated. Then the impact b is selected at random between 0 and $\mu\Gamma^{c-c}/(\hbar g N_S)$. If $b < b_{max}$, Θ is calculated from (1) and both carriers are scattered. If $b > b_{max}$, both carriers remain unscattered. Our simulation gives the same plasma thermalization for any $2b_{max} > d$ (we have tested $2b_{max}$ up to 1000 d). Thus the divergency of (3) for $b_{max} = b(\Theta = 0, g) \to \infty$ has no real meaning (scattering angles Θ for $b > d/2$ are too close to zero to affect the thermalization).

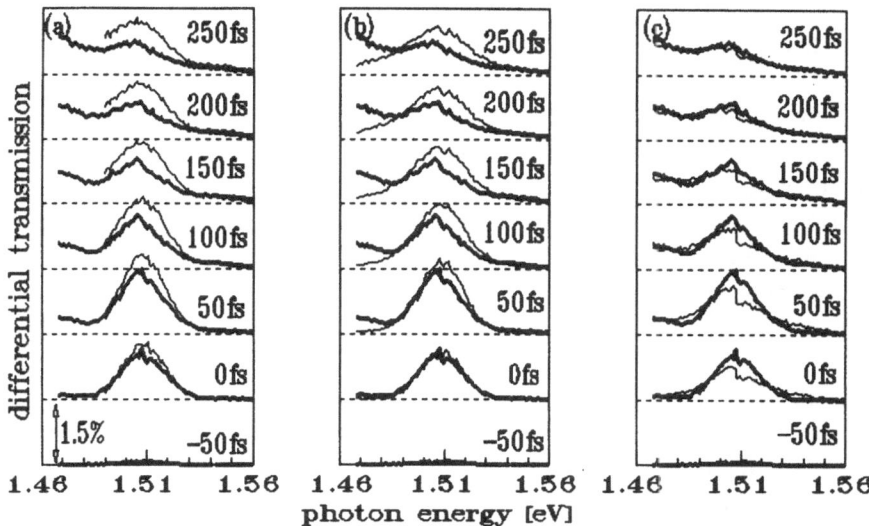

Figure 1. DT for $E_g = 1.4675eV$ (thick lines) compared (a) with the result for $E_g = 1.489eV$ (thin lines), (b) with the result without optic phonon scattering (thin lines), and (c) with the result which includes collisional broadening (thin lines).

The laser excitation into state k is described by the carrier generation rate $g(\mathbf{k}, t) = \int d\omega\, I(\omega, t)\alpha(\omega, t)L_W/\omega$, where $I(\omega, t)$ is the Gaussian pump spectrum with central frequency $\hbar\omega_0 = 1.509eV$, half width of $10meV$, and time spread of $100\,fs$. The band–to–band absorption $\alpha(\omega, t)$ reads

$$\alpha(\omega, t) = \alpha(\omega) \sum_{\mathbf{k}} (1 - f_e(\mathbf{k}, t) - f_h(\mathbf{k}, t)) \frac{\hbar^2}{\mu} \delta\left(\frac{\hbar^2 k^2}{2\mu} + E_g - \hbar\omega\right) \quad , \quad (4)$$

$$\alpha(\omega) = H(\hbar\omega - E_g) \times 2\left(1 + exp\{-2\pi|\frac{Ry}{E_g - \hbar\omega}|^{1/2}\}\right)^{-1} \times \alpha_0 \quad , \quad (5)$$

where $\alpha(\omega)$ is the absorption of unpumped sample, $H(x)$ the Heavyside step function, E_g the energy gap, Ry the Rydberg energy, and $\alpha_0 = 2.7 \times 10^5 m^{-1}$ [6]. The differential transmission $DT = N_w L_w(\alpha(\omega) - \alpha(\omega, t))$, where L_w is the well width and N_w the number of the wells.

Fig.1a shows the DT for experimental conditions of Ref.1. In Ref.1 the excess energy $\hbar\omega_0 - E_g$ was estimated to be less than the optic phonon energy $\hbar\omega_{LO} = 36.8meV$. The value of $20meV$, which corresponds to $E_g = 1.489eV$, was specified in later papers [2, 7, 8]. In Fig.1a the DT for $E_g = 1.489eV$ shows much slower thermalization than experiment [1]. We estimate from the heavy–hole exciton peak ($E_{HH} = 1.4585eV$[6]) that E_g is at most $E_{HH} + 4Ry \approx 1.477eV$, and more precisely [6] only $1.4675eV$. For $E_g = 1.4675eV$ Fig.1 shows much faster relaxation due to the enhancement of the optic phonon emission ($\hbar\omega_0 - E_g = 41.5meV > \hbar\omega_{LO}$). Unlike

Ref. 1 we do not think that numerical integration of DT spectra [1] indicates little interaction with lattice. First, cooling of electrons with energies $> \hbar\omega_{LO}$ is partly compensated by the escape of cold holes (via optic phonon absorption) from the spectral hole region [8]. Second, near the band edge the DT spectrum merges into the exciton peak, which obscures the presence of cold electrons. *The optic phonon scattering strongly contributes to the DT spectra relaxation*, as illustrated in Fig.1b. Without optic phonons only the 2D c–c scattering is operative (thin curves in Fig.1b) and the thermalization is too slow compared to the experiment [1]. *We conclude that the spectral hole burning measurements [1, 2] do not isolate the 2D c–c scattering due to too high $\hbar\omega_0 - E_g$.* In Fig.1a, the DT spectra for $E_g = 1.4675eV$ show almost unchanged occupancy of near–band edge states at time delays $> 150fs$. This should lead to the 150-fs excitonic absorption saturation, which is indeed observed [1]. The unrelaxed central peak at times $> 100fs$ is in contrast with experiment [1], which indicates presence of some other broadening mechanisms. In Fig.1c we illustrate the effect of collisional broadening of optically coupled electron–hole states, incorporated into the generation rate $g(\mathbf{k}, t)$ by replacing the δ–function in (4) by Lorentzian of width $\hbar\lambda_{tot}(\mathbf{k})$, where $\lambda_{tot}(\mathbf{k})$ is the sum of total electron and hole scattering rates. Inhomogenous broadening and broadening due to finit duration of the probe pulse, ignored in present work, would further relax the central peak.

References

1. Knox, W.H., Hirliman, C., Miller, D.A.B., Shah, J., Chemla, D.S., and Shank, C.V. (1986) Femtosecond excitation of nonthermal carrier population in GaAs quantum wells, *Phys. Rev. Lett.* **56**, 1191–1193.
2. Knox, W.H., Chemla, D.S., Livescu, G., Cunningham, J.E., and Henry, J.E. (1988) Femtosecond carrier thermalization in dense Fermi sea, *Phys. Rev. Lett.* **61**, 1290-1293.
3. Moško, M., Mošková, A., and Cambel, V. (1995) Carrier–carrier scattering in photoexcited intrinsic GaAs quantum wells and its effect on femtosecond plasma thermalization, *Phys. Rev.* **B 51**, 16860–16866.
4. Balescu, R. (1975) Equilibrium and nonequilibrium statistical mechanics, John Willey and Sons, N.Y.
5. Mošková, A., and Moško, M. (1994) Exchange carrier-carrier scattering of photoexcited spin-polarized carriers in GaAs quantum wells: Monte Carlo study, *Phys. Rev.* **B 49**, 7443–7449
6. Chemla, D.S., Miller, D.A.B., Smith, P.W., Gossard, A.C., Wiegmann, W. (1984) Room temperature excitonic nonlinear absorption and refraction in GaAs/AlGaAs multiple quantum wells., *IEEE J. Quantum Electron.* **20**, 265–275.
7. Goodnick, S.M., Lugli, P., Knox, W.H., and Chemla, D.S. (1989) Monte Carlo simulation of femtosecond spectroscopy in semiconductor heterostructures, *Solid State Electron.* **32**, 1737–1741.
8. Goodnick, S.M., and Lugli, P. (1988) Influence of electron–hole scattering on subpicosecond carrier relaxation in $Al_xGa_{1-x}As/GaAs$ quantum wells, *Phys. Rev.* **B 38**, 10135-10138.

CARRIER CAPTURE DUE TO CARRIER-CARRIER
INTERACTION IN QUANTUM WELLS

K. KÁLNA AND M. MOŠKO

Institute of Electrical Engineering, Slovak Academy of Sciences
Dúbravská cesta 9, SK-842 39 Bratislava, Slovakia

Abstract. Electron capture times in a separate confinement quantum well (QW) with finite electron density are calculated for electron-electron (e-e) and electron-polar optic phonon (e-pop) scattering. In both cases the capture time versus QW width oscillates with the same period, but with the quite different amplitude. For electron density of $10^{11} cm^{-2}$ the e-e capture time is $10^1 - 10^3$ times larger than the e-pop capture time except for QW widths near resonance minima. Near the resonances it is only $2 - 3$ times larger and at high enough electron densities even smaller. Thus the resonant capture can be optimized by varying the carrier density.

The electron capture in a quantum well plays an important role in optimizing the performance of separate confinement heterostructure quantum well (SCHQW) lasers. Quantum calculations [1] of pop emission induced capture in $GaAs$ QW predicted oscillations of the capture time versus the QW width, which have been observed . The minima of the oscillations provide the optimum well and barrier width for an optimized capture efficiency. At high electron densities the e-e scattering induced capture is expected to play an important role. In this work the e-e and e-pop scattering induced capture times are calculated and their relative importance is discussed for typical SCHQW structures [2, 3].

The analyzed SCHQW consists of the $Al_xGa_{1-x}As/GaAs/Al_xGa_{1-x}As$ QW with 500 Å $Al_xGa_{1-x}As$ barriers, embedded between two semiinfinite $AlAs$ layers. The e-e scattering is treated following Ref. 4. When two electrons in subbands i, j with wave vectors $\mathbf{k}$ and $\mathbf{k}_0$ are scattered to subbands m, n with wave vectors $\mathbf{k}'$ and $\mathbf{k}'_0$, the e-e scattering rate of an electron with

J. Novák and A. Schlachetzki (eds.), Heterostructure Epitaxy and Devices, 79–82.

wave vector $\mathbf{k}$ from subband i to subband m is given by [4]

$$\lambda_{im}(\mathbf{k}) = \frac{1}{N_S A} \sum_{j,n,\mathbf{k}_0} f_j(\mathbf{k}_0) \lambda_{ijmn}(g) \quad , \tag{1}$$

where $g = |\mathbf{k} - \mathbf{k}_0|$,

$$\lambda_{ijmn}(g) = \frac{N_S m^* e^4}{16\pi \hbar^3 \kappa^2} \int_0^{2\pi} d\theta \, \frac{F_{ijmn}^2(q)}{q^2 \, \epsilon^2(q)} \quad , \tag{2}$$

$$q = \frac{1}{2} \left[2g^2 + \frac{4m^*}{\hbar^2} E_S - 2g \left(g^2 + \frac{4m^*}{\hbar^2} E_S \right)^{1/2} \cos\theta \right]^{1/2} , \tag{3}$$

$$F_{ijmn}(q) = \int_{-\infty}^{\infty} dz \int_{-\infty}^{\infty} dz_0 \, \chi_i(z) \, \chi_j(z_0) \, e^{-q|z-z_0|} \, \chi_m(z) \, \chi_n(z_0). \tag{4}$$

$E_S = E_i + E_j - E_m - E_n$, the summation over $\mathbf{k}_0$ includes both spin orientations, m^* is the electron effective mass in $GaAs$, κ the static permittivity, A the normalization area, E_j the subband energy and $f_j(\mathbf{k}_0)$ the electron distribution in subband j. Wave functions χ_i in the QW are obtained assuming the x-dependent, flat Γ-band with parabolic energy dispersion. To deal with the 0.3-eV $GaAs$ QW [3] we take $x = 0.305$. The e-e capture time $\tau_{e-e} = \sum_{i,\mathbf{k}} f_i(\mathbf{k}) / \sum_{i,\mathbf{k},m} f_i(\mathbf{k}) \lambda_{i,m}(\mathbf{k})$, where the summation over i (m) includes the subbands above (below) the $AlGaAs$ barrier, and summation over j, n in (1) involves the subbands below the $AlGaAs$ barrier. $f_j(\mathbf{k}_0)$ is the Fermi function taken at temperature 8K and for an electron density $N_S = 10^{11} cm^{-2}$. $\epsilon(q) = 1 + (q_S/q) F_{1111}(q) f_1(\mathbf{k}_0 = 0)$ is the static screening function due to the electrons in the lowest subband [4], where $q_S = e^2 m^* / (2\pi \kappa \hbar^2)$. Fig. 1 shows τ_{e-e} versus the QW width for $f_i(\mathbf{k})$ taken as a constant distribution up to 36.8 meV above the $AlGaAs$ barrier, which roughly models the injected "barrier" distribution after a rapid phonon cooling [1, 3]. In the inset our calculation is compared with the result (crosses) of Ref. 3. Both τ_{e-e} curves oscillate with the QW width and reach a resonant minimum, whenever a new bound state merges into the QW. However, our τ_{e-e} is two-to-three orders of magnitude larger. The difference of a factor of 4 is due to the missing factor of 1/4 in the e-e scattering rate of Ref. 3 (see Ref. 4 for details). When the τ_{e-e} values from Ref. 3 are multiplied by a factor of 4, our τ_{e-e} is still ~ 100 times larger. We believe that Ref. 3 predicts much smaller τ_{e-e} due to a numerical error.

The e-pop scattering rate of an electron with wave vector $\mathbf{k}$ from subband i to subband m reads [5] (for spontaneous phonon emission only)

$$\lambda_{im}(\mathbf{k}) = \frac{e^2 \omega m^*}{8\pi \hbar^2} \left(\frac{1}{\kappa_\infty} - \frac{1}{\kappa} \right) \int_0^{2\pi} d\theta \, \frac{F_{iimm}(q)}{q \, \epsilon(q)}, \tag{5}$$

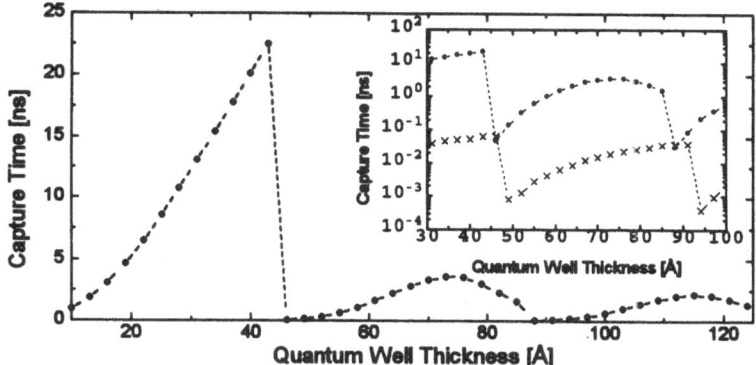

Figure 1. E-e capture time τ_{e-e} vs. the QW thickness for $N_S = 10^{11} cm^{-2}$ and for $f_i(k)$ taken as a constant up to 36.8 meV above the barrier. In the inset this result is compared with the data (crosses) from Ref. 3, calculated for the same SCHQW.

$$q = \left[2k^2 + \frac{2m^*}{\hbar^2}P - 2k\left(k^2 + \frac{2m^*}{\hbar^2}P\right)^{1/2} \cos\theta \right]^{1/2}, \qquad (6)$$

where $P = E_i - E_m - \hbar\omega$, $\hbar\omega$ is the pop energy and κ_∞ is the high frequency permittivity. We calculate the e-pop scattering induced capture time τ_{e-pop} by averaging (5) as discussed for τ_{e-e}. Figure 2 compares τ_{e-pop} with τ_{e-e} for the parameters and the constant distribution $f_i(k)$ from Fig. 1.

The τ_{e-pop} data shown by empty circles are calculated using the same static screening $\epsilon(q)$ as for the e-e scattering, empty squares show τ_{e-pop} for $\epsilon(q) = 1$. τ_{e-e} is one-to-three orders larger than τ_{e-pop} except for QW widths near the resonance minima. This finding differs from previous analysis [3] which predicts nearly the same oscillation amplitude in both cases. Ref. 3 also predicts that in the SCHQW lasers with a QW width below 40 Å the e-e capture causes excess carrier heating in the QW. Figure 2 does not support this conclusion, because the e-e capture is negligible. It is easy to asses the dependence of both capture times on the electron density N_S. For $N_S \geq 10^{11} cm^{-2}$ and temperature 8 K the static screening $\epsilon(q)$ is independent on N_S, because $f_1(0) \simeq 1$. Therefore, the τ_{e-pop} values in Fig. 2 would be the same also for higher N_S and the τ_{e-e} values would decrease approximately like N_S^{-1} for each QW width. In Fig. 2 we show τ_{e-e} for $N_S = 2.8 \times 10^{11} cm^{-2}, 5 \times 10^{11} cm^{-2}$ and $10^{12} cm^{-2}$ only at QW widths of 43 Å and 46 Å. At 43 Å τ_{e-e} is much larger than τ_{e-pop} even for $N_S = 10^{12} cm^{-2}$ due to the absence of resonance. At 46 Å, when the first excited subband merges into the QW, τ_{e-e} resonantly decreases about 500 times and becomes smaller than τ_{e-pop} when $N_S \simeq 5 \times 10^{11} cm^{-2}$. When $N_S = 10^{12} cm^{-2}$, the total capture time $\tau_{e-e}\tau_{e-pop}/(\tau_{e-e} + \tau_{e-pop})$ is 3.8 ps for the unscreened e-pop capture ($\tau_{e-pop} = 11$ ps) and 4.3 ps for

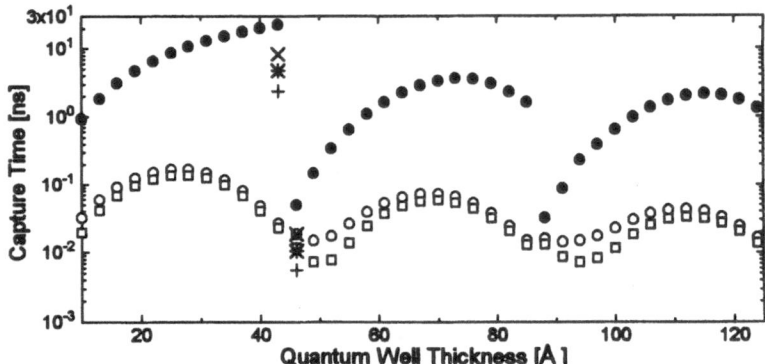

Figure 2. E-pop capture time τ_{e-pop} and e-e capture time τ_{e-e} vs. the QW thickness for $N_S = 10^{11} cm^{-2}$. Open circles show τ_{e-pop} for the statically screened e-pop interaction, open squares show τ_{e-pop} for the unscreened e-pop interaction and full circles are the τ_{e-e} data from Fig. 1. Crosses, asterisks and pluses at 43 Å and 46 Å show the τ_{e-e} data for $N_S = 2.8 \times 10^{11} cm^{-2}, 5 \times 10^{11} cm^{-2}$ and $10^{12} cm^{-2}$, respectively.

the screened e-pop capture ($\tau_{e-pop} = 18$ ps). Thus, compared to the case $\tau_{e-e}^{-1} = 0$ the capture efficiency of the QW with the optimized (resonant) width can be improved with a factor $2.9 - 4.2$ by increasing N_S to $10^{12} cm^{-2}$.

In summary, the e-e and e-pop capture times in the SCHQW oscillate with the same period, but with very different amplitude. The e-e capture time is much larger than the e-pop capture time except for the QW widths near resonances where it can be even smaller for electron densities close to $10^{12} cm^{-2}$ which leads to an improved capture efficiency of the QW. An inefficient e-pop capture in the SCHQW laser should not lead to excess carrier heating [3] due to e-e capture, because away from the resonance it is still much stronger than the e-e capture.

References

1. Brum, J.A. and Bastard, G. (1986) Resonant carrier capture by semiconductor quantum wells, *Phys. Rev.* **B 33**, 1420–1423.
2. Blom, P.W.M., Smit, C., Haverkort, J.E.M., and Wolter, J.H. (1993) Optimization of barrier thickness for efficient carrier capture in graded-index and separate-confinement multiple quantum well lasers, *Phys. Rev.* **B 47**, 2072–2081.
3. Blom, P.W.M., Haverkort, J.E.M., van Hall, P. J., and Wolter, J. H. (1993) Carrier-carrier scattering induced capture in quantum well lasers, *Appl. Phys. Lett.* **62**, 1490–1492.
4. Moško, M., Mošková, A., and Cambel, V. (1995) Carrier-carrier scattering in photoexcited intrinsic GaAs quantum wells and its effect on femtosecond plasma thermalization, *Phys. Rev.* **B 51**, 16860–16865.
5. Goodnick, S. M. and Lugli, P. (1992) Hot-carrier relaxation in quasi-2D systems, in J. Shah (ed.), *Hot Carriers in Semiconductor Nanostructures*, Academic, New York, pp. 191–234.

OPTICAL AND THEORETICAL STUDY OF GAAS QUANTUM WELLS EMBEDDED IN GAAS/ALAS SUPERLATTICES

V.DONCHEV, I.IVANOV AND K.GERMANOVA
Faculty of Physics, Sofia University
5 J.Bourchier blvd., Sofia-1126, Bulgaria

The replacement of the AlGaAs alloy barriers of a GaAs quantum well (QW) by short period GaAs/AlGaAs superlattices (SLs) is beneficial for several reasons [1].

The aim of this work is to study the electronic states and the optical properties of GaAs QWs embedded in GaAs/AlAs SLs in order to determine to what extent they are influenced by the presence of a SL in the barrier region.

Two types of samples representing planar structures grown by MBE on (100) GaAs substrates were studied. Samples type A consist of a 5 nm thick GaAs QW sandwiched between 26 and 20 periods of a $(GaAs)_8/(AlAs)_4$ SL. Samples type B are similar but contain a GaAs buffer layer of 400 nm and a second QW of 12 nm situated bellow that of 5 nm. The barrier regions in this case represent 5, 24 and 20 periods of SL situated bellow the 12 nm QW, between the QWs and above the 5 nm QW, respectively. The cap layer is $Al_{0.33}Ga_{0.67}As$ of 50 and 30 nm in samples A and B, respectively.

Low temperature photoluminescence (PL) measurements were performed with a resolution of 0.2nm using a He-Ne laser, a SPEX double monochromator and a GaAs photomultiplier.

Theoretical assessment was made using two approaches:

1) The effective mass approximation (EMA) [2] applied to a single GaAs QW (5 or 12 nm) having uniform $Al_xGa_{1-x}As$ alloy barriers with x equal to the mean Al content in the SL ($x=4/(8+4)=0.33$). One-band model is used taking into account the nonparabolicity of CB and light hole (lh) band.

2) The empirical tight binding (ETB) approach with the surface Green function matching method. The algorithm is described in [3,4]. In the present calculations the inhomogeneous region was a finite $(GaAs)_8/(AlAs)_4$ SL (10 or 20 periods) with an embedded centered GaAs QW (18 or 42 MLs).

J. Novák and A. Schlachetzki (eds.), Heterostructure Epitaxy and Devices, 83–86.
© 1996 *Kluwer Academic Publishers.*

An ideal finite SL and a single QW with $Al_{0.33}Ga_{0.67}As$ barriers were also studied for comparison.

Table I shows that the EMA and ETB approaches give similar energies for e1-hh1 and e1-lh1 transitions of a single QW. When the $Al_{0.33}Ga_{0.67}As$ barrier is replaced by a $(GaAs)_8/(AlAs)_4$ SL these energies increase which can be explained by an increase of the effective potential barrier. The effect is stronger for the narrower QW where the confinement energies are larger.

TABLE 1. Calculated and experimental transition energies (in meV) for GaAs QWs of 5 and 12 nm. The values in parentheses in the 7th column are the exciton binding energies. The e1 state in all cases is localized in the QW.

QW width	Transition	Hole localization	Single QW		QW embedded in a $(GaAs)_8/(AlAs)_4$ SL		
			EMA	ETB	ETB	ETB-E_{exc}	experiment (sample)
12 nm	e1-hh1	QW	1551	1570	1572	(-9) 1563	1552 (B)
	e1-lh1	QW	1561	1579	1581	(-11) 1570	1564 (B)
	e1-hh1	QW	1625	1624	1639	(-12) 1627	1625 (A) 1620 (B)
	e1-lh1	QW	1653	1648	1670	(-15) 1655	1643 (A) 1637 (B)
	e1-hh2	SL			1701	1690	
	e1-hh3	SL			1703	1692	1689 (A)
5 nm	e1-hh4	SL			1705	1694	1670 (B)
	e1-hh5	SL			1708	(-11) 1697	
	e1-hh6	SL			1710	1699	1693 (A)
	e1-hh7	QW			1714	1703	(PLE [6])
	e1-lh2	SL			1743	1732	
	e1-lh3	SL			1753	(-11) 1742	1736 (A)
	e1-lh4	SL			1755	1744	(PLE [6])

The comparison with experiment is done after subtracting the exciton binding energies E_{exc} from the calculated transition energies (column 7 in Table 1). In the case of e1-hh1 and e1-lh1 transitions which are localized in the QW we use as a first approximation the values of E_{exc} corresponding to a single QW and calculated by an interpolation formula [5].

The QW of 12 nm is manifested in the PL spectrum (Fig.2) by the sharp peaks P1 and P2. They are accounted for by the e1-hh1 and e1-lh1 transitions according to the calculated transition energies.

In the case of the 5 nm QW the PL structures at energies close to the calculated energies for the transitions e1-hh1 and e1-lh1 have very small amplitude (see Table 1 and Figs.1,2). This indicates that an other recombination channel is dominant.

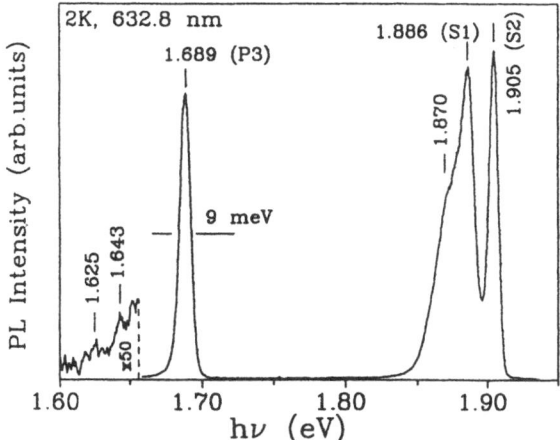

Figure 1. PL spectrum of a sample with one QW of 5 nm (sample type A). The excitation power is 0.5 W/cm².

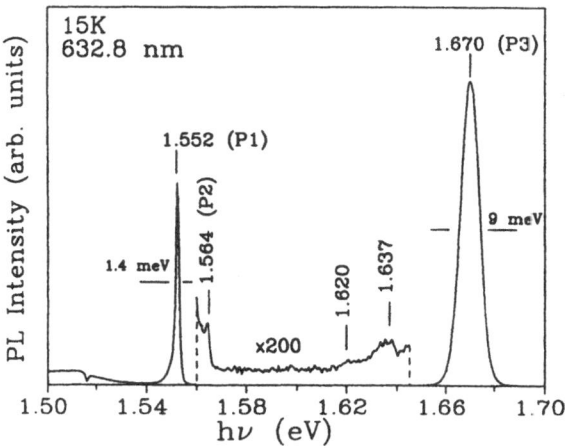

Figure 2. PL spectrum of a sample with two QWs of 12 and 5 nm (sample type B). The excitation power is 10 W/cm².

It can be seen from Figs.1 and 2 and Table 1 that the big peak P3 cannot fit the calculated energies of a single QW nor of an ideal SL. Therefore, it is connected with the 5 nm QW embedded in the SL. The group of transitions e1-hh2 to e1-hh7 which have very close energies do not exist in the spectrum of the ideal SL, nor of the single QW. The states e1 and hh7 are localized mainly in the 5 nm QW. The states hh2 to hh6 correspond to the SL hh mini-band. However they have non-zero spectral strength in the QW. So, we suppose that this group of transitions gives rise to the large PL peak P3 in Figs.1 and 2. A support of this assumption is found by PLE spectroscopy of a sample type A [6]. The value for E_{exc} in this case is found from temperature dependent PL of a sample type B.

To explain the origin of P3 we assume random well-width fluctuations in the SL which cause hh localization in the growth direction as discussed in [7]. At low temperatures this disorder can reduce considerably the hh transport to the QW. The sharp doublet around 1.89 eV in Fig.1 is typical for a SL with well-width fluctuations [7]. The energy distance between the peaks roughly corresponds to 1 ML fluctuations of a 8 MLs QW. So, we suppose that much more electrons than holes arrive to the 5 nm QW where they thermalise to e1 state and recombine with hh from the SL, thus giving rise to the peak P3. As to the 12 nm QW, it is very close to the buffer layer and the hh generated there can reach it and give rise to the peaks P1 and P2 in Fig.2.

In conclusion, the replacement of the AlGaAs alloy barriers by a GaAs/AlAs SL modifies the electronic structure of the QW and can give rise to optical transitions which do not exist in the spectrum of a single QW, nor of an ideal SL. These effects are more pronounced for narrower QWs. It is found that some of these new transitions are dominant in the emission spectrum of the structures considered.

We are grateful to Dr. St.Vlaev for performing ETB calculations and fruitful discussions, to prof. F.K.Reinhart and D.Martin for supplying the samples, to prof. P.Tronc and prof. A.Miller for helpful discussions and to D.Batovski for technical help. This work was supported by the Bulgarian National Science Fund.

References

1. Sakaki, H., Tsuchiya, M. and Yoshino, J. (1985) Energy levels and electron wave functions in semiconductor quantum wells having superlattice alloylike material (0.9nm GaAs/0.9nm AlGaAs) as barrier layers, *Appl.Phys.Lett.* 47, 295-297.
2. Herman,M.A., Bimberg,D. and Christen,J. (1991) Heterointerfaces in QWs and epitaxial growth processes: Evaluation by luminescence techniques, *J.Appl.Phys.* 70, R1-R51.
3. Vlaev, S., Velasco, V.R., and Garcia-Moliner,F. (1994) Electronic states in graded-composition heterostructures, *Phys.Rev.B* 49, 11222-11229; (1994) Tight-binding calculation of electronic states in a triangular symmetrical quantum well, *Phys.Rev.B* 50, 4577-4580; (1995) Tight-binding calculation of electronic states in an inverse parabolic quantum well, *Phys.Rev.B* 51, 7321-7324.
4. Vlaev, S. (1995) A tight-binding calculation of transition energies in triangular symmetrical QWs, in M.Balkanski and I.Yanchev (eds.), *Fabrication Properties and Applications of Low-Dimensional Semiconductors*, Kluwer Academic Publishers, Dordrecht, pp.141-142.
5. Oelgart, G. *et al.* (1994) Experimental and theoretical study of excitonic transition energies in GaAs/AlGaAs quantum wells, *Phys.Rev.B* 49, 10456-10465.
6. Donchev, V., Ivanov, I. and Germanova, K., (1996) Optical and theoretical assessment of GaAs quantum wells having superlattices as barrier layers, in M.Balkanski (ed.), Devices based on Low-Dimensional Semiconductor Structures, Kluwer Academic Publishers, Dordrecht, (in press)
7. Fujiwara, K., Tsukada, N. and Nakayama, T., (1989) Linear polarization effects on PL properties of GaAs/AlAs QW heterostructures, *Solid State Commun.* 69, 63-66.

MOCVD GROWTH AND CHARACTERISATION OF InAs/GaAs SUPERLATTICES

M.ČERNIANSKY, J.KOVÁČ, V.GOTTSCHALCH*

Slovak Technical University, Department of Microelectronics
Ilkovičova 3, SK 812 19 Bratislava , Slovakia
*Fakultät für Chemie und Mineralogie, Universität Leipzig
Linnéstrasse 3-5, 04103 Leipzig, Germany

1. Introduction

A growth of highly strained InAs/GaAs heterostructures on GaAs substrates has attracted much interest because of their potential application in optoelectronics. InAs/GaAs superlattice-based high quality lasers have been already reported [1]. The 7.16% lattice mismatch in these structures can lead to various effects. Self-organised growth of InAs quantum dots and formation of wire-like InAs islands at the initial stages of the growth are the most interesting ones [2, 3].

Most of the reports on the growth of InAs/GaAs heterostructures, however, have been done on structures grown by MBE. Just very few papers deal with MOCVD growth of InAs/GaAs heterostructures, probably because of lower controllability of the growth process by this method. We report here on the MOCVD growth of high quality InAs/GaAs superlattices. Optical properties of the superlattices and also the influence of substrate misorientation were investigated by photoluminescence.

2. MOCVD growth of InAs/GaAs superlattices

Superlattices on variously oriented GaAs substrates have been grown by AIX-200 low pressure MOCVD. The reactor was equipped with a rotating substrate holder and standard precursors (TMGa, TMIn, TMAl and AsH₃) were used. Quality of the InAs layers was examined by high-resolution transmission electron microscopy, double-crystal x-ray diffraction and atomic force microscopy. Abrupt and smooth interfaces and mirror-like surfaces of InAs

J. Novák and A. Schlachetzki (eds.), Heterostructure Epitaxy and Devices, 87–90.
© 1996 *Kluwer Academic Publishers.*

monolayers have been observed. Very good quality of the InAs monolayers has been confirmed also by photoluminescence [4]. Our PL measurements agree well with the data published by other authors [5], so just negligible segregation of In atoms is supposed.

Seven different sets of superlattices were investigated. Details of the structures are given in table 1. Each superlattice is embedded between thick AlGaAs layers, so the whole structure constitutes GRIN SCHS laser. The structures have been grown on (100) and vicinal GaAs substrates to investigate effect of substrate misorientation. Substrates intentionally misoriented by 0.5°, 2° and 6° towards [010] or [01$\bar{1}$] directions were used. Each set of samples with differently oriented substrates was grown in one epitaxial run to ensure the best possible comparability of the samples.

TABLE 1. Parameters of the superlattices - number of periods, composition of one period and used substrates. Denotation of the substrates identifies angle and direction of misorientation.

Sample	Superlattice	Substrates				
		(100)	0.5° [010]	2° [010]	6° [010]	6°[01$\bar{1}$]
A	6x 0.5ML InAs/3ML GaAs	+		+	+	
B, F	6x 1 ML InAs/ 6 ML GaAs	+		+	+	
C, G	6x 1 ML InAs/ 4 ML GaAs		+	+	+	+
D	8x 1 ML InAs/ 4 ML GaAs		+	+	+	+
E	11x 1 ML InAs/4ML GaAs		+	+	+	+

3. Optical properties

Temperature dependent and excitation power dependent PL measurements were performed to analyse optical properties of the superlattices. The samples were mounted in closed-cycle He cryostat and the 488nm line of an argon laser was used for excitation. Photoluminescence was detected using 0.5m-focal-length monochromator and cooled Ge photodiode.

Low temperature PL spectra of five different superlattices are shown in fig.1. Very intense luminescence measured for all our samples is nearly two orders of magnitude higher compared to the luminescence of GaAs cap layers. Integrated PL intensity is weakly temperature dependent. Just very small decrease of PL intensity has been observed. FWHM of the PL peaks of exactly oriented and 0.5° misoriented structures is rather small (about 20meV), except for the sample denoted B. This again demonstrates good quality of the superlattices.

The influence of substrate misorientation on the optical properties of the superlattices is illustrated in fig.2.It must be pointed out that the presence of two

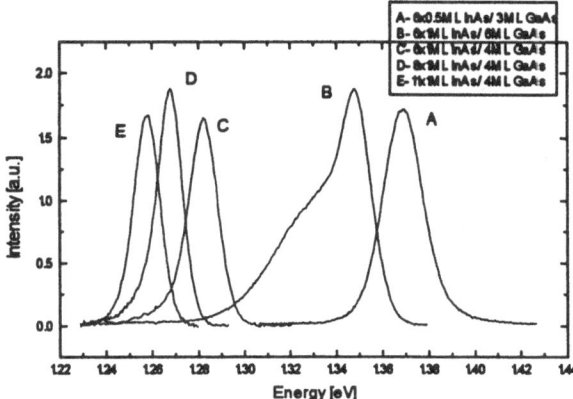

Figure 1. Low temperature PL spectra of five different types of superlattices grown on exactly oriented substrates or substrates misoriented by 0.5° . Very high intensity and rather small FWHM (about 20meV) indicate good quality of the superlattices.

clearly resolved peaks for the 2° misoriented structure does not depend on the substrate misorientation. The shape of the spectra of 6° misoriented superlattices indicates that two electronic states are involved in the radiative recombination. Also by increasing excitation power or temperature the second peak can appear.

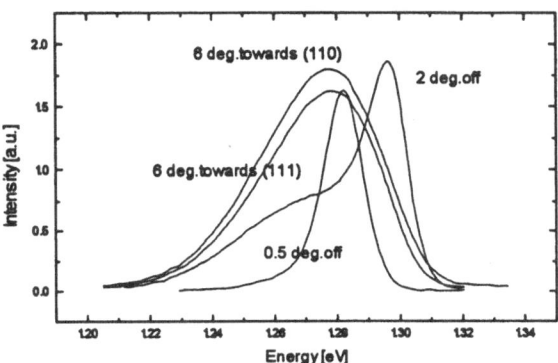

Figure 2. A comparison of the low temperature PL spectra of the samples grown on differently misoriented substrates. All samples were grown in the same epitaxial run.

What is common for all sets of the samples is a shift of the main peak in the case of 2° misoriented samples and higher integrated luminescence for 6° misoriented structures. The first fact is rather surprising and we have no explanation for it.

A difference between structures grown on substrates tilted towards different directions could be expected. It has been found out by some authors that the initial stage of the growth of InAs on GaAs is highly anisotropic [3] and this could lead to differences in the optical properties for structures grown on variously oriented substrates, but we have observed no such effect.

The origin of the second radiative transition is not completely understood. It can be deduced from the temperature dependence and the excitation power dependence of the luminescence that the second peak is excitonic in nature. We believe that this effect is probably connected with microscopic details of the InAs layers. This assumption is also supported by the optical properties of the superlattice A, in which just submonolayers of InAs have been deposited. In this case only one peak has been observed.

Broad area and ridge-waveguide lasers have been fabricated from the structures. A threshold current density lower than $300A \cdot cm^{-2}$ was achieved for laser with $600\mu m$ cavity length. Rather surprising fact has been discovered. It has been found that different lasers made from the same wafer can emit at two wavelengths which correspond to the peaks observed in the PL spectra. This fact also confirms excitonic origin of both transitions. However, no correlation with cavity length or current density has been found.

4. Summary

High quality InAs/GaAs superlattices have been grown by MOCVD. Two close electronic states exist in the superlattices as can by seen from PL and stimulated emission spectra. These states are probably excitonic in nature and they might be connected with microscopic details of the InAs layers. Substrate misorientation seems to affect optical properties of the superlattices, but no evident correlation with the existence of two electronic states was found.

5. References

1. Dutta, N.K., Chand, N., Lopata, J. and Wetzel, R. (1993) Temperature characteristics of InAs/GaAs short-period superlattices quantum well laser, *Appl.Phys.Lett.* **62**, 2018-2020
2. Marzin, J.-Y., Gérard, J.-M., Izraël, A., Barrier, D. and Bastard, G. (1994) Photoluminescence of single InAs quantum dots obtained by self-organized growth on GaAs, *Phys.Rev. Lett.* **73**, 716-719
3. Bressler-Hill,V., Lorke,A., Varma,S., Petroff,P.M., Pond, K. and Weinberg,W.H. (1994) Initial stages of InAs epitaxy on vicinal GaAs (001)-(2x4), *Phys.Rev.B* **50**, 8479-8487
4. Schwabe, R., Pietag, F., Faulkner, M., Lassen, S., Gottschalch, V., Franzheld, R., Bitz, A. and Staehli, J.L. (1994) Optical investigations on isovalent δ-layers in III-V semiconductor compounds, *submitted to J.Appl.Phys.*
5. Brandt, O., Lage, H. and Ploog, K. (1992) Heavy and light-hole character of optical transitions in InAs/GaAs single-monolayer quantum wells, *Phys.Rev.B* **45**, 4217-4220

ELECTRICAL CHARACTERISTICS OF EPITAXIAL Al/Al$_x$Ga$_{1-x}$As/n-Al$_{0.25}$Ga$_{0.75}$As HETEROSTRUCTURES

ZS. J. HORVÁTH AND L. DOZSA
Research Institute for Technical Physics
of the Hungarian Academy of Sciences,
Budapest, Újpest 1, P.O.Box 76, H-1325 Hungary

1. Introduction

In this work the current-voltage (I-V) and capacitance-voltage (C-V) characteristics of MBE grown epitaxial Al/undoped Al$_x$Ga$_{1-x}$As/n-Al$_{0.25}$Ga$_{0.75}$As junctions have been studied in a wide temperature range as a function of the AlAs mole fraction of the 4 nm thick Al$_x$Ga$_{1-x}$As cap layer. Several anomalous behaviours have been observed, that are mainly due to the influence of DX centres present in AlGaAs.

2. Experimental

Five different samples have been prepared. The 2 μm thick Al$_{0.25}$Ga$_{0.75}$As active epitaxial layer was grown on an n$^+$ GaAs buffer layer and n$^+$ GaAs substrate by MBE. The active layer was Si-doped with nominal doping of 2x10^{17} cm^{-3}. A 4 nm thick undoped Al$_x$Ga$_{1-x}$As cap layer was then grown in four of the samples with x=0, 0.5, 0.75 and 1, respectively. (The sample without cap layer was considered having a cap with x=0.25.) The 300 nm thick epitaxial Al layer was grown in-situ. Circular dots with diameters of 300 μm were patterned by conventional photolithography. The In that bonded the sample to the holder during the epitaxial growth, provided the backside ohmic contact. Details see in [1].

The I-V and 1 MHz C-V characteristics have been measured in the dark as a function of temperature.

3. Results and discussion

3.1. I-V CHARACTERISTICS

Evaluating the measured I-V characteristics by the thermionic emission (TE) theory, we obtained temperature dependent ideality factor and barrier height values: the ideality factor increased, while the apparent barrier height decreased with decreasing temperature [1]. This phenomenon may be connected either with the lateral inhomogeneity of the Schottky barrier height [2], or with the role of the recombination current mechanism [3], or with anomalously high

J. Novák and A. Schlachetzki (eds.), Heterostructure Epitaxy and Devices, 91–94.
© 1996 *Kluwer Academic Publishers.*

Table 1: The Schottky barrier heights of the studied $Al/i-Al_xGa_{1-x}As/n-Al_{0.25}Ga_{0.75}As$ junctions evaluated from the I-V characteristics for the termionic emission (ϕ_{b0TE}) and for the termionic-field emission (ϕ_{b0TFE}) theories, and from the C-V characteristics (ϕ_{b0C}), and those expected for $Al/n-Al_xGa_{1-x}As$ junctions (ϕ_{b0th}) [10], as a function of the Al mole fraction x

x	ϕ_{b0TE}, V	ϕ_{b0TFE}, V	ϕ_{b0C}, V	ϕ_{b0th}, V
0	0.84	1.05	1.05	0.84
0.25	0.92	1.10	1.02	1.01
0.5	1.02	1.17	1.10	1.09
0.75	1.01	1.24	1.12	1.01
1	0.97	1.14	0.98	0.95

thermionic-field emission (TFE) [4]. The Schottky barrier height values evaluated from the room temperature I-V characteristics for the TE theory, are presented in Table 1.

We also evaluated the I-V characteristics by the TFE theory [1,5], and obtained very good agreement between the experimental I-V curves and the theoretical ones calculated for TFE by using the evaluated parameters [1]. The Schottky barrier height values evaluated from the temperature dependence of the I-V characteristics for the TFE theory, are also presented in Table 1.

The evaluated characteristic energy values (E_{00}) did not exhibit systematic dependence on the composition of the cap layer, but they were about 1.5-2 times higher than their theoretical value [1]. This anomaly, in general, may be due to any mechanism, which enhances the electric field or the density of states at the semiconductor surface [4,5]. In our junctions the obtained anomaly may be partly due to the electric field enhancement at the periphery of the diodes. Another possible mechanism may be the multistep tunneling through deep levels and/or DX centres [6].

We also obtained anomalous excess currents at low forward biases and low temperatures in some of the diodes studied, similar to those obtained by Drápal and Hampl for Ti/n-GaAs diodes [7]. However in our case the excess current depended on the reverse bias applied to the diode before measuring the forward I-V characteristics. The higher was the applied reverse bias the higher was the forward excess current. This phenomenon also did not depend on the cap layer, its measure was different for different diodes on the same wafer. The origin of this phenomenon is not clear yet. Perhaps it is connected with the ionization of DX centres.

3.2. C-V CHARACTERISTICS

The typical Schottky-Mott (C^{-2}-V) plots obtained were not entirely linear, they showed a small curvature with decreasing slope at higher reverse biases. The slopes of the Schottky-Mott plots decreased with increasing temperature for all the samples, indicating the effect of DX centres, which are not fully ionized even at 360 K.

In the most cases the voltage intercepts decreased with increasing temperature, as it can be expected for n-type Schottky junctions, due to the temperature dependence of the barrier height [8], while in other cases it exhibited a maximum at about 300-310 K. Therefore, in these cases a competitive mechanism existed, which may also be connected with recharging the DX centres.

The barrier height values obtained from the C-V measurements for 300 K, are also presented in Table 1.

In some cases we obtained increasing capacitance values with increasing reverse bias at high reverse biases and low temperatures. Detailed experiments have shown that this phenomenon was connected with the pre-breakdown ionization of DX centres [9].

3.3. EFFECT OF THE CAP COMPOSITION ON THE BARRIER HEIGHT

The barrier height values obtained from the different measurements and evaluations, are presented in Table 1 as a function of the Al content of the cap layer (x). Our earlier analysis of the experimental Schottky barrier height values obtained in AlGaAs Schottky junctions showed that - independent of the Al content - the Schottky barrier height is due to the pinning of the Fermi-level at the AlGaAs surface at 1.15±0.10 eV below the L-band minimum [10]. The corresponding barrier height values expected for $Al/n-Al_xGa_{1-x}As$ junctions are also presented in Table 1 for the same Al contents.

The barrier heights obtained from the different measurement techniques and evaluations, are different. Those obtained for TFE are higher by a factor of 1.15-1.25, than the values obtained for TE. The barrier heights evaluated from the C-V measurements, are between those obtained for TE and TFE. All of them exhibit a maximum at x=0.5 or 0.75. The barrier heights obtained for TFE, are influenced by the temperature dependence of barrier height neglected during the evaluation. If this dependence is taken into account, lower barrier heights are expected [1]. On the other hand, the barrier height values evaluated from the C-V characteristics, may be strongly effected by the charge being in the DX centres near the Al/AlGaAs interface [11,12].

The comparison of the evaluated barrier heights with the expected ones shows a good agreement for TE at x=0, 0.75 and 1, and for C-V at x=0.25, 0.5 and 1. The general tendency of the composition dependence suggests that the barrier height in the studied $Al/i-Al_xGa_{1-x}As/n-Al_{0.25}Ga_{0.75}As$ junctions is mainly determined by the interface states at the $Al/i-Al_xGa_{1-x}As$ interface.

4. Acknowledgements

The authors thank A. Bosacchi, S. Franchi, E. Gombia, and R. Mosca (MASPEC Institute, Parma, Italy) for their contribution. This work has been supported in part by the COST and COPERNICUS

94

programs of the Commission of the European Communities under contracts N⁰ CIPA3510CT923133 and N⁰ CP941180, respectively.

5. References

1. Horváth, Zs.J., Bosacchi, A., Franchi, S., Gombia, E., Mosca, R., and Biondelli, D. (1995) Electrical behaviour of epitaxial Al/n-Al$_{0.25}$Ga$_{0.75}$As junctions: effect of the composition of undoped Al$_x$Ga$_{1-x}$As cap layer, *Vacuum* **46**, 959–961.
2. Sullivan, J.R., Tung, R.T., Pinto, M.R. and Graham, W.R. (1991) Electron transport of inhomogeneous Schottky barriers: A numerical study, *J. Appl. Phys.* **70**, 7403–7424.
3. Donoval, D., Barus, M., and Zdimal, M. (1991) Analysis of I–V measurements on PtSi–Si Schottky structures in a wide temperature range, *Solid-State Electron.* **34**, 1365–1373.
4. Horváth, Zs.J. (1992) A new approach to temperature dependent idality factors in Schottky contacts, *Mat. Res. Soc. Symp. Proc.* **260**, 359–366.
5. Horváth, Zs.J. (1995) Comment on "Analysis of I–V measurements on CrSi$_2$–Si Schottky structures in a wide temperature range", *Solid-State Electron.* **38**, in press.
6. Eizenberg, M., Heiblum, M, Nathan, M.I., Braslau, N. and Mooney, P.M. (1987) *J. Appl. Phys.* **61**, 1516–1522.
7. Drápal, S. and Hampl, J. (1993) Anomalies in Schottky diode I–V characteristics induced by annealing, *Solid-State Electron.* **36**, 1639–1640.
8. Revva, P., Langer, J.M., Missous, M., Peaker, A.R. (1993) Temperature dependence of the Schottky barrier in Al/AlGaAs metal–semiconductor junctions, *J. Appl. Phys.* **74**, 416–425.
9. Horváth, Zs.J., Gombia, E., Mosca, R., and Motta, A. (1994) Capacitance–voltage anomaly in Al/AlGaAs Schottky junctions due to DX centers, *Workshop on New Developments in Semiconductor Physics, Sept. 25–30, 1994, Balatonfüred, Hungary,* Abstract p.28.
10. Horváth, Zs.J. (1993) Fermi-level pinning at metal/AlGaAs and GaAs/AlGaAs interfaces, in *Proc. 16th Annual Semiconductor Conf. CAS'93, Oct. 12–17, 1993, Sinaia, Romania,* Research Institute for Electronic Components, Romanian Academy of Sciences, Bucharest, 1993, pp. 311–313.
11. Horváth, Zs.J. (1988) Comment on "Engineered Schottky barrier diodes for the modification and control of Schottky barrier heights" [J. Appl. Phys. 61, 5159 (1987)] *J. Appl. Phys.* **64**, 443–444.
12. Horváth, Zs.J. (1992) Effect of near-interface concentration change on barrier height in ion-bombarded and heat-treated GaAs Schottky contacts, *Mat. Res. Soc. Symp. Proc.* **260**, 441–446.

INVESTIGATION OF A GaAs HETEROSTRUCTURE WITH AN AlAs POTENTIAL BARRIER BY DLTS MEASUREMENTS

Ľ. STUCHLÍKOVÁ, L. HARMATHA, J. KOVÁČ,
A. KOVÁČIK
*Slovak Technical University, Faculty of Electrical Engineering
and Information Technology, Department of Microelectronics,
Ilkovičova 3, 812 19 Bratislava, Slovakia*
V. GOTTSCHALCH
*Fakultät für Chemie und Mineralogie, University of Leipzig,
D-04103 Leipzig, Germany*
B. RHEINLÄNDER
*Fakultät für Physik und Geowissenschaften, University of Leipzig,
D-04103 Leipzig, Germany*

1. Introduction

The emission and capture processes of charge carriers in a tunnelling photodetector based on a GaAs NIN heterostructure with an AlAs barrier [1] were investigated.

These processes can be characterized by the DLTS (Deep Level Transient Spectroscopy) measuring method. DLTS methods are commonly applied on structures with a space charge region (pn-junction, Schottky barrier, MOSFET).

A pn-junction has been created by diffusion into the top layer of the tunnelling photodetector, so as to be able to use a DLTS calculation model for identifying parameters.

Using the standard DLTS and its modifications (MCTS and MCTS&CLR), parameters of deep levels in a GaAs NIN heterostructure with an AlAs barrier have been detected. The measured results have been verified by measurements on the same structure with the pn-junction created on the top layer.

2. Samples

We used the structures shown in Table 1. The basic structure NIN 5 - a tunnelling photodetector based on modulation of the potential barrier height was grown by MOCVD [2]. These structures have a shielding AlAs barrier layer

95

J. Novák and A. Schlachetzki (eds.), Heterostructure Epitaxy and Devices, 95–98.
© 1996 *Kluwer Academic Publishers.*

placed asymmetrically in the intrinsic region of GaAs betwen the two n$^+$ contacts. On this structure a pn-junction has been formed by a diffusion of ZnO (DNIN 5, NIN 5-421). Standard device processing with MESA area 0.02371 mm^2 [3] was used for structures NIN 5 and DNIN 5. Structure NIN 5-421 has an area of 0.325 mm^2.

TABLE 1. Layer composition of the heterostructures

NIN5				\ DNIN5 NIN5-421
	nm	cm^{-3}	material	PN junction
contact	500	5x10^{17}	GaAs	diff ZnO 200 - 300 nm
spacer	500	i	GaAs	
barrier	8.7	i	AlAs	
spacer	70	i	GaAs	
buffer	50	5x10^{17}	GaAs	
substrate	n+	n+	GaAs	

3. Experiment

Standard DLTS, MCTS and MCTS &CLR measuring methods were used. The initial measuring conditions for these methods were determined from CV [4] and IV [5] measurements realized in darkness and under illumination at various temperatures. DLTS measurements have been performed using a Polaron DLTS spectrometer with a pulsed GaAs laser source. This spectrometer uses a boxcar detection system for acquiring the DLTS output signal. The laser excitation unit used for charge carriers excitation is capable of 100 mW maximum power, the wavelength at peak intensity being 850 nm.

4. Results and discussion

The measured DLTS spectra exhibit a strong deviation from an exponential dependence. This can be caused both by the presence of several mutually influencing deep energy levels and by contributions of emission and capture of charge carriers in the barrier region.

Wide ranges of reverse voltage biases (-3.5 V, 0.2 V), temperatures (80 to 400 K) and times of filling (0.04 to 11 ms) have been used as the initial measuring condition. The MCTS measurements used an optical pulse of a very low intensity, since the signal from minority traps rises and is non-exponential when rising the pulse intensity. We assumed that it was caused by retained holes beyond the retaining barrier since light generates electron-hole pairs in the 50 nm spacer under illumination and the electric field separates them.

Parameters of deep levels in both types of structures were determined by an experimental regression. They are listed in Table 2 along with results of other authors obtained by standard DLTS measurements [6].

TABLE 2. Characteristic parameters of deep levels in both types of structures

DNIN5 NIN5-421	Standard DLTS		MCTS		MCTS & CLR		Similar traps in lit. [6]		
Deep Level	ΔE_T [eV]	σ_T[m²]	ΔE_T [eV]	σ_T[m²]	ΔE_T [eV]	σ_T[m²]	ΔE_T [eV]	σ_T[m²]	Deep Level
HT1			0.235	1.87E-21			0.220	1.48E-20	EH3
HT2	0.257	1.24E-20	0.283	1.01E-20			0.300	4.61E-20	EH4
HT3	0.408	7.37E-19			0.361	3.89E-19	0.380	8.16E-19	EH5
HT4	0.446	2.14E-16	0.421	3.91E-17			0.420	1.00E-17	EL5
HT5	0.411	7.94E-19					0.470	1.52E-17	EL4
HT6	0.410	4.77E-20			0.479	1.28E-19	0.440	4.35E-18	EI1
HT7	0.192	8.83E-20	0.289	3.81E-20	0.282	5.73E-19			
HT8	0.461	6.20E-20			0.479	1.28E-19			
HT9	0.562	1.19E-17					0.575	1.25E-17	EL3

NIN5	DLTS		MCTS		MCTS & CLR		Similar traps in lit. [6]		
Deep Level	ΔE_T [eV]	σ_T[m²]	ΔE_T [eV]	σ_T[m²]	ΔE_T [eV]	σ_T[m²]	ΔE_T [eV]	σ_T[m²]	Deep Level
HT2	0.279	3.69E-21					0.300	4.61E-20	EH4
HT3	0.396	1.43E-19			0.369	5.31E-20	0.380	8.16E-19	EH5
HT5	0.465	2.79E-18					0.470	1.52E-17	EL4
HT6	0.508	3.93E-18	0.489	6.21E-20	0.426	7.47E-20	0.440	4.35E-18	EI1
HT7	0.190	4.61E-23	0.279	4.47E-20	0.245	9.76E-19			
HT8					0.481	1.16E-19			
HT9	0.576	4.38E-17	0.588	1.78E-17			0.575	1.25E-17	EL3
HT10	0.729	1.99E-17					0.78	4.91E-16	EL12

The values of electron traps activation enthalpies ΔE_T and capture cross sections σ_T were calculated from an Arrhenius diagram using the known equation

$$\ln e_n = \ln (k_m \, \sigma \, T^2) - \Delta E_n / k_B T, \qquad (1)$$

where e_n is the emission rate, k_m a material constant, T the absolute temperature, k_B the Boltzmann constant.

On both structures we detected HT1-HT10 majority deep levels. Parameters of deep level HT1 approximately correspond to EH3 in literature, HT2 to EH4, HT3 to EH5, HT4 to EL5, HT5 to EL4, HT6 to EL1, HT9 to EL3 and HT10 to

EL12. We summed up deep levels with enthalpy of ionization 0.192-0.289 eV into a deep level HT7 and about 0.4 eV into HT8.

After a comparison of results received by measuring methods applied on both structures we can say that traps HT2, HT3, HT5, HT6, HT7, HT8, HT9 were detected in both structures. This occurrence of traps in both types of structure validates the use of DLTS methods for investigation NIN structure. Trap HT10 was detected only in NIN5. Traps HT1 and HT4 were detected only in DNIN5, NIN5-421. The trap HT7 was detected by all measuring methods in both structures. The theoretical background for explaining this deep level has not been sufficient, but we can assume that they can be associated with the presence of a barrier in the structure. On the structure NIN5, the trap HT6 was detected by all measuring methods. This trap was dominant.

No method was able to detect all deep energy levels.

5. Conclusion

The results achieved show that our procedure - to measure the NIN structure with a barrier and a pn-junction and then to apply these measuring methods to a NIN structure with a barrier was right.

We verified the application of the standard DLTS method and its modification for investigation of NIN structures with barriers and detected 10 majority deep levels. We assumed that trap HT7 ≈ 0.2 eV stems from the presence of the barrier in the structure. Modifications of our mathematical model aiming at getting the parameters of deep levels from the measured DLTS spectra and finding the influence of the barrier are the topic of research at present.

References

1. Redhammer, R., Kováč, J., Németh, Š., Gottschalch, V., Rheinländer, B., Kováčik, A., Jakabovič, J., Tomaška, M. and Škriniarová, J. (1994) Charge retaining tunnelling structures, *Proceedings EDS'94 II*, Brno, 289-293
2. Redhammer, R. and Allsopp, D. W. E. (1992) Self-consistent modelling of charge effects in resonant tunnel diodes, *Proc. of ESSDERC'92, Microelectronics Engineering*, Leuven, Belgium , 899-902
3. Gregušová, D., Eliáš, P., Malacký, L., Kúdela ,R. and Škriniarová, J. (1995), Wet Chemical MESA Etching of InGaP and GaAs with Solution Based on HCl, CH_3COOH, H_2O_2, *Phys. Stat Sol.(a)* **151**, 113-118
4. Botka,V., Csabay,O.,Harmatha,L. (1992) A complex working set-up for MOS structure measurement and evaluation , *Czechos. J. of Pfysics* **42**, 46-56 (in Slovak)
5. Donoval, D., Barus, M., Bedlek, M., Nágl, V., Racko, J., Sitar, P., Kulak, D. (1992) Schottky structures with higher breakdown voltage, *J. Elect .Engineering* **43**, 265-269
6. Milnes, A. G. (1983) Impurity and defect levels (Experimental) in Gallium Arsenide, *Advances in Electronics and Electron Physics* **61**, 63-160

EFFECT OF HETEROBARRIERS ON THE DX CENTER IN ALGAASSB AND IN GAALAS

L. DÓZSA
Research Institute for Technical Physics of the
Hungarian Academy of Sciences
H-1325 Budapest P.O. Box 76.

P. HUBIK, J. KRISTOFIK AND N. TERZIEV
Institute of Physics of the Czech Academy of Sciences
Cukrovarnicka 10, 162 00 PRAHA 6

Abstract
DX levels were investigated in GaAlAs, and in GaAlAsSb. The deep level transient spectroscopy results are in agreement with the literature data. The deep level transients above 200 K show new features of the DX centers. An influence of the heterobarrier in the structure was found significant. These properties are explained by a modification in the large lattice relaxation model.

Introduction
The DX centers are basic defects in III-V semiconductors. The main properties are explained by large lattice relaxation [1]. The level was found in most of the III-V compound as in GaSb [2] and in InP under high hydrostatic pressure [3]. The main characteristics are summarized in review papers [4]. The DX concentration was found to depend on the cooling rate at low temperatures [5]. The most accepted model is a transition of the substitutional donor to interstitial site theoretically justified for Si in GaAlAs [6].
In this paper we present results of our investigations of DX centers in different materials. Based on these results we suggest that in a general model the large lattice relaxation of DX is a recombination enhanced defect reaction, explaining why it is not very sensitive to the material and dopant properties.

J. Novák and A. Schlachetzki (eds.), Heterostructure Epitaxy and Devices, 99–102.
© *1996 Kluwer Academic Publishers.*

Material and measurement

The DLTS measurements were carried out with a Polaron S4600 set up, which included a three-gate boxcar and a capacitance bridge working at 1 MHz frequency. The recombination properties in p-n junction structures were investigated by the method described in [7] with a DLS82 system. The fast capture and emission transients were measured by a method decribed in [8].

The GaInAsSb/AlGaAsSb mesa structure avalanche photodiodes and 180 μm p-n junctions were grown in the Ioffe Institute in Sankt Petersburg in Russia by liquid phase epitaxy on Te-doped GaSb substrates. The composition of the investigated layer was $Al_{0.3}Ga_{0.7}As_{0.02}Sb_{0.98}$ doped by Te to $10^{17}/cm^3$.

The QW lasers were grown by MOCVD at the University of British Columbia in Vancouver, Canada. The 80 A GaAs QW, and the graded Al content optical waveguide layers were undoped, while the n- and p- type cladding layers were Si doped $Al_{0.6}Ga_{0.4}As$. Our results refer to DX in the 0.6 Al content cladding layer. LPE GaAlAs laser structures were grown with 0.3 Al content.

Results

In GaAlAsSb photodiode the effect of the heterobarriers onto the impedance is shown in the Cole-Cole diagrams in Fig.1 and Fig.2 .

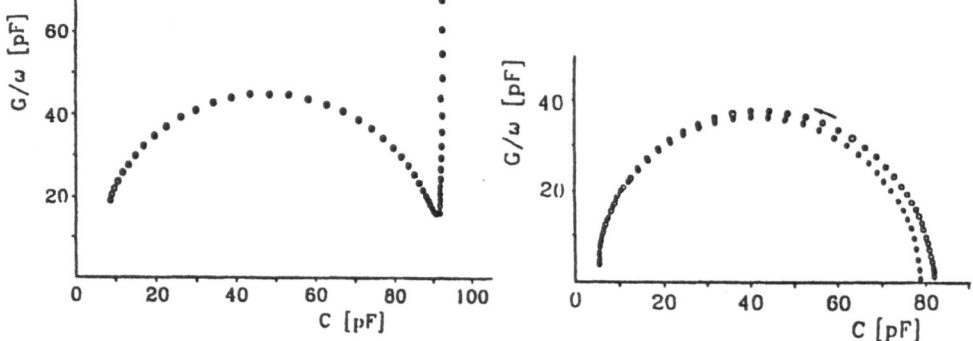

Fig.1 Cole-Cole diagram of the photodiode at 300 K

Fig.2 Cole-Cole diagram of the photodiode at 77 K before (o) and after (●) illumination

The GaAlAs LPE laser structure show similar effect. The capacitance of these samples as a function of temperature show that the effect is influenced by the free carrier distribution at the interface during cooling [5].

The capture barrier was determined from the pulse width giving half the saturation amplitude in DLTS. The emission barrier was determined from the temperature shift of the DLTS peak maxima. A capture barrier of E_B=0.33 eV, and an emission barrier of 0.306 eV were found in the $Al_{0.3}Ga_{0.7}As_{0.02}Sb_{0.98}$ samples. In the MOCVD GaAlAs samples a capture barrier of 0.31 eV, and an emission barrier of 0.38 eV were determined by DLTS. The Sn doped LPE samples have shown double DLTS peak with 0.18 and 0.28 eV emission energies. These values agree with the literature data. The time dependence of the capture process was strongly non-exponential in all cases. Illumination of the samples at 77K has revealed persistent photoconductivity.

The DX centers have shown similar recombination properties in all these materials. The recombination spectra measured in the GaAlAsSb p-n junctions is shown in Fig.3. The activation energy of the recombination agreed with the capture barrier of the DX levels in all cases, showing that the recombination

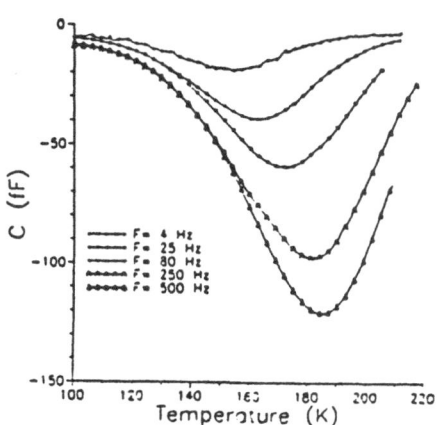

Fig.3 *The recombination spectra of the DX level in the GaAlAsSb samples*

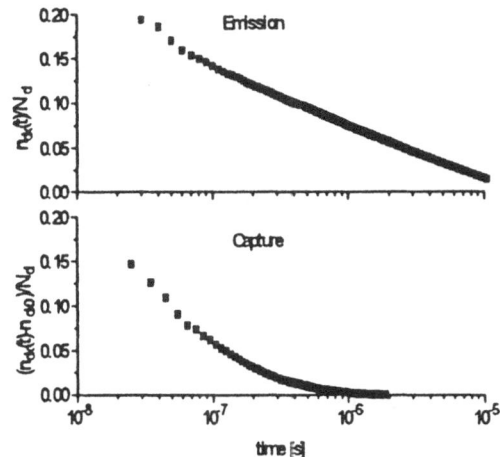

Fig.4 *The capture (lower plot) and the emission (upper plot) in LPE GaAlAs samle at 300 K.*

efficiency is limited by the electron capture. It indicates that recombination process is connected to the shallow-deep transition of DX centers.

The room temperature capture and emission transients of the the LPE GaAlAs samples are shown in Fig.4 on logarithmic scale. The signal is nonexponential from the nanoseconds range, and an extrapolation of the DLTS data does not fit to them. The capture signal changes around 200 K. At lower temperatures it is deep level like, while above it takes place in the depleted layer.

The following simple model is suggested to explain the DX related phenomena.

The two configuration states of DX are characterised by approximately equal total energic E_1 and E_2. The two states are separated by a barrier over 1 eV. The transition is possible only in the presence of free electrons, when the captured electron can supply the energy for the transition. It is analogous to a recombination enhanced defect reaction process. The occupancy of the two states in equilibrium is determined by the energy difference $\Delta E = E_1 - E_2$

$$n_{dx} = n_{ddx} \exp(\Delta E / kT)$$

where n_{dx} is the concentration in the shallow and n_{ddx} that in the deep state.

In this model the capcitance DLTS peaks describe the transition between the two states in the presence of free electrons. It means that the DLTS measures the excited state of the DX. This model can explain the usual DX phenoma, explains the similarity of DX centers in different materials and with different dopants, and it relates DX to the stability of the crystalline structure.

102

Conclusion

The DX centers were investigated in GaAlAs and GaAlAsSb samples grown by different techniques and in different structures. The values measured by DLTS agree well with the literature data. The persistent photoconductivity was observed in all materials. The temperature and the frequency dependence of the capacitance was shown to be sensitive to the heterobarriers. The recombination properties were dominated in all samples by the electron capture of the DX center. The results suggest that the recombination is important in the transition between the two DX configuration states. The capture and emission properties of the center around room temperature are markedly different from those extrapolated from low temperatures capacitance DLTS measurements. Based on these results we suggest a model, where the two configurations of the DX center are coupled by a recombination enhanced defect reaction.

Acknowledgements .

This work has been supported by the Hungarian National Research Fund (OTKA) under grant number 14114 and by the Grant Agency of the Czech Republic, grant number 202/93/1160.

References
1. D.V. Lang and R.A. Logan Phys. Rev. B 19,1015 (1979)
2. B. Stepanek, Sestakova, P. Hubik, V. Smid and V. Charvat
   J. Cryst. Growth 126, 617 (1993)
3. J.A. Wolk,W. Walukiewitz, M.L. Thewalt, and E.E. Haller
   Phys. Rev. Lett. 68, 3619 (1992)
4. P. M. Mooney J. Appl. Phys. 67, No. 3,R1 (1990)
5. C. Ghezzi, E. Gombia and R. Mosca   J. Appl. Phys. 70, 215 (1991)
6. D.J. Chadi and K.J. Chang Phys Rev. B39,10366(1989)
7. L. Dózsa Solid-State Electronics 29, 861 (1986)
8. L. Dózsa Solid-State Electronics 35,228 (1992)

MECHANICAL STUDY OF THE STRAINED $In_xGa_{1-x}As$/GaAs HETEROSTRUCTURES

Á.NEMCSICS, J.SZABÓ, S.GURBÁN and L.CSONTOS
Research Institute for Technical Physics of the Hungarian Academy of Sciences; P.O.Box 76 H-1325 Budapest, Hungary

1. Introduction

Heterostructures of compound semiconductors grown on the GaAs substrate have most importance in fabricating optical and high speed devices. The GaAs(001) surface is of great technological importance because it serves as a good substrate for production of electronic devices. Layers of $In_xGa_{1-x}As$ on GaAs because of mismatch in lattice parameter can be strained or relaxed, depending on growth technology. The molecular beam epitaxy (MBE) is an appropriate method for growing thin and controlled layers. The surface quality and composition of epitaxial layers can be best monitored in-situ with reflection high energy electron diffraction (RHEED). The wafer with strained epitaxial layer as expected is bended. The deviation of crystal planes can be investigated with X-ray topography, and the flatness of the surface by optical methods.

2. Experimental

The sample and its preparation. The growth of $In_xGa_{1-x}As$ layers and the measurements were performed in a special MBE equipment. Supplemented computer control was applied for good reproducibility of technological parameters [1]. For growth there were used two different kinds of GaAs wafers. One of both kinds was cut from (001) oriented (accuracy ±0.1°) Zn-doped crystals. Other subtrates were used undoped semiinsulating (SI) and oriented 2° off the (001) plane towards the [110] direction. The samples were mounted on sampleholders by liquid In. The sizes of the samples were about 7x9mm^2. After cleaning the sample, first pure GaAs was grown under suitable growth condition (Fig.1.). We have determined the composition of the $In_xGa_{1-x}As$ layer during growth from the time period of the RHEED-oscillations, taking the deviation in lattice parameter into consideration [2]:

$$x = \frac{\tau_{GaAs} - b^2\tau_{InGaAs}(x)}{\tau_{GaAs}} \tag{1}$$

103

J. Novák and A. Schlachetzki (eds.), Heterostructure Epitaxy and Devices, 103–106.
© 1996 *Kluwer Academic Publishers.*

where τ is the growth time period and $b = a_{InGaAs}(x)/a_{GaAs}$ is the ratio of the lattice parameters. $In_xGa_{1-x}As$ can be grown with good surface quality until high In composition only at special growth condition (Fig.1) [2].

The curvature measurement using X-ray diffraction. The curvatures of the samples were measured by double crystal X-ray topography. With this method, there is no limitation for measuring the curvature of a single crystal, but with an epitaxial layer overlapping the crystal, the method got new limits. The total intensity of the X-ray reflection will add-up from the intensity of the components. In case, if both intensity components are large enough and separable, isolated images can be made from the substrate and the epitaxial layer. But if any of them is neglegible, a decision has to be made about which is the dominant. When the substrate intensity is dominant (the layer is very thin), the curvature can be measured (Fig.2.) on the same way as measuring would happen to a single crystal. The principle of the method is, that only a part of the whole sample is in reflecting position. Turning the sample round, another part of the sample will get into reflecting position. The layer intensity component has a too wide range of reflection angle, so the whole surface of the sample will reflect, that`s why this intensity component is unsuitable for curvature measuring.

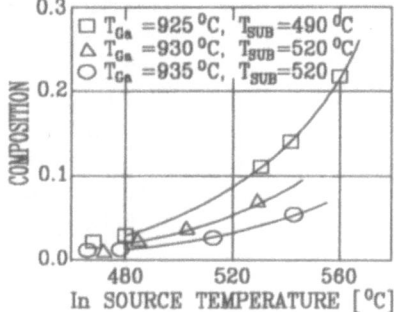

Figure 1. Composition x as function of the In-source temperature

Figure 2. Turning the sample 15 at the z-axis around by 0.0018° results a 5mm move of the reflection image

Chemical treatment of samples. For optical measurement a smooth back surface of sample is needed. The In used for sample sticking have etched the back surface of wafer roughly. First the back sides of samples were treated with 80°C Ga, and with alcoholic wool almost all In-Ga eutectic were removed at room temperature. Afterwards the front sides of samples were stuck to a glass with wax to protect from etching. The polishing process was carried out in three etch steps (in HCl 3-5min., in 1% HF 5-10hours and in $3HNO_3 + 1HF + nH_2O$ where n=0-5). The accurate etch parameters of different samples were adjusted individually. (Sample 08 was etched with too concentrated HF.)

Wafer flatness measurements using Makyoh method. According to the literature we have developed a Makyoh equipment [3] for determination of the radii of curvatures of semiconductor wafers [4]. Our designed equipment is shown by Figure 3. A flat mirror and the contour images the radius of curvature of the sample can be issued. We prepared a set of calibration mirrors by evaporating gold on plano-convex and plano-concave lenses with known radii of curvatures. The results of curves are shown by Figure 4. The Figure 5 shows a typical image from our optical flatness measurement. The radius of curvature was in this case about 10m.

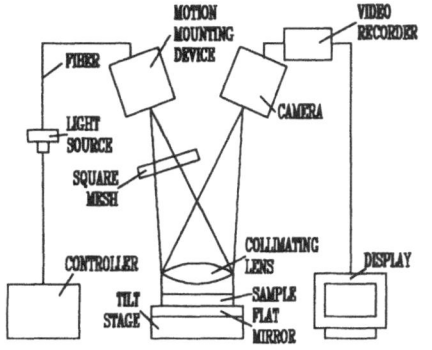

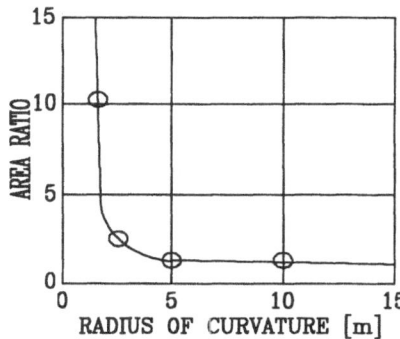

Figure 3. The scheme of our Makyoh apparatus

Figure 4. Area ratio of the Makyoh and contour image

3. Results and discussion

We have compared the results of both different measurements with the results of calculation. We have used a simple mechanical model. Lattice fitness results in the epitaxial layer compressive stress, in substrate at interface tensile stress and at the back side compressive stress again. The mechanical constants of our materials to be found in [5]. The lattice parameter is a linear function of In composition:

$$a_{InGaAs}(x) = a_{GaAs} + 0.4050 \cdot x \qquad (2)$$

We have supposed with calculation of curvature radius, that thickness of epitaxial layer can be neglected beside the thickness of substrate (300μm). With same ignored factors the radius R of curvature in meter:

$$R = \frac{62550 \cdot 1 + 250 \cdot 5 + b \cdot t_e}{t_e(1-b)} \qquad (3)$$

Where b is identical as in eq. (1) and t_e is the thickness of the epitaxial layer in μm. The results are shown in Figure 6. The dotted line shows the critical layer thickness [6].

TABLE 1. Measured and calculated results

Sample No	06	07	08	09	10	13	15	22
Substrat	:Zn	:Zn	:Zn:	:Zn	SI	SI	SI	SI
Thickness [μm]	5	0.02	0.02	1	0.01	0.02	1	0.2
Composition	0.03	0.1	0.3	0.05	0.4	0.1	0.05	0.05
R (X-ray) [m]	11.5	-	-	11.5	>80	-	15.9	-
R (Makyoh) [m]	>10	>>10	-	>10	>10	-8	≅10	≅10
R (comp.) [m]	6	440	150	18	225	440	18	87

106

The table shows the parameters of samples and the radius of curvature as result of both different measurement and calculation. Between results correlation can be found. By evaluation it must be taken into consideration that the optical measurement near wafer edge shows higher curvature because of polishing. The samples are not always being equally strained. For example, at the sample 09 an unevenness of the curvature was detected. One half of the sample had about 11.5m radius of curvature and the other half section showed nearly a perfect flat reflecting surface. So the sample 09 curving is not homogeneous. The sample 13 is concave.

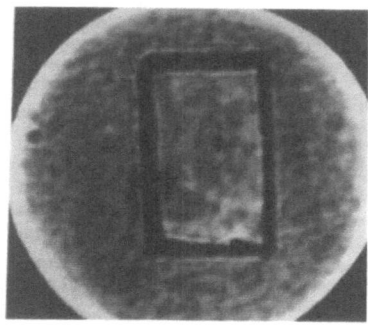

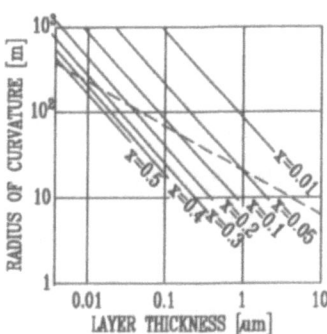

Figure 5. This Figure shows a 10m convex curvature sample 15. The light part shows the Makyoh-, dark part the contour-image.

Figure 6. Radius of curvature as function of the layer thickness at different composition

Acknowledgments

The preparation of samples was supported for Á.N. in CAU Kiel by DAAD (Deutsche Akademischer Austauschdienst) and by the Hungarian Foundation Széchenyi. This work was supported still for J.Sz. by OTKA No.14094 (Hungarian National Scientific Research Fund).

References:

1. Geyer,M., Nemcsics,Á., Olde,J.,Reshöft,K., Manzke,R. and Skibowski, M. (1992) Computergesteuerte Optimierung der Molekularstrahlepitaxie mittels RHEED-Oszillationen, Verhandl. DPG(VI) **27** 513
2. Schnurpfeil,R., Reshöft,K., Müller,A., Nemcsics,Á., Manzke,R. and Skibowski,M.(1994) The electronic structure of InGaAs(001) (2x4) surfaces, J.El. Spectr. **68** 175-183
3. Szabó,J.,Makai,J.(1993) Mirror like surface investigation using Makyoh method (in Hungarian), El. Technol-Mikrotechn. **32** 15-18
4. Szabó,J., Riesz,F.and Szentpáli,B. (1995) Wafer curvature and flatness measurements using the magic mirror (Makyoh) method, (to be published)
5. Adachi, S.(1992) Physical Properties of III-V semiconductor compounds; John Wiley New York
6. Ordens,P.J. and Usher,B.F. (1987) Determination of critical layer thickness in InGaAs/GaAs heterostructures by X-ray diffraction, Appl. Phys. Lett. **50** 980-982

HOT ELECTRONS AT SEMICONDUCTOR HETEROJUNCTIONS

M. HORÁK
Dept. of Microelectronics, Fac. Electr. Eng. Comp. Sci., TU Brno
Údolní 53, 602 00 Brno, Czech Republic

Abstract

The model of electron transport across abrupt Np-heterojunctions based on the solution of the balance equations and thermionic-field emission through the interface potential barrier is presented.

1. Introduction

Theoretical discussions of N-p heterojunctions have been presented by many authors, e.g. [1], [2]. Since a metal-semiconductor junction is a special kind of heterojunction, investigations of the injection mechanisms at the heterojunction have been performed in similar ways to those used in the Schottky barrier case. In modern submicrometer structures the carrier transport is determined by various nonlinear and nonlocal effects. The simplest hydrodynamic model based on the drift-diffusion approximation for carrier currents fails and significant attention is recently devoted to the balance equations [3]. This method was applied to high-field transport in the depletion layer of the metal-semiconductor contact [4] and in this paper we apply the same method to the abrupt N-p-heterojunction.

2. Heterojunction model

We consider the conductivity band edge of the $(N\text{-}Al_xGa_{1-x}As/p\text{-}GaAs)$-type, parabolical spherical bands and neglect the satellite conduction band minima and the hole current.

J. Novák and A. Schlachetzki (eds.), Heterostructure Epitaxy and Devices, 107–110.
© 1996 *Kluwer Academic Publishers.*

For steady state and one-dimensional electron transport the system of balance equations is [3]

$$p\frac{\partial v}{\partial x} = enE - \frac{\partial(nkT_{el})}{\partial x} - \frac{p}{\tau_m} ,$$

(1)

$$\frac{\partial(vW)}{\partial x} = jE - j\frac{k}{e}\frac{\partial T_{el}}{\partial x} - \frac{W - W_0}{\tau_E} .$$

(2)

In these equations p is the momentum density, v the electron velocity, j the electron current density, n the electron concentration, E the electric field intensity, T_{el} the electron temperature, W the mean kinetic energy density, τ_m and τ_E the momentum and energy relaxation times, W_0 the equilibrium energy related to the lattice temperature T, $W_0=3nkT/2$. The current density and momentum density are related by $j = env$, $p = mnv$, thus for steady state both are constant. The mean energy density W and the mean carrier kinetic energy w are defined as $W = nw = 3nkT_{el}/2 + mnv^2/2$.

As the concept of electron temperature is related to sufficiently high electron concentration, we consider only the forward bias of the heterojunction. The strong electric field in the depletion layer of the N-region $(-x_n< x < 0)$ and its small thickness for donor concentrations above approx. 10^{16} cm^{-3} (thus the possibility of quasi-ballistic transport) give reasons for application of (1) and (2). Hot electrons are injected above the barrier to the p-region. If the acceptor concentration is at least one order of magnitude higher than the donor concentration, the influence of the space-charge layer in the p-region $(0 < x < x_p)$ should not be considered. We assumme zero electric field in the neutral p-region $(x_p < x < x_C)$. To avoid problems with correct description of hot electron diffusion, we again apply eqs. (1), (2). Using the method of integration factor applied in [4] to the metal-semiconductor contact, in our case we obtain the following expression for the current density

$$j = \frac{ek}{m}\frac{n(b)T_{el}(b)\exp\left[\int_a^b \frac{eE}{kT_{el}}dx'\right] - n(a)T_{el}(a)}{\int_a^b\left(\frac{1}{\tau_m} + \frac{\partial v}{\partial x}\right)\exp\left[\int_a^x \frac{eE}{kT_{el}}dx'\right]dx} ,$$

(3)

electron concentration profile (4)

$$n(x) = n(b)\frac{T_{el}(b)}{T_{el}(x)}\exp\left[\int\limits_x^b \frac{eE}{kT_{el}}dx'\right] - \frac{jm}{ekT_{el}(x)}\int\limits_x^b\left(\frac{1}{\tau_m}+\frac{\partial v}{\partial x'}\right)\exp\left[\int\limits_x^{x'}\frac{eE}{kT_{el}}dx''\right]dx' ,$$

and mean carrier kinetic energy profile

$$w(x) = w_0 + \int\limits_x^b\left(eE + \frac{\partial(kT_{el})}{\partial x'}\right)\exp\left[\int\limits_x^{x'}\frac{dx''}{v\tau_E}\right]dx' . \tag{5}$$

where for the depletion layer in N-region we substitute $a=0$, $b=-x_n$, and for the neutral p-region $a=x_C$ (contact), $b=x_p$.

The above described transport model makes implicit use of a displaced Maxwell distribution function with electron temperature. At the heterointerface we consider the injection of high-energy electrons above the barier and tunneling through the barrier. Using the standard approach we calculate the electron current density at the heterointerface as the difference of flows from N to p and from p to N , thus

$$j = e\left[n(0)v_N - n(x_p)v_p\exp\left(\frac{e(U_{D2}-U_2)-\Delta E_C}{kT_{el}}\right)\right]F(\xi)(1+\delta) , \tag{6}$$

where v_N and v_p are mean electron thermal velocities, ξ is related to the shift of the Maxwellian, $F(\xi) = \exp(-\xi^2) - \sqrt{\pi}\,\xi\,erfc(\xi)$ is the emission function, δ characterizes the tunneling,

$$\delta = \frac{1}{kT_{el}}\int\limits_{E_{\perp min}}^{E_C(0_-)} t(E_\perp)\exp\left(\frac{E_C(0_-)-E_\perp}{kT_{el}}\right)dE_\perp , \tag{7}$$

and $t(E_\perp)$ is the transmission probability.

The boundary conditions at the heterointerface can be formulated as follows: (i) if the applied voltage $U=0$, then the current (6) accross the hetrointerface $x=0$ vanishes, this gives the relation between v_N and v_p; (ii) at $x=-x_n$ and $x=x_p$ the current transmitted above and through the barrier (6) is equal to the hydrodynamic current (3), this enables to calculate $n(0)$ and $n(x_p)$.

As a result we obtain the following expression for the current density: (8)

$$
j = \cfrac{ev_N(1-\exp(-\frac{eU_2}{kT_{el}}))F(\xi)(1+\delta)n(-x_n)T_{el}(-x_n)\exp\left[\int\limits_{0}^{-x_n}\frac{eE(x')}{kT_{el}(x')}dx'\right]}{T_{el}(0_-)+\frac{m_N v_N}{k}(1-\exp(-\frac{eU_2}{kT_{el}})F(\xi)(1+\delta)\int\limits_{0}^{-x_n}\left(\frac{1}{\tau_m}+\frac{\partial v}{\partial x}\right)\exp\left[\int\limits_{0}^{x}\frac{eE(x')}{kT_{el}(x')}dx'\right]dx}
$$

This is an analytical expression for the current density, but as the electron temperature is not known, it is necessary to solve the whole transport problem numerically.

The numerical algorithm can be described as follows:
(i) zero approximation: electron concentration profile is given by the potential inside the depletion layer, electron temperature is equal to lattice temperature; (ii) momentum and energy relaxation times calculation: various scattering processes should be considered (acoustic phonons, polar and nonpolar optical phonons and for momentum also ionized impurity scattering) and combined in accordance with Matthiessen's rule, formulas can be found in [5]; (iii) current density calculation (8); (iv) electron concentration profile (4) and velocity profile $v=j/(en)$; (v) electron energy (5) and temperature; (vi) the process is repeated with modified concentration and temperature starting with (ii) until good accuracy is reached.

The described model is quite general, based on the balance equation and shifted Maxwellian with electron temperature. It includes some special transport models as quasi-ballistic limit, strictly local hydrodynamic model and drift-diffusion model; the discussion of these special cases can folow [4].

References

1. Stettler, M.A., Lundstrom, M.S. (1994) A detailed investigation of heterojunction transport using a rigorous solution to the Boltzmann equation. *IEEE Trans. ED* **41**, 592-599.
2. Hjelmgren, H., Tang, T.W (1994) Thermionic emission in a hydrodynamic model for heterojunction structures. *Solid State Electronics* **37**, 1649-1657.
3. Bloetekjaer, K. (1970) Transport equations fo electrons in two-valley semiconductors. *IEEE Trans. ED* **17**, 38-47.
4. Darling, R.B. (1988) High-field nonlinear electron transport in lightly doped Schottky-barrier diodes. *Solid State Electronics* **31**, 1031-1047.
5. Ridley, B.K. (1982) *Quantum processes in semiconductors.* Clarendon Press, Oxford.

III/V-COMPOUND SEMICONDUCTORS ON SILICON

A. SCHLACHETZKI
Institut für Halbleitertechnik, Technische Universität Braunschweig
Hans-Sommer-Str. 66, D-38106 Braunschweig, Germany

1. Introduction

Optical communication systems with glass fibres operating at data rates of 2.5 Gbit/s are presently introduced into practical use. Systems for 10 Gbit/s and beyond are in development. Optoelectronic devices and electronic integrated circuits are available for this purpose, however both components not integrated together in a monolithic form.

Personal computers which are presently in use are limited in their system's frequency to around 100 MHz. Higher frequencies require the careful consideration of signal delay as well as distortion in the case of larger distances and a precise matching of the impedance of the connecting lines. Coaxial cables reach their limits because of the ever progressing miniaturization of electronic circuits.

For both fields of application cited above, a viable solution would be the monolithic integration of light transmitter and receiver with the electronic circuits which latter will preferably be based on silicon for the foreseeable future. Already now silicon can be used in integrated circuits, fabricated by well-established production techniques in industry, at data rates of 15 to more than 30 Gbit/s depending on the special application. An increase of speed by a factor of 2 to 3 can be expected for the next years [1]. If the monolithic integration of electronic circuits in Si with optoelectronic devices in direct-bandgap III/V-compound semiconductors could be accomplished, a number of advantages of Si substrates can be exploited:

- an optical transmitter can be placed directly on the Si substrate;
- the large size of available Si (as compared to GaAs substrates) can accommodate complex, electronic circuits;
- the mechanical strength of Si offers new functions for micro-electromechanical systems;
- the large thermal conductivity of Si leads to reduced temperature sensitivity;
- Si is available at low cost.

Additionally, the envisaged integration scheme would alleviate the problem of electromagnetic interference which will become more pressing with increased systems complexity. However, several problems arise from the mismatch in lattice constant of Si and most III/V-compounds, from their difference in thermal-expansion coefficient, and

111

J. Novák and A. Schlachetzki (eds.), Heterostructure Epitaxy and Devices, 111–124.
© *1996 Kluwer Academic Publishers.*

from the nonpolar character of Si. One might expect the same problems from wide-bandgap semiconductors, like GaN and others, which are presently developed on alien substrates, like sapphire and SiC, for green and blue emitting diodes. They will find their use with displays and compact data storage.

It is the intention of this paper to review the present status of III/V-compound semiconductors on Si substrates since a large volume of relevant data is available. Moreover, a comparison with wide-bandgap semiconductors which show several similar features might be useful for a better understanding of epitaxial growth in lattice-mismatched composites.

2. Devices in III/V-Compounds on Si

Almost a decade ago the interest in lattice-mismatched epitaxial growth was strongly revived leading to the demonstration of nearly any kind of device in III/V-compound semiconductors on Si [2]. Most convincing was the development of a multi-quantum-well laser (MQW) on Si emitting at a wavelength of 1.5 μm for 8000 h [3]. More recently a double-heterostructure laser of basically the same layer sequence was built for 1.3 μm operation; the aging test so far extended for 800 h with only 10% degradation [4]. The performance of these lasers is essentially the same as devices of equal structure, but processed simultaneously on InP substrates in lots of 30 devices. The key parameters threshold current and differential quantum efficiency are 36 mA and 0.19 W/A, respectively, for cw operation.

At the receiving end metal-semiconductor-metal photodiodes (MSM) are available which suit to the lasers and are based on the same technology. They compare favourably with test devices on InP substrate as far as the dark current and the quantum efficiency are concerned [5]. However, the fall time of pulses is very long, a frequently observed phenomenon which might be caused by undepleted regions at sharp interfaces between epitaxial layers. At the cost of a demanding electron-beam lithography, results much improved in this respect were obtained with patterned substrates [6]. More on the research side is the concept of a tandem wavelength-division photosensor where the shorter-wavelength signal is detected by an AlGaAs structure while an underlying pn junction in Si is employed for the signal component in the 0.8 to 1.0 μm wavelength range [7].

Whereas lasers are most sensitive to crystal defects, majority-carrier devices are not affected by high-density dislocations. This was experimentally proven by investigations with microwave metal-semiconductor field-effect transistors (MESFET) [8] and with high-electron-mobility transistors (HEMT) [9]. In both cases a comparison was made with test devices on native substrates. The MESFETs were around 10% within the attainable goal. For HEMTs the standard deviation of the threshold voltage is the most prominent parameter for industrial application. No significant influence of the substrate is observed.

Parallel to the lattice-mismatched heteroepitaxy, several variations of a grafting technique were developed during the last decade where a III/V-compound device or epitaxial layer is transferred to the Si substrate in a hybrid or semi-monolithic manner. An example is the integration of a laser diode with a Si substrate carrying a planar silica optical waveguide [10]. The precision alignment of the two optical components is achieved by a ridge on the laser chip which locks into a V-groove in the Si motherboard. This latter requires some micromachining because of the alignment V-groove and because of a specifically designed recess to house the upside-down mounted laser with its contact. In spite of a geometrical precision of around 1.5 μm, a coupling loss of 14 dB from laser to waveguide was encountered.

The alternative to this hybrid integration is the transfer of a III/V epitaxial layer on the Si substrate. The active device which is to be transferred (laser, photodiode, MESFET or others) can be processed either while the epitaxial layer is still on the III/V substrate (preprocessing) or after it has been transferred to the Si substrate (postprocessing). The second case promises easier alignment of the III/V device relative to the Si substrate. An important precondition is the simple and precise severing of the epitaxial film from the III/V substrate. This involves a sacrificial layer between the substrate and the device layers which can be dissolved by chemical etching to remove the III/V substrate.

To provide sufficient mechanical stability the device layers are covered by thick Apiezon W wax at the beginning of the epitaxial lift-off (ELO). The wax is removed once the device layers have been bonded to the Si [11]. Generally, the Si substrate is prepared by evaporated Pd to ensure a tight bonding by the formation of an intermetallic compound between the Pd and the III/V layers. For safe bonding the Si substrate must be smooth and planar, e.g. better than 50 nm across distances of 10 μm. Such stringent requirements are not easily met by Si substrates containing electronic circuitry connecting to the grafted III/V device. Techniques are suggested to solve this problem [11].

The ELO process was used to provide up to 4 cm² of a 5" Si wafer with postprocessed GaAs LEDs [11]. However, doubts were raised whether postprocessed devices can perform as well as preprocessed devices since device fabrication normally requires heat cycles. Therefore the difference in thermal expansivity between III/V layer and Si which is a major concern with lattice-mismatched heteroepitaxy comes into play. This is exemplified by postprocessed pin photodetectors showing substantially enhanced dark currents [12]. Although the prospects of precise lateral alignment are lost, a number of preprocessed devices were transferred by ELO. Among them are a surface-emitting laser [13], a resonant-cavity photodetector and LED [14], and an MSM photodetector [15].

To circumvent the problems of difference in thermal expansion coefficient and the ensuing residual thermal stress, attempts were made to prevent threading dislocations from penetration from the Si interface to the active region of the device. This was accomplished by epitaxially growing a GaP/GaAs buffer layer, an InAlGaAs stepgraded layer, and an InP cover layer on the Si substrate [16]. On the InP substrate an InGaAs/InP-MQW structure plus InP cladding layers were deposited which were separated from the InP substrate after bonding by way of a sacrificial layer. This procedure very much resembles the ELO process with the addition of the epitaxial layers on the Si substrate. The result is the very low etch-pits density (EPD) of around 10^4cm^{-2}.

3. Lattice-Mismatched MOVPE Growth

As any other epitaxial growth-technique, also the metal-organic vapour-phase epitaxy (MOVPE) has to encounter four main factors: (1) the match in crystal structure and lattice constant between substrate and epitaxial film; (2) the congruence of the thermal expansion coefficients; (3) the polar or nonpolar character of the constituents of the composite system affecting the bonds at the substrate/film interface; (4) a substrate-surface finish free of work damage. This last point which was already of great importance during the development of silicon-on-sapphire structures is also of concern with GaN on sapphire [19]. It is as difficult to quantify as the third point.

Tab. 1 Lattice properties of substrate and layer materials as selected from [17, 18]. The lattice constants for the hexagonal structures apply to the basal plane (a) and to the hexagonal axis (c). If known, the thermal expansion coefficient (perpendicular to the c-axis) is given for low temperatures (low value) up to high temperatures near the crystal-growth temperature (large value).

Material	structure	lattice constant a / c (nm)	thermal expansion coeff. $\perp$c(K^{-1})
Si	diamond	0.543095	3.2 ... 4.2 × 10^{-6}
sapphire (α-Al$_2$O$_3$)	· corundum	0.4765 / 1.3001	4.5 ... 5.4 × 10^{-6}
6H-SiC	(hexagonal)	0.30806 / 1.51173	4.2 ... 5.4 × 10^{-6}
GaAs	zincblende	0.565325	6.5 ... 7.5 × 10^{-6}
InP	zincblende	0.58687	3.5 ... 6.5 × 10^{-6}
AlN	wurtzite	0.311 / 0.498	5.27 × 10^{-6}
GaN	wurtzite	0.3186 / 0.5176	3.17 ... 7.75 × 10^{-6}

Tab.1 lists some data on the lattice constant and on the thermal expansivity of substrates (first group) and of semiconductors (second group) which are of interest in the present context. We note, however, that a closer match of the lattice constant not necessarily leads to an improved epitaxial growth. For comparison in this section we augmented Tab. 1 by nitrides which are presently very actively studied for short-wavelength-laser applications. The nitrides are commonly grown, by MOVPE on sapphire of (0001) surface with the c-axis of nitride and substrate in parallel. The hexagonal basal plane of the film is rotated by 30° with respect to the sapphire resulting in a mismatch of the relevant crystal planes of 15% for GaN [20, 21]. Because of this large value, the polytype 6H of the numerous modifications of SiC is considered as an alternative substrate. 6H-SiC which originates from a particular stacking sequence of the Si-C tetrahedron in a hexagonal unit cell [17] differs by 3.4% in its lattice constant from GaN if both basal planes are oriented alike and if the c-axes coincide [22].

However, as mentioned previously the reduced lattice mismatch alone does not guarantee superior epitaxial growth [20]. We emphasize in the following the MOVPE growth of GaAs and InP on exactly (100) oriented Si substrates with a lattice mismatch of 4.1 and 8.1%, respectively $((a_{film}-a_{sub})/a_{sub})$. Frequently, Si substrates misoriented by a few degrees off towards the [011] direction are used. But they are not in accordance with the exact orientation which is well introduced in the electronics industry.

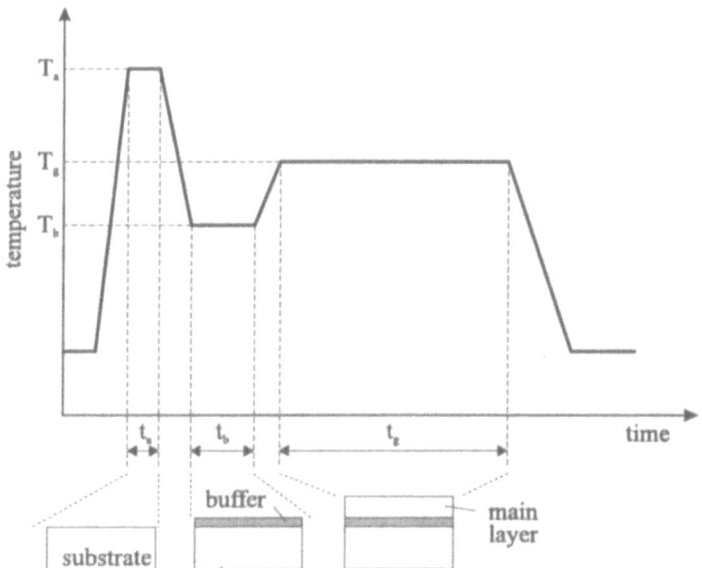

Figure 1. Schematic representation of lattice-mismatched growth by MOVPE.

Fig. 1 schematically depicts the course of the temperature during MOVPE growth of lattice-mismatched films. As a pretreatment the substrate is annealed *in situ* at a temperature T_a for a time interval t_a of around 10 min. In the case of a Si substrate this phase serves to deoxidize the surface and to saturate open bonds, e.g. bonds of Si surface atoms by As provided by an AsH_3 preflow [23, 24].

This prepares the substrate for the crucial step of buffer-layer growth at a much reduced temperature T_b [19-25]. The buffer grown first in the 3-dimensional mode and eventually in the 2-dimensional mode during time intervals of around 10 min accommodates the misfit dislocations resulting from the lattice mismatch and serves as a nucleus of the subsequently deposited main layer at the elevated temperature T_g. The growth time t_g depends on the intended main-layer thickness. It can be determined from the growth rate which is typically around 1 μm/h. It can be expected that the main layer is essentially stress-free at T_g. However, while the crystal is cooled down to room temperature after the completed growth mechanical tensions build up due to the difference in thermal expansions of film and substrate (cf. Tab. 1). The following Tab. 2 gives some indications on the parameters encountered during the low-pressure MOVPE as outlined before.

Tab. 2 Approximate values of essential parameters during low-pressure MOVPE of lattice-mismatched composites.

Film/substrate	T_a (°C)	T_b (°C)	buffer thick-ness (nm)	T_g (°C)
GaAs/Si	≈ 950	≈ 400	10 ... 20	≈ 700
InP/Si	≈ 950	≈ 400	40 ... 80	600 ... 650
GaN/sapphire	900 ... 1000	400 ... 600	20 ... 50	1020 ... 1050

Successful MOVPE growth leads to specular surfaces of the epitaxial layer (rms roughness around 1 nm). This is also true of area-selective growth which will be employed in the case of lasers on Si circuits. For area-selective epitaxy the growth procedure is interrupted after the buffer layer has been deposited. The buffer is then photolithographically patterned. After the substrate has been reintroduced in the MOVPE reactor the main layer grows only in those areas of the Si substrate where the buffer layer was not removed [26]. This growth technique is compatible with the fabrication of MOS technology. MOS circuits in Si perform satisfactorily after the high-temperature steps of the III/V-crystal epitaxy [27].

4. Crystal Defects in Epitaxial Films

In the following discussion we concentrate on III/V-compound semiconductors on Si. However, the main results apply to a range of heteroepitaxial systems differing widely in lattice constant, thermal expansion, and/or ionicity of binding [28].

If an atom impinging on the substrate from the gas phase is trapped at the surface it diffuses along it until it is more tightly bound to an epitaxial nucleus. It might also be desorbed back into the gas phase. This process is the more likely the higher the temperature. Therefore, the low temperature T_b for the buffer growth (cf. Fig. 1) is favourable for the formation of an epitaxial film. In other words, at low T_b the growth of nuclei is stimulated. In addition their critical radius beyond which they can grow is smaller. Also the nucleation rate of epitaxially oriented nuclei is higher since there are fewer broken bonds at the interface between substrate and epitaxial nucleus. Misoriented nuclei are more likely to collapse since their atoms are more easily desorbed into the gas phase. Finally, epitaxial nuclei grow faster laterally than

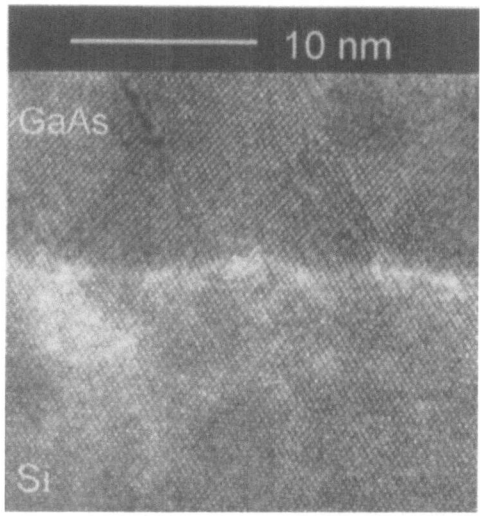

Figure 2. High-resolution TEM microphoto-graph of epitaxially grown GaAs on Si.

misoriented nuclei since the nucleus/substrate interfacial energy is smaller in the former case. Consequently less energy is stored in the interface and the total system assumes a lower energy. Briefly, at optimum growth conditions (= sufficiently low T_b) more epitaxially oriented nuclei grow with sufficiently large radii so that they are able to

quickly grow laterally. Once they coalesce and cover the total substrate surface the buffer layer growth is finished. The buffer layer then serves as the "quasi substrate" for the nearly homoepitaxial growth of the main layer.

To accommodate the lattice mismatch, misfit dislocations form already in the epitaxial nuclei along the nucleus/ substrate interface. As an example Fig. 2 shows a high-resolution transmission-electron-microscope microphotograph (TEM) of a GaAs film on Si clearly demonstrating epitaxial growth [29], but also dislocation lines penetrating the picture plane vertically at the horizontal separation between GaAs and Si (middle of Fig. 2).

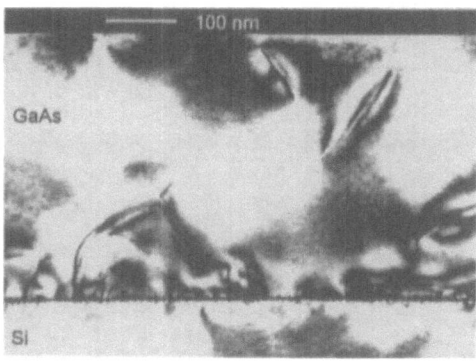

Figure 3. Cross-section of GaAs on Si. TEM microphotograph. (Kindly provided by G.-P. Tang, H.-H. Wehmann, and M. Seibt).

118

epitaxial growth [29], but also dislocation lines penetrating the picture plane vertically at the horizontal separation between GaAs and Si (middle of Fig. 2). They are around 10 to 13 nm apart. This period is also found from the black, wavy line in the lower part of Fig. 3 representing the misfit dislocation in the GaAs buffer on Si. The main layer can be grown at the elevated temperature T_g (cf. Fig. 1) since the buffer presents a clean, native surface for twodimensional layer growth. Threading dislocations visible as dark curves in the GaAs film of Fig. 3 penetrate the main layer.

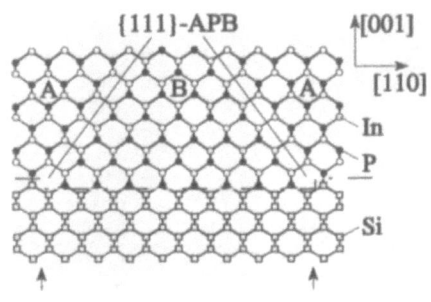

Figure 4. Schematic of an InP epitaxial layer on a Si substrate. The arrows point to the position where monatomic steps on the Si surface (broken line) are located.

The misfit-dislocation period as given above compares well with the theoretically expected Vernier period of 10 nm calculated for the lattice-mismatch between GaAs and Si. For InP/Si both these figures are 5 nm. In the case of AlN on sapphire with its lattice mismatch of 12.5% also good agreement at 2 nm is found [21].

Antiphase domains (APD) are defects which can be found in compound semiconductors. APD boundaries (APB) can be interpreted as planar defects originating from monatomic steps in the substrate surface. They are indicated by arrows pointing towards the steps in the broken line in Fig. 4. This line marks the surface of the Si substrate on which InP is deposited. The direction of the bonds of the surface atoms between the two steps is in the plane of Fig. 4 whereas those to the right and to the left are vertically oriented. If the epitaxial growth begins with a P monolayer the result is a film containing a domain B whose sequence of atoms is shifted relative to the surrounding domains A. For the growth mode of Fig. 4 APBs in {111} planes result which can be understood as stacking faults. The same crystal structure is found when prismatic nuclei growing on the lower and the upper level of the monatomic step merge to form the epitaxial film [28]. Due to the particular orientation of the bonds of the Si surface-atoms these nuclei are rotated by $\pi/2$ with respect to each other.

APDs appear as irregularly shaped islands on the [001] surface of the InP layer as demonstrated by Fig. 5 [24]. They are visible by microscopic inspection (as grown film, upper-left part of Fig. 5). If an

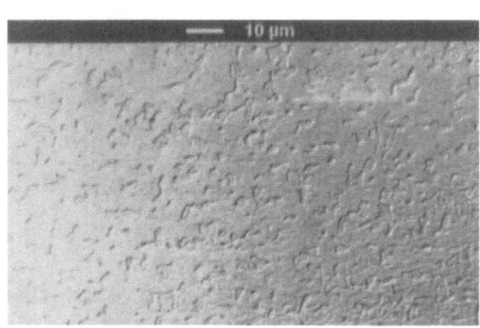

Figure 5. Antiphase domains. Plan view of InP on Si.

anisotropic etching is applied clearly the vertical etching patterns within the APDs and the horizontal features in the surrounding matrix emerge (lower-right part of Fig. 5).

APDs can be avoided by either improving the growth process or by using vicinal substrates where diatomic steps are prevailing [24].

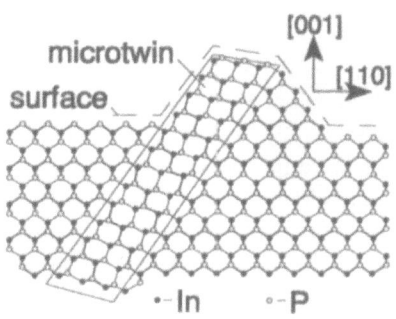

As another consequence of the hypothesis of the threedimensionality of the first steps in the epitaxial growth twins are frequently observed. We use the common ABCABC... notation to set up the hexagonal close-packed lattice (like zincblende) in the [111] direction. The prismatic nuclei grow by sequential deposition of crystal planes on the {111} surfaces. If a stacking fault occurs and if the stacking sequence is subsequently reversed a twin with lattice planes in the

Figure 6. Schematic representation of a twin in a zincblende lattice.

order ABCA/CBACBA... results. After coalescence of neighbouring nuclei a continuous film appears comprising twin lammellae as indicated in Fig. 6. The ensueing lattice configuration can also be understood as the reflection of the lattice on the (111) plane as the coherent twin boundary, whose trace appears as lines tilted at an angle of 54°, plus

a stacking fault at this plane. Since the orientation of the epitaxial film is changed within the microtwin its growth velocity is also changed. As depicted in Fig. 6, we expect a growth irregularity extending on the film surface along the [$\bar{1}$10] direction. This expectation is borne out by experiment as exemplified by the epitaxial InP layer on Si shown in Fig. 7 in cross-section. Twin lamellae block the movement of threading dislocations. Therefore, the trace of a twin on the epitaxial film surface is delineated by pits after dislocation etching. To obtain high-quality films it is important to avoid the

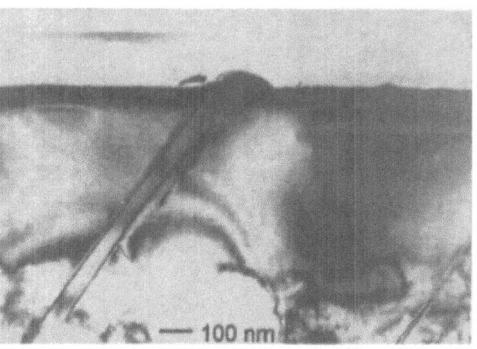

Figure 7. Twin lamellae in cross-section. InP on Si. TEM micrograph. (Kindly provided by G.-P. Tang, H.-H. Wehmann, E. Bugiel.)

growth of twins, e.g. by an underlying GaAs layer in the case of InP on Si [30].

120

5. Characterization Methods

The main incentive to study lattice-mismatched heterostructures is the potential for new devices. Consequently, a large group of publications is concerned with basic characteristics of device performance, e.g. [4-9]. Rather separate from this, the other large group investigates the epitaxial layers proper. We quoted a number of the second group in our discussion of GaN. In this case as well as with GaAs and InP, the main characterization methods are photoluminescence (PL) [22, 25], x-ray diffractometry (XRD) [19, 22], the measurement of carrier mobility μ [19], and TEM [20, 21, 28]. Atomic force microscopy is finding more and more interest [20]. Relatively few results are available on the optical absorption [25], although as proven by spectroscopic ellipsometry [31] a distinct effect can be expected. As shown in the contribution by Peiner et al., the observation of electron-channelling patterns can be advantageously employed to determine the crystalline quality and the mechanical stress in the epitaxial layer.

Very few investigations combine the characteristics of a layer with the performance of the device made thereof, as is done as an example for GaN-based LEDs in [32]. The main result is that even very large numbers of dislocations in GaN, which are of pure edge-type and penetrate the epitaxial film along the c-axis [33], do not affect the device efficiency as adversely as with other III/V compounds. The dislocations appear to originate from small-angle boundaries forming upon coalescence of the growth nuclei, which we discussed in the preceding section.

Only limited information on the material characteristics in spatial resolution is available. In most cases either capacitance-voltage measurements (CV) are performed to obtain the electrically-active carrier concentration [34, 35] or secondary-ion mass spectroscopy to find the in-depth composition [36]. A better understanding of the film/substrate interface and of the effect of the dislocations is gained if such measurements are combined with the profile of the mobility μ. As an example, Fig. 8 shows measurements taken with an InP layer on (100) Si [37].

The mobility μ and the electron concentration n were obtained by differential Hall measurements in the van-der-Pauw configuration after sequential

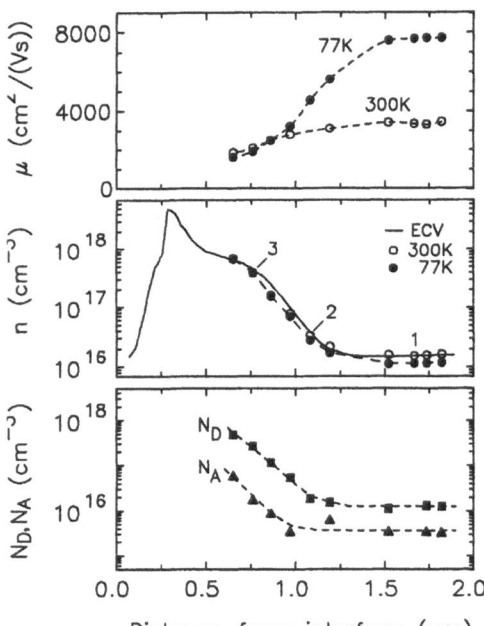

Figure 8. Mobility μ, electron concentration n, and donor and acceptor concentration N_D and N_A, respectively, In a 1.8 μm thick InP film in dependence on the distance from the Si-substrate surface.

removal of the epitaxial layer by chemical etching. The n profile is convincingly confirmed by electrochemical CV analysis (ECV). If the temperature dependence of n is evaluated the donor (N_D) and acceptor concentration (N_A) result as displayed in Fig. 8c. Similarly, the scattering mechanisms can be found from the temperature dependence of μ. They are ionized impurities, polaroptical phonons, space charges, and - what is of interest here - dislocations. This allows to determine the dislocation density within the epitaxial layer. The result are circles in Fig. 9 which is supplemented by XRD measurements of the etch-pits density (EPD), and by a theoretical curve [37].

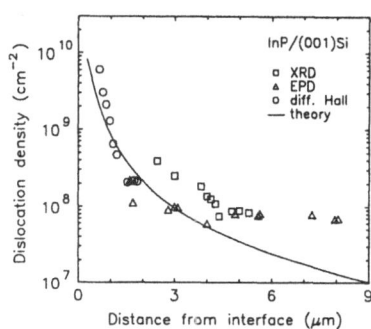

Figure 9. Dislocation density in InP on Si in dependence on the distance from the film/substrate interface.

We conclude that the dislocation density rapidly increases within the lattice-mismatched epitaxial film when approaching the film/substrate interface (Fig. 9). Related with it is the increase of N_A whereas N_D results from Si outdiffusing into the film (Fig. 8c). N_D saturates around $2 \times 10^{19} cm^{-3}$ at the interface. We speculate that the dislocation density, i.e. N_A, saturates at the same level leading to decreasing n by compensation (Fig. 8b). At these high levels dislocations become more and more effective as scatterers adversely affecting μ (Fig. 8a). It is fortunate that μ deteriorates only at very large dislocation densities ($10^8 cm^{-2}$ and more) [38]. Thus, at least majority-carrier devices are not much influenced by dislocations [9].

Dislocations are introduced into lattice-mismatched composites at growth temperature in order to relieve mechanical strain. This does not mean that the system is stress-free at room temperature since due to the incongruence of the thermal expansion coefficient (cf. Tab. 1) residual strain remains at low temperatures. In the case of InP on Si we expect a biaxial strain which is compressive within the plane of the epitaxial layer and as a consequence tensile in the normal direction . We expect a reduced bandgap with tension and an increased bandgap with compression leading to an ellipsoidal Fermi-body. This is indeed reflected in very fundamental properties, as for example in the effective mass m_c. Fig. 10 shows the results of Shubnikov-de Haas measurements taken with the magnetic field varying from the

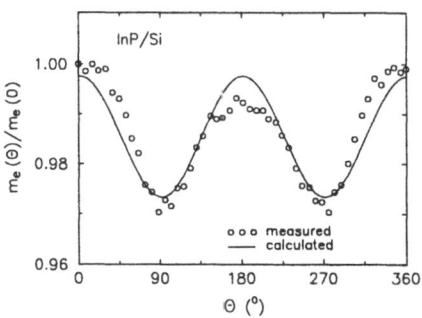

Figure 10. Effective mass m_c in dependence on the angle θ relative to the normal of an epitaxial InP layer on Si. Result of Shubnikov-de Haas measurements.

direction of the epitaxial InP-layer normal (θ = 0 or 180°) to the in-plane direction (θ = 90° or 270°) [39]. At field directions 0 or 180° when the electrons are probing the in-plane compression m_e is increased. At 90° or 270° when the cyclotron rotation is in the vertical plane tension is experienced leading to the minima of m_e. The calculation shown as solid curve in Fig. 10 and based on a Fermi ellipsoid supports this interpretation.

6. Conclusion

After the successful introduction of silicon-on-sapphire in integrated circuits the lattice-mismatched heteroepitaxy was employed for GaAs- and InP-based devices on Si as well as with GaN on sapphire. Whereas in the latter case blue-emitting LEDs are in the centre of interest almost any kind of device was demonstrated in the former case. There are many similarities in the MOVPE growth of both systems. One is the importance of the buffer layer which determines the quality of the main layer [20, 21, 23-25, 31]. In both cases misfit dislocations are incorporated to accommodate the lattice mismatch. However, dislocations in GaN seem not to be as detrimental to device performance as in InP-based alloys or particularly in GaAs-based alloys. Efficient GaN LEDs can be made even at dislocation-density levels of up to 10^{11} cm^{-2} [32]. The dislocations in GaN are of purely-edge type oriented along the c-axis. A consequence might be that they are difficult to suppress [33].

One of the techniques to avoid the adverse effects of dislocations on devices is the introduction of strained superlattices with InP- or GaAs-based compounds [4, 5, 40]. Others are annealing steps [4, 7, 16, 36] or the stepwise reduction of the lattice mismatch by interposing a GaAs layer between Si and the InP-based alloy [5]. The growth process is simplified by use of epitaxial Si on the Si substrate [4, 36]. APD-free material is obtained if the growth is adequately controlled [24]. Finally, patterned substrates promise to improve layer and device quality [6, 41].

It appears that sufficient knowledge about lattice-mismatched heteroepitaxy is available to develop useful devices. Whether they will be introduced into wide use depends on the urgency of a particular application.

Acknowledgement

Many helpful discussions with E. Peiner and H.-H. Wehmann are gratefully acknowledged.

References

1. Rein, H.-M. (1994) as quoted in *Mikroelektronik* **8**, 256
2. Schlachetzki, A., and Wehmann, H.-H. (1991) Mechanical Strain Relaxation During Lattice-Mismatched

Epitaxial Growth, *Solid-State Phenomena* **19 & 20**, 551-562

3. Sugo, M., Mori, H., Itoh, Y., and Sakai, Y. (1992) InP-based optical devices on Si substrates, Ext. Abstract 1992 *Int. Conf. Solid State Dev. and Mat.*, 656-658

4. Yamada, T., Tachikawa, M., Sasaki, T., Mori, H., Kadota, Y., and Yamamoto, M. (1995) Stable CW operation of 1.3 μm double-heterostructure laser heteroepitaxially grown on Si, *Electronics Lett.* **31**, 455-457

5. Sasaki, T., Enoki, T., Tachikawa, M., Sugo, M., and Mori, H. (1994) InAlAs/InGaAs metal-semiconductor-metal photodiodes heteroepitaxially grown on Si substrates, *Appl. Phys. Lett.* **64**, 751-753

6. Dröge, E., Schnabel, R.F., Böttcher, E.H., Grundmann, M., Krost, A., and Bimberg, D. (1994) High-speed InGaAs on Si metal-semiconductor-metal photodetectors, *Electronics Lett.* **30**, 1348-1350

7. Yang, M., Umeno, M., Jimbo, T., Shinmizu, H., Soga, T., Egawa, T., and Azuma, Y. (1994) Integrated wavelength-division photosensor using GaAs on Si, *Sensors and Actuators* A **40**, 121-123

8. Georgakilas, A., Halkias, G., Christou, A., Papavassiliou, C., Perantinos, G., Konstantinidis, G., and Panayotatos, P.N. (1993) Microwave Performance of GaAs-on-Si MESFET's with Si Buffers, *IEEE Trans. Electron. Dev.* **40**, 507-512

9. Aigo, T., Jono, A., Tachikawa, A., Hiratsuka, R., and Moritani, A. (1994) High uniformity of threshold voltage for GaAs/AlGaAs high electron mobility transistors grown on a Si substrate, *Appl. Phys. Lett.* **64**, 3127-3129

10. Jones, C.A., Cooper, K., Nield, M.W., Rush, J.D., Waller, R.G., Collins, J.V., and Fiddyment, P.J. (1994) Hybrid integration of a laser diode with a planar silica waveguide, *Electronics Lett.* **30**, 215-216

11. Ersen, A., Schnitzer, J., Yablonowitch, E., and Gmitter, T. (1993) Direct Bonding of GaAs Films on Silicon Circuits by Epitaxial Liftoff, *Solid St. Electron.* **36**, 1731-1739

12. Justice, J., Corbett, B., Walsh, S., Cosidine, L., and Kelly, W.M. (1995) Dark curents in pin photodetectors fabricated by preprocessing and postprocessing techniques of epitaxial liftoff, *Electronics Lett.* **31**, 1382-1383

13. Fatollahnejad, H., Mathine, D.L., Droopad, R., Maracas, G.N., and Daryanani, S. (1994) Vertical-cavity surface-emitting lasers integrated onto silicon by PdGe contacts, *ibid.* **30**, 1235-1236

14. Salvador, A., Huang, F., Sverdlov, B., Botchkarev, A.E., and Morkoc, H. (1994) InP/InGaAs resonant cavity enhanced photodetector and light emitting diode with external mirrors on Si, *ibid.* **30**, 1527-1529

15. Herrscher, M., Grundmann, M., Dröge, E., Kollakowski, St., Böttcher, E.H., and Bimberg, D. (1995) Epitaxial liftoff InGaAs/InP MSM photodetectors on Si, *ibid.*31, 1383-1384

16. Mori, K., Tokutome, K., Nishi, K., and Sugou, S. (1994) High quality InGaAs/InP muliquantum-well structure on Si fabricated by direct bonding, *ibid.* **30**, 1008-1009

17. Landolt-Börnstein, (K.-H. Hellwege, ed.) (1982) Numerical Data and Functional Relationships in Science and Technology, vol. 17a, Springer, Berlin etc.

18. Cullen, G.W., and Wang, C.C., eds. (1978) Heteroepitaxial Semiconductors for Electronic Devices, Springer, New York etc.

19. Doverspike, K., Rowland, L.B., Gaskill, D.K., and Freitas, J.A. (1995) The Effect of GaN and AlN Buffer Layers on GaN Film Properties Grown on Both C-Plane and A-Plane Sapphire, *J. Electron. Mat.* **24**, 269-273

20. George, T., Pike, W.T., Kan, M.A., Kuznia, J.N., and Chan-Chien, P. (1995) A Microstructural Comparison of the Initial Growth of AlN and GaN Layer on Basal Plane Sapphire and SiC Substrates by Low Pressure Metalorganic Chemical Vapor Deposition, *ibid.* **24**, 241-247

21. Ponce, F.A., Major, J.S., Plano, W.E., and Welch, D.F. (1994) Crystalline structure of AlGaN epitaxy on sapphire using AlN buffer layers, *Appl. Phys. Lett.* **65**, 2302-2304

22. Wetzel, C., Volm, D., Meyer, B.K., Pressel, K., Nilsson, S., Mokhov, E.N., and Baranov, P.G. (1994) GaN epitaxial layers on 6H-SiC by the sublimation sandwich technique, *ibid.* **65**, 1033-1035

23. Lubnow, A., Tang, G.-P., Wehmann, H.-H., Schlachetzki, A., Bugiel, E., and Zaumseil, P., (1992) The Influence of a Hydride Preflow on the Crystalline Quality of InP Grown on Exactly Oriented (100) Si, *J. Electron. Mat.* **21**, 1141-1146

24. Tang, G.-P., Lubnow, A., Wehmann, H.-H., Zwinge, G., and Schlachetzki, A. (1992) Antiphase-Domain-

124

Free InP on (100) Si, *Japan. J. Appl. Phys.* Pt. 2, **31** L 1126-L 1128

25. Akasaki, I., and Amano, H. (1994) Widegap Column-III Nitride Semiconductors for UV/Blue Light-Emitting Devices, *J. Electrochem. Soc.* **141**, 2226-2271

26. Tang, G.-P., Wehmann, H.-H., Peiner, E. Lubnow, A., Zwinge, G., Schlachetzki, A., and Hergeth, J. (1992) A New Maskless Selective Growth Process for InP on (100) Si, *J. Appl. Phys.* **72**, 4366-4368

27. Lubnow, A., Tang, G.-P., Wehmann, H.-H., Peiner, E., and Schlachetzki, A. (1994) The Effect of III/V-Compound Epitaxy on Si Metal-Oxide-Semiconductor Circuits, *Japan. J. Appl. Phys.* **33**, 3628-3634

28. Pirouz, P., Ernst, F. , and Cher, T.T. (1988) Heteroepitaxy on (001) Silicon: Growth Mechanisms and Defect Formation, *Mat. Res. Soc. Symp.* Proc. **16**, 57-70

29. Wehmann, H.-H., Tang, G.-P., Koch, A., Seibt, M., and Schlachetzki, A. (1995) Twin Formation during Epitaxial Growth of InP on Si, *Solid-State Phenomena* **47-48**, 547-552

30. Wehmann, H.-H., Tang, G.-P., Bartels, A., Iber, H., Dettmer, K., Schlachetzki, A. (1995) High Quality $In_{0.53}Ga_{0.47}As$ Layers on (001)Si, *6th European Workshop on Metal-Organic Vapour Phase Epitaxy and Related Growth Techniques*, Gent, Belgium June 25-28

31. Zwinge, G., Ziegenmeyer, I., Wehmann, H.-H., Tang, G.-P., and Schlachetzki, A. (1993) InP on Si substrates characterized by spetroscopic ellipsometry, *J. Appl. Phys.* **74**, 5889-5891

32. Lester, S.D., Ponce, F.A., Craford, M.G., and Steigerwald, D.A. (1995) High dislocation densities in high efficiency GaN-based light-emitting diodes, *Appl. Phys. Lett.* **66**, 1249-1251

33. Quian, W., Skowronski M., De Graef, M., Doverspike, K., Rowland, L.B., and Gaskill, D.K. (1995) Microstructural characterization of α-GaN films grown on sapphire by organometallic vapor phase epitaxy, *ibid.* **66**, 1252-1254

34. Uchida, Y., Kakibayashi, H., and Goto, S. (1993) Electrical and structural properties of dislocations confined in a InGaAs/GaAs heterostructure, *J. Appl. Phys.* **74**, 6720-6725

35. Schnabel, R.F., Krost, A., Grundmann, M., Heinrichsdorff, F., Bimberg, D., Pilatzek, M., and Harde, P. (1993) Epitaxy of high resistivity InP on Si, *Appl. Phys. Lett.* **63**, 3607-3609

36. Mori, H., Tachikawa, M., Sugo, M. and Itoh, Y. (1993) GaAs heteroepitaxy on an epitaxial Si surface with a low-temperature process, *ibid.* **63**, 1963-1965

37. Bartels, A., Peiner, E., and Schlachetzki, A. (1995) The effect of dislocations on the transport properties of III/V-compound semiconductors on Si, *J. Appl. Phys.* **78** *(in print)*

38. Ohori, T., Ohkubo, S., Kasai, K., and Komeno, J. (1994) Effect of threading dislocations on mobility in selectively doped heterostructures grown on Si substrates, *ibid.* **75**, 3681-3683

39. Schneider, D., Himstedt, E., Schlachetzki, A., and Tang, G.-P. *(will be published)*

40. Kawai, T., Yonezu, H., Ogasawara, Y., Saito, D., and Pak, K. (1993) Suppression of threading dislocation generation in highly lattice mismatched heteroepitaxies by strained short-period superlattices, *Appl. Phys. Lett.* **63**, 2067-2069

41. Knall, J., Romano, L.T., Biegelsen, D.K., Bringans, R.D., Chin, H.C., Harris, J.S., Treat, D.W., and Bour, D.P. (1994) The use of graded InGaAs layers and patterned substrates to remove threading dislocations from GaAs on Si, *J. Appl. Phys.* **76**, 2697-2702

CRYSTAL GROWTH OF COLUMN III NITRIDES BY OMVPE

I. AKASAKI AND H. AMANO
Meijo University, Department of Electrical and Electronic Engineering,
1-501 Shiogamaguchi, Tempaku-ku, Nagoya 468, Japan

Abstract

Column III nitrides are one of the most promising materials for the applications to short wavelength light emitters as well as high-temperature electronic devices. Recent development of the technology and understanding of the growth mode in the heteroepitaxial growth of nitrides on highly-mismatched substrates enable us to grow high-quality nitride films. Conductivity control of both n-type and p-type nitrides has also been achieved. This paper reports the recent progress of crystal growth of column III nitrides by OMVPE. Electrical and luminescence properties of these nitrides are also described.

1. Introduction

Aluminum nitride (AlN), gallium nitride (GaN), indium nitride (InN) and their alloys AlGaInN posses the direct transition type band structure with the bandgap energy ranges from 1.9 eV for InN, 3.4 eV for GaN [1], and to 6.2 eV for AlN at room temperature (RT). Moreover, they have superior physical and chemical properties e.g. high thermal conductivity [2], high electron saturation velocity, and physical and chemical stability. Therefore they are promising for the fabrication of high-temperature electronic devices as well as short wavelength light emitters such as light emitting diode (LED) and laser diode (LD) in the blue to ultraviolet (UV) regions. The high-temperature electronic devices have many desirable electrical properties that are useful for the current high-performance engines and the future electronic components operated in a harsh environment. The short wavelength light emitters enable us to produce all solid state flat-panel full-color displays, high-performance optical storage and printing systems, and small medical apparatus.

To realize such devices, it is essential to grow high-quality column III nitride films and to control their electrical conductivity. During the two decades before 1985, many pioneering work on these nitrides was conducted on crystal growth and doping [1, 3-9], on device fabrication and physics [10] and on physical properties [11].

On the contrary to other III-V compounds such as GaAs and InP, however, it had been quite difficult to grow high-quality epitaxial nitride films with a flat surface free from cracks. This is mainly due to the lack of substrate materials with the lattice constant and thermal expansion coefficient close to those of GaN and nitride alloys. Furthermore, it has

125

J. Novák and A. Schlachetzki (eds.), Heterostructure Epitaxy and Devices, 125–134.
© *1996 Kluwer Academic Publishers.*

been well known that undoped nitrides were of strong n-type conductivity, and p-type nitrides had not been realized until recently. These problems had prevented from their applications to devices for a long time.

Yoshida et al. [12] reported that cathodoluminescence intensity from GaN can be increased by using AlN single crystal as an intermediate layer on sapphire substrate in molecular beam epitaxy (MBE) of GaN. However, the residual electron concentration was still on the order of 10^{19} cm^{-3} and the electron mobility was around from 30 to 40 cm^{2}V^{-1}s^{-1}.

In 1986, we proposed a new process [13] and succeeded in overcoming the problems mentioned above and in growing high-quality GaN with a specular surface free from cracks, for the first time, by the low-temperature-deposition of a very thin AlN buffer layer in organometallic vapor phase epitaxy (OMVPE) of GaN. The crystalline quality as well as the electrical and optical properties of GaN can be remarkably improved at the same time. The concept and the structure of the low-temperature-deposited AlN buffer layer [14], are essentially different from those of the single-crystalline AlN intermediate layer ever reported. Details of the AlN buffer layer will be described later. Immediately (1986), bright MIS-type blue LED with a brightness of about 200 mcd at a forward current of 10 mA, was developed by using this GaN. Also, the RT operation of UV stimulated emission was achieved from such a high-quality GaN film by optical pumping [15], for the first time. Khan et al. achieved the first surface-mode UV stimulated emission from optically pumped GaN [16] and developed GaN-based electronic devices [17], which were grown by this low-temperature-deposited buffer layer technique. Nakamura [18] reported the low- temperature-deposited GaN buffer layer method for the growth of GaN, the concept of which is essentially the same as that of the AlN buffer layer technique.

Conductivity control of n-type nitrides has been achieved by us using silane (SiH$_4$) as a doping gas and high-quality nitrides grown with the buffer layer [19].

In 1989, Amano et al. [20] succeeded in producing, for the first time in the world, distinct p-type GaN with low resistivity by low-energy electron beam irradiation (LEEBI) treatment of such a high-quality GaN doped with magnesium (Mg).

In this paper, details for the growth of GaN and nitride alloys using the low-temperature-deposited AlN buffer layer, and their electrical and luminescence properties are reported.

2. Growth of GaN by OMVPE Using AlN Buffer Layer

A dual-flow channel horizontal type OMVPE reactor [21] operated at an atmospheric pressure was used for the growth of GaN. Trimethylgallium (TMGa), trimethylaluminum (TMAl), trimethylindium (TMIn) and ammonia (NH$_3$) were used as source gases and hydrogen (H$_2$) as a carrier gas. In this system, the column III source gases and column V source gas (NH$_3$) were mixed just before reaching the substrate to maintain the laminar flow and to suppress the formation of adducts. Figure 1 shows the sequence of the process. The ordinates are the substrate temperature and the flow rates of organometallic compounds and ambient gases. Polished and etched sapphire C-faces were used as substrates. The misorientation was less than 1°. Until 1985, conventionally, GaN was grown directly on the substrate. While, in the new process, before GaN growth, the thin

AlN layer about 50 nm thick was deposited at around 600℃ by feeding TMAl and NH₃ diluted with H₂ as shown in Fig. 1. Then the substrate temperature was raised to about 1000℃, and single crystalline GaN several microns thick was grown in the same way as that for conventional OMVPE.

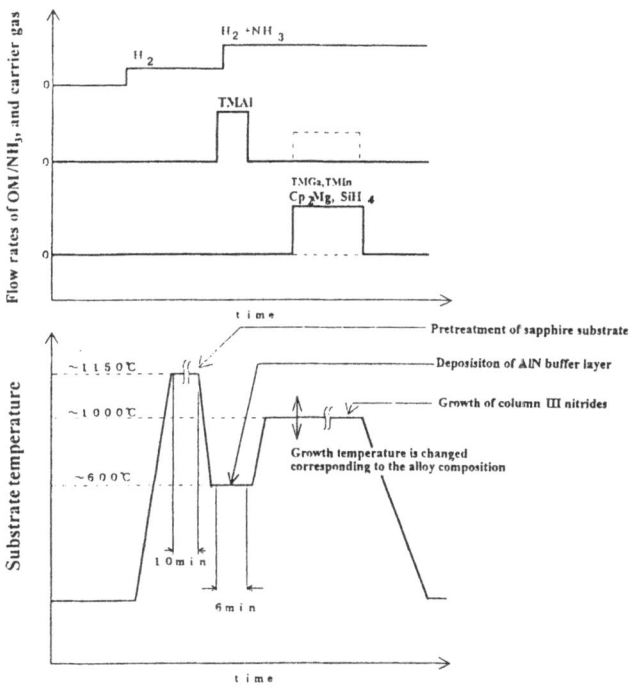

Figure 1. Sequence for the growth of GaN, and nitride alloys. The growth temperature is changed corresponding to the alloy composition.

The surface morphology of GaN film can be drastically improved by using the AlN buffer layer. The above-mentioned condition for AlN deposition was optimum for the growth of high-quality GaN with a flat surface free from cracks. In the case that the thickness of the AlN buffer layer exceeded 100 nm, polycrystalline GaN was obtained.When the deposition temperature of the AlN buffer layer was higher than 1000℃, at which single-crystalline AlN can be grown, the surface of GaN film became rough [14, 22].

The full-width at half-maximum (FWHM) of the double-crystal X-ray rocking curve (XRC) obtained from the GaN grown with the AlN buffer layer under the optimum condition was about 110 arcsec, which is the narrowest to date for this material [14], as shown in Fig. 2.

In the 4.2 K photoluminescence (PL) spectrum of undoped GaN grown with the AlN buffer layer, the sharp and strong neutral donor-bound exciton (I₂) line and sometimes free exciton (E_X) line appeared, while emission bands in the long wavelength region, which may be due to deep-level defects, were scarcely observed. On the other hand, emission bands at long wavelengths dominated the spectrum of GaN grown by conventional OMVPE.

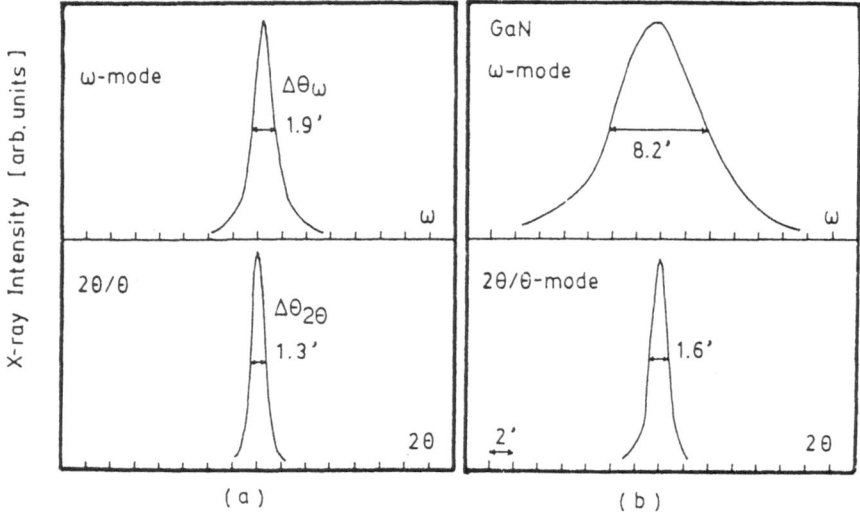

Figure 2. Double crystal X-ray diffraction profiles obtained from GaN film grown (a) with and (b) without the AlN buffer layer. The top shows ω-mode and the bottom 2θ/θ mode.

The GaN grown with the AlN buffer layer had n-type conductivity with an electron concentration of about 1×10^{17} cm^{-3} and recently on the order of 10^{16} cm^{-3} at RT, which was two to four orders of magnitude lower than those of conventional samples [14]. The electron mobility at RT was about 600 cm^{2}V^{-1}s^{-1} or higher, which is at least one order of magnitude higher than those of conventional ones.

The temperature dependence of the mobility suggested that the mobility is limited by the optical phonon scattering at high temperatures and by the ionized impurity scattering at low temperatures.

All these results (surface morphology, XRC, PL and electrical properties) clearly showed that by the preceding deposition of the AlN buffer layer, the electrical and optical properties as well as the crystalline quality of GaN can be drastically improved.

Transmission electron microscopy (TEM) as well as scanning electron microscopy (SEM) were carried out to clarify the growth mechanism of GaN grown by OMVPE on a sapphire substrate with the AlN buffer layer. The cross-sectional TEM has revealed that

GaN has many defects near the GaN/AlN interface, composed of columnar fine crystals and trapezoid crystals, but the defect density decreased abruptly for the layer of GaN thicker than about 300 nm [23]. From the fine structure of the GaN/AlN interface, a new growth mechanism using the AlN buffer layer which is important to obtain a uniform and high-quality GaN layer has been found. It consists of the following stages: (1) high-density nucleation of GaN, (2) geometric selection arranging the crystallographic direction of the GaN columnar crystals and (3) highly lateral growth velocity of the trapezoid islands. The growth process is schematically shown in Fig. 3 [23].

As we reported elsewhere [14, 22], the essential roles of the AlN buffer layer are both the supply of nucleation centers having the same orientation as the substrate and the promotion of lateral growth of the GaN film due to the decrease in interfacial free energy between the film and the substrate.

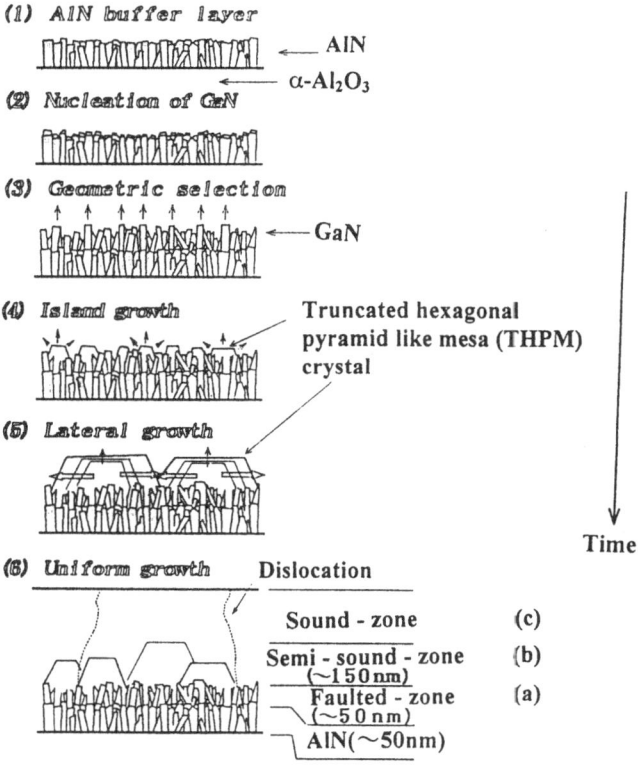

Figure 3. Schematic diagram showing the growth process of GaN on the AlN buffer layer as the cross sectional views.

3. Growth of AlGaN and GaInN Alloys

As mentioned before, in the dual-flow channel reactor, the column III source gases and column V source gas (NH_3) were mixed just before reaching the substrate to maintain the laminar flow and to suppress the formation of adducts. Therefore we can grow AlGaN and GaInN having desired alloy composition. Thus, single-crystalline AlGaN alloys with AlN molar fraction less than 0.4 have been successfully grown in the same way as that for GaN by supplying TMAl together with TMGa during growth [24].

GaInN alloys with InN molar fraction less than 0.15 have also been successfully grown by supplying TMIn, together with TMGa during growth, and by changing the growth temperature corresponding the alloy composition [25]. In the case of growth of GaInN, however, the InN molar fraction depended strongly on the growth temperature, because the vapor pressure of N_2 over InN is very high. The growth-temperature dependence of the InN molar fraction Y in GaInN alloys obtained experimentally agreed with the calculated result which was obtained as follows.

$$Y = \frac{(\alpha \cdot P_{In}) \cdot [TMIn]}{(\alpha \cdot P_{In}) \cdot [TMIn] + [TMGa]}$$

where $[TMIn]$ and $[TMGa]$ are flow rate of TMIn and TMGa, respectively, and P_{In} is vapor pressure of metallic In, and α fitting parameter [26].

The effectiveness of the AlN buffer layer on the improvement of crystalline quality of AlGaN or GaInN alloys has also been proved [25, 26].

4. Conductivity Control of n-Type Nitrides

Epitaxial nitride films grown by the method mentioned above have low residual electron concentration of about $10^{15} \sim 10^{17}$ cm^{-3}. It was found that SiH_4 is suitable for Si doping[19]. The electron concentration and the resistivity of n-type nitrides changed linearly with the flow rate of SiH_4 for GaN, AlGaN and GaInN. The electron concentration has been controlled from the undoped level up to near 10^{19} cm^{-3} for these nitrides, without deterioration of crystalline quality and surface flatness, by using the buffer layer technique.

5. Realization of p-Type Nitrides and Their Electrical Properties

We used, for the first time, biscyclopentadienyl magnesium (bis-Cp_2Mg) as a Mg source gas in OMVPE of GaN [27], and later for AlGaN [28, 29] and for GaInN [30]. Mg concentration in GaN changed linearly with the flow rate of bis-Cp_2Mg, and this relationship was almost independent of the substrate temperature in the temperature range from 850 to 1100°C. RHEED [23] and SEM [27] observations showed that single crystalline GaN having a flat surface with Mg concentration up to about 10^{20} cm^{-3}, can be grown by using the AlN buffer layer. Generally speaking, as-grown Mg-doped GaN (or nitride alloys) were highly resistive. In 1989, we [20] realized, for the first time, a distinct p-type GaN with low resistivity by the LEEBI treatment of such a high-quality GaN doped with Mg grown with buffer layer. The treatment condition was reported elsewhere [19].

Hole concentration up to 2×10^{18} cm^{-3}, and resistivity of 0.2 Ω cm at RT have been achieved for GaN [29]. P-type AlGaN [25, 28, 29] and p-type GaInN [30] were also achieved in the same way as that for p-type GaN . Figure 4 shows the temperature dependence of hole concentration in Mg-doped GaN, AlGaN [29] and GaInN [30].

For three kinds of nitrides (GaN, AlGaN and GaInN), the hole concentration did not saturate at high temperatures, suggesting the possibility of fabrication of nitride-based devices operating at high temperatures.

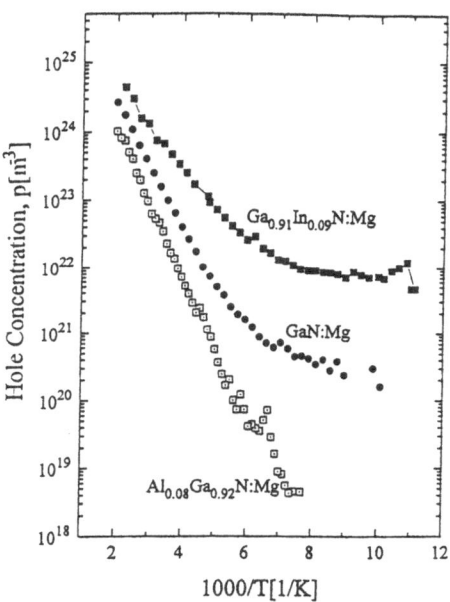

Figure 4. Temperature dependence of the hole concentration in GaN, and Al$_{0.08}$Ga$_{0.92}$N and Ga$_{0.91}$In$_{0.09}$N doped with Mg.

6. Recent Progress of Crystal Growth

Shown in Fig. 5 is the PL spectrum of undoped GaN (full line curve) grown on sapphire using the AlN buffer layer [31]. The PL is dominated by the intrinsic free exction (FE) with peaks at about 3.488, 3.495 and 3.506 eV, (labelled A, B and C), corresponding to the transition between valence band A, B and C, respectively. The intensity of main exciton peak FE(A) is about 2 orders of magnitude higher than those of impurity-bound excitons as they are donated ABE and DBE in the figure. The PL FWHM for the FE(A) is about 3 meV. Such a sharp and strong FE emission indicates the much improved crystalline quality.

Figure 6(a) shows high-resolution X-ray diffraction pattern obtained from GaN/GaInN multi quantum well (MQW) grown on sapphire, whose structure is depicted schematically in Fig. 6(c). The satellite peaks characteristic of MQW structure are clearly shown. This fine MQW structure is also verified by the In concentration profile by SIMS as shown in Fig. 6(b) [32].

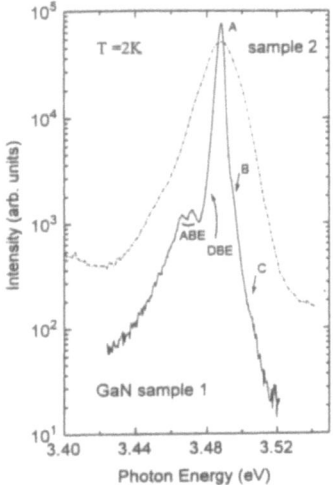

Figure 5. Photoluminescence spectra of GaN showing the free excition emissions dominate.

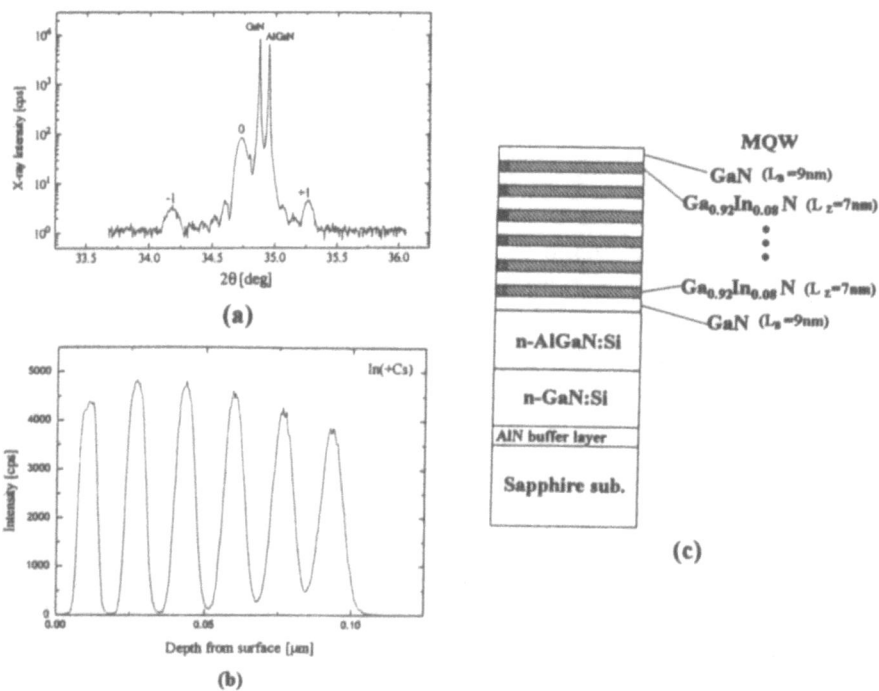

Figure 6. (a) High-resolution X-ray diffraction pattern. (b) In concentration profile by SIMS. (c) Schematic of GaN/GaInN multi quantum well.

Shown in Fig. 7 is the dependence of the PL band edge emission intensity at 77 K on GaInN well layer thickness in GaN/GaInN MQW. The PL intensity increases with the decrease of the well thickness. The intensity increases remarkably with the reduction of well layer thickness [32].

The cross-sectional TEM showed that dislocation density in the MQW is about 2×10^8 cm^{-2}, which is at least an order of magnitude lower that those ever reported [33]. These results indicate that crystalline quality of nitrides has been steadily improving.

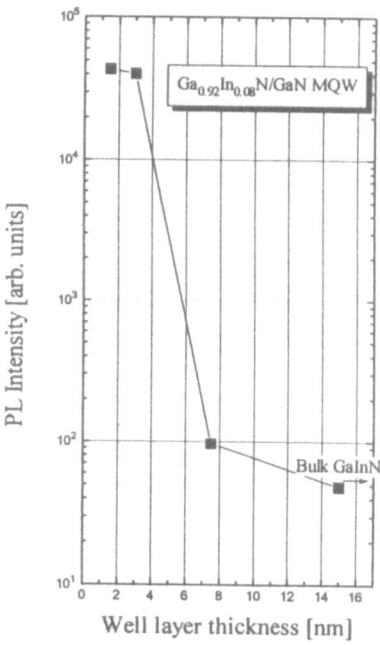

Figure 7. Dependence of photoluminescence band edge intensity on GaInN well layer thickness.

7. Summary

By OMVPE using the AlN buffer layer, the electrical and luminescence properties as well as crystalline quality of GaN, AlGaN and GaInN have been remarkably improved. Utilizing such high-quality nitrides grown by the AlN buffer layer technique, conductivity control for n-type GaN, AlGaN and GaInN has been achieved. By combining this buffer layer technique with the LEEBI treatment, GaN, AlGaN and GaInN with distinct p-type conduction have been realized, for the first time. The crystalline quality and luminescence property of GaN/GaInN multi quantum well structure are drastically improved by OMVPE using the AlN buffer layer technique.

134

Acknowledgments

The authors are grateful to H. Sakai, S. Sota, M. Koike, (Toyoda Gosei Co.) and T. Tanaka (Pioneer Electronic Corp.) for help in measuring the electrical and optical properties.

This work was partly supported by the Ministry Education, Science, Sports and Culture of Japan (contract nos. 06452114, 07505012 and 07650025), Research Foundation, of Electrotechnology of Chubu, Nissan Science Foundation, Hoso Bunka Foundation, Iketani Science and Technology Foundation and Daiko Foundation.

References

1. Maruska, H. P. and Tietjen, J. J. (1969) Appl. Phys. Lett., **15**, 327.
2. Sichel, E. K. and Pankove, J. I. (1977) J. Phys. Chem. Solids., **38**, 330.
3. Pankove, J. I., Maruska, H. P. and Berkeyheiser, J. E. (1970) Appl. Phys. Lett., **17**, 197.
4. Dingle, R., Shaklee, K. L., Leheny R. F. and Zetterstrom, R. B. (1971) Appl. Phys. Lett., **19**, 5.
5. Manasevit, H. M., Erdmann F. M. and Simpson, W. I. (1971) J. Electrochem. Soc., **118**, 1864.
6. Tansley T. L. and Foley, C. P. (1986) J. Appl. Phys., **59**, 3241.
7. Koide, Y., Itoh, H., H. Khan, M. R., Hiramatsu, K., Sawaki N. and Akasaki, I. (1987) J. Appl. Phys., **61**, 4540.
8. Yoshimoto, N., Matsuoka, T., Sasaki T. and Katsui, A. (1991) Appl. Phys. Lett., **59**, 2251.
9. Maruska, H. P., Rhines W. C. and Stevenson, D. A. (1972) Mat. Res. Bull., **7**, 777, Monemar B. and Lagerstedt, O. (1979) J. Appl. Phys., **50**(10), 6480.
10. for example, Pankove, J. I., Miller E. A. and Berkeyheiser, J. E. (1971) RCA Review, **32**, 383, Maruska, H. P., Stevenson D. A. and Pankove, J. I. (1973) Appl. Phys. Lett., **22**, No.6 303, Pankove J. I. and Lampert, M. A. (1974) Phys. Rev. Lett., **33**, No. 6, 361, Ohki, Y., Toyoda, Y., Kobayashi K. and Akasaki, I. (1981) Inst.Phys.Conf.Ser., **63**, 479.
11. Grimmeiss H. G. and Monemar, B. (1970) J. Appl. Phys., **41**, 4054.
12. Yoshida, S., Misawa S. and Gonda, S. (1983) Appl. Phys. Lett., **42**, 427.
13. Amano, H., Sawaki, N., Akasaki I. and Toyoda, Y. (1986) Appl. Phys. Lett., **48**, 353.
14. Akasaki, I., Amano, H., Koide, Y., Hiramatsu K. and Sawaki, N. (1989) J. Crystal Growth, **98**, 209.
15. Amano, H., Asahi T. and Akasaki, I. (1990) Jpn. J. Appl. Phys., **29**, L205.
16. Khan, M. A., Olson, D. T., Van Hove J. M. and Kuznia, J. N. (1991) Appl. Phys. Lett., **58**, 1515.
17. Khan, M. A., Bhattarai, A., Kuznia J. N. and Olson, D. T. (1993) Appl. Phys. Lett., **63**, 1214.
18. Nakamura, S. (1991) Jpn. J. Appl. Phys., **30**, L1705.
19. Amano H. and Akasaki, I. (1990) Ext. Abs. Mat. Res. Soc. Fall Meeting, **21**, 165.
20. Amano, H., Kito, M., Hiramatsu K. and Akasaki, I. (1989) Jpn. J. Appl. Phys., **28**, L2112.
21. Hirosawa, K., Hiramatsu, K., Sawaki N. and Akasaki, I. (1993) Jpn. J. Appl. Phys., **32**, L1039.
22. Amano, H., Akasaki, I., Hiramatsu, K., Koide N. and Sawaki, N. (1988) Thin Solid Films, **63**, 415.
23. Hiramatsu, K., Itoh, S., Amano, H., Akasaki, I., Kuwano, N., Shiraishi T. and Oki, K. (1991) J. Crystal Growth, **115**, 628.
24. Koide, Y., Itoh, H., Sawaki, N., Akasaki I. and Hashimoto, M. (1986) J. Electrochem. Soc., **133**, 1956.
25. Akasaki I. and Amano, H. (1994) Mat. Res. Soc. Symp. Proc., **339**, 443.
26. Akasaki I. and Amano, H. (1995) J. Crystal Growth, **146**, 455.
27. Amano, H., Kitoh, M., Hiramatsu K. and Akasaki, I. (1990) J. Electrochem. Soc., **137**, 1639.
28. Akasaki I. and Amano, H. (1992) Mat. Res. Soc. Symp. Proc., **242**, 383.
29. Tanaka, T., Watanabe, A., Amano, H., Kobayashi, Y., Akasaki, I., Yamazaki, S. and Koike, M. (1994) Appl. Phys. Lett., **65**, 593.
30. Yamasaki, S., Asami, S., Shibata, N., Koike, M., Manabe, K., Tanaka, T., Amano H. and Akasaki, I. (1995) Appl. Phys. Lett., **66**, 1112.
31. Harris, C. I., Monemar, B., Amano H. and Akasaki, I. (1995) Appl. Phys. Lett., **67**, 840.
32. Akasaki, I., Amano H. and Suemune, I. accepted for publication in Proc, ICSCRM' 95.
33. Lester, S. D., Ponce, F. A., Craford, M. G. and Steigerwald, D. A. (1995) Appl. Phys. Lett., **66**, 249.

GaSb dots grown on GaAs surface by MOCVD

R. Bożek, J.M. Baranowski, R. Stępniewski
Institute of Experimental Physics, Warsaw University
Hoża 69, 00-681 Warsaw, Poland
J. Wróbel
Institute of Physics, Polish Academy of Sciences
Al. Lotników 32/46, Warsaw Poland

ABSTRACT   We report on the MOCVD growth of GaSb islands on GaAs substrate at the temperature of 610 °C. The islands are ordered along the [110] direction. Their lateral dimensions are independent of the amount of the deposited material.

Introduction

In recent years self-organised structures formed during heteroepitaxy of highly mismatched semiconductor systems attract much attention, because they may potentially be used to obtain quantum dots and wires. So far most often investigated systems are germanium on silicon and indium (-gallium) arsenide on gallium arsenide. Only a few papers are devoted to gallium antimonide grown on gallium arsenide. In this case most results were obtained for MBE [1-3] or MOCVD growth at relatively low temperatures not exceeding 560°C [4]. However it is known that higher temperatures should enhance formation of islands. In our paper we present results obtained for GaSb layers grown on a GaAs substrate at temperature of 610°C.

The lattice constants of GaAs and GaSb are 5.65Å and 6.09Å respectively. The resulting mismatch of the system is 7.8%. Another fact that should be pointed out is that GaSb is grown at temperature relatively close to its melting point 708°C.

Experimental

The samples were grown in a horizontal, RF heated, atmospheric pressure MOCVD reactor, using trimethylgallium (TMG) and trimethylantimony (TMSb). Substrates were [001] epi-ready, vertical gradient freeze (VGF) N^+ GaAs:Si wafers, without disorientation. Their high quality was confirmed by X-ray measurements. The

J. Novák and A. Schlachetzki (eds.), Heterostructure Epitaxy and Devices, 135–138.

136

growth procedure was as follows. After bake out of the substrates (10 min., 800°C), 0.5μm thick GaAs buffer layer was grown at the temperature of 675°C. Then the TMG line was closed and the substrate cooled down to 610°C in an arsine ambient. The AsH₃ line was closed and the TMSb line immediately opened 10 seconds before letting TMG into the reactor in order to flush the remaining arsine. After deposition of the GaSb layer the sample was cooled down under flow of TMSb. The deposition time was changed in the range from 1.5 to 10.5 seconds. This nominally should correspond to thickness from 2 to 14 monolayers, assuming the same growth rate (1.5 μm/h) as calibrated for a 2 μm thick layer.

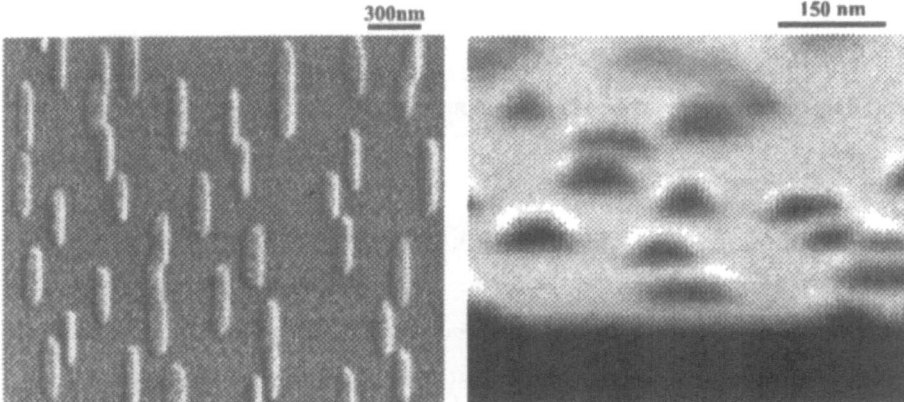

Fig. 1. SEM micrographs of the GaSb islands grown on the GaAs substrate taken from the top (left) and in the direction of the long axis (right).

The surface of the samples was investigated using scanning electron microscope (SEM) type JEOL 6400. The results of the observations are summarised in the Table 1.

TABLE 1. Summary of islands characteristics for the samples grown at 610°C.

Nominal thickness	Average		
	Length [nm]	Width [nm]	Length/Width
2	No islands were observed		
4	No islands were observed		
6	300± 90	90 ± 22	3.3± 0.9
8	290± 50	115± 15	2.6± 0.5
10	300± 90	95 ± 20	3.1± 0.9
12	290± 80	95± 15	3.1± 0.8
14	320± 90	105± 20	3.1± 1.0

We could not detect any islands in case of the layers with nominal thickness of 2 and 4 ML. For the thicker layers we observed elongated islands with their long axes parallel to the [110] direction. The typical pictures of the islands, taken for the 14 ML sample, are presented in the Figure 1. The lateral dimensions were independent of the time of the growth. The average length and width are respectively 300 nm and 100 nm. The length distribution is distinctly broader than the width distribution. We could observe islands with an aspect ratio up to 8:1, while the average aspect is close to 3:1. The density of the islands slightly increases from $3.7*10^8$ cm^{-2} in case of 6 ML to $6.1*10^8$ cm^{-2} in case of 14 ML (Fig. 2). For the 14 ML sample we found the height of islands as close to 30 nm. This result in conjunction with the lateral dimensions and the average density corresponds very well with the nominal amount of the deposited material. This shows that the growth after formation of an island continues in the vertical direction.

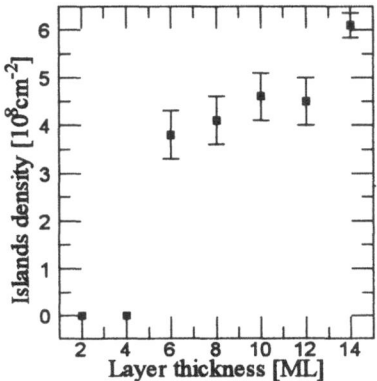

Fig.2 Dependence of the islands density on the nominal layer thickness.

Conclusions

Our observations are generally consistent with the Stranski-Krastanov growth mechanism in which planar growth of few monolayers is followed by formation of 3D islands, allowing the system to lower its elastic energy. Observation of the islands only for layers thicker than 4 ML may be explained by an initially 2D growth mode. This result is in agreement with paper [2], in which the authors report the formation of islands for layers thicker than about 1.2 nm grown in the MBE at 470°C.

138

Reports devoted to the kinetics of islands formation in much lower temperatures describe two stages [5]. The first one is the formation of coherent (strained) islands with highly homogenous dimensions and the next their growth accompanied by processes of relaxation. In our case the formation of large islands (probably relaxed) is observed, which are growing mainly vertically. This is probably a result of much more efficient surface diffusion of the adatoms.

The separate problem is the orientation of the islands, which we found to be different, than usually reported orientation for the epitaxy of III-V semiconductors e.g. [1,3]. Graham et al. observed a change of the orientation GaSb islands with orientation from [110] to [-110] at the temperature of 530°C, which they related to the change of the surface reconstruction. However this result is inconsistent with our findings and the observation on all MBE samples, which are usually grown at temperature below 500°C.

It seems that the vertical growth mode and the increased elongation during growth performed at more elevated temperatures could be used for preparation of self-organised quantum wires.

Acknowledgement

This work was partially supported by Polish Committee for Scientific Research under Grant no. PBZ-101-01. Authors wish also to express their gratitude to the Foundation for Polish Science for supplying funds for the MOCVD scrubbing system.

References

1. Brar and D. Leonard, Spiral growth of GaSb on (001) GaAs using molecular beam epitaxy, *Appl. Phys. Lett.* **66**, (1995), 463-465
2. F. Hatami et al.., Radiative recombination in type-II GaSb/GaAs quantum dots, *Appl. Phys. Lett.* **67**, (1995), 656-658
3. J.M. Kang, Suk-Ki Min, A. Rocher, Asymmetric tilt interface induced by 60 misfit dislocation arrays in GaSb/GaAs (001), *Appl. Phys. Lett.* **65**, (1994), 2954-2956
4. R.M. Graham, A.C. Jones, N.J. Mason, S. Rushworth, L. Smith, P.J. Walker, Growth of GaSb on GaAs substrates, *J. Crystal Growth* **145** (1994), 363-370
5. D. Leonard, K.Pond and P.M. Petroff, Critical layer thickness for self-assembled InAs islands on GaAs, *Phys.Rev.* **B50**, (1994) 11687-11692

CRYSTALLOGRAPHIC TILTING IN LATTICE-MISMATCHED HETEROEPITAXY: A KINETIC APPROACH

Ferenc RIESZ
Research Institute for Technical Physics of the Hungarian Academy of Sciences, P. O. Box 76, H-1325 Budapest, Hungary
E-mail: Rie7883@helka.iif.hu

1. Introduction

Crystallographic tilting (that is, the small misorientation of corresponding epilayer and substrate lattice planes) is commonly observed in (100) oriented vicinal lattice-mismatched heteroepitaxial systems [1-6]. This tilt is due to the asymmetry of the strain relaxation process induced by the substrate miscut.

In this paper, this tilting is analysed theoretically. The need for a kinetic relaxation model is pointed out, and tilting is calculated for a model system using the kinetic models of Matthews and Dodson and Tsao (DT). We show an intimate connection between strain relaxation mechanism and tilt. Similarities and dissimilarities between the two models are pointed out.

2. Strain Relaxation Models and Extension to Vicinal Epitaxy

In (100) zinc blende semiconductors, strain relaxation proceeds by the glide of 60° MDs. These dislocations have in-plane (misfit) and out-of-plane (tilt) Burgers vector components. In a vicinal system, the substrate miscut changes the dislocations' orientations, thus causing a preference of certain slip systems, inducing a net, overall tilt. To calculate the tilt, we must calculate the number of the MDs having different-orientation slip systems.

The earliest models of strain relaxation are equilibrium ones. They based either on an energy minimization or force-balance approach (for review, see Ref. [7]). These equilibrium models usually predict a critical layer thickness (CLT) proportional to $\cos\lambda$, where λ is the angle between the Burgers vector and that line in the epilayer plane which is perpendicular to the intersection of the glide plane and the surface. Equilibrium models thus predict that only MD arrays with the smallest λ (and CLT) will be present in the epilayer. However, transmission electron microscopy studies and tilt data clearly indicate that other MDs with larger CLT are present in vicinal systems as well. That is, there must be some kinetic limitation that delays the introduction of the MDs having the smallest CLT until the others, with larger CLT, appear. The need for a kinetic strain relaxation model is recognized long ago in heteroepitaxy [7]. Matthews' kinetic model [8] incorporates a frictional force on the moving dislocation,

139

J. Novák and A. Schlachetzki (eds.), Heterostructure Epitaxy and Devices, 139-142.
© 1996 *Kluwer Academic Publishers.*

while a more recent, DT approach [9] includes also multiplication of MDs. The original formulation of these two models differs, since the DT model is a phenomenological one. In order to compare the results, we use an alternative formulation of the Matthews model that conforms to the DT model. The generation rate of MDs for the Matthews and DT models is given by the following equations respectively [7,9]:

$$\frac{\partial \rho}{\partial t} = C_M \frac{\sigma_{ex}}{G} \rho_0 \text{ and } \frac{\partial \rho}{\partial t} = C_{DT} \left(\frac{\sigma_{ex}}{G} \right)^2 [\rho(t)+\rho_0] \text{ (for } \sigma_{ex}>0), \quad (1)$$

where ρ is the linear density of MDs, ρ_0 is the initial defect density that initiates the strain relaxation, G is the shear modulus; the rate constants are thermally activated: $C_{DT,M} = C_{0\,DT,M} \exp(-E_a / kT)$ with dislocation glide activation energy E_a. The excess stress σ_{ex} is the difference between the misfit stress acting on the threading arm and the line tension of the misfit segment:

$$\sigma_{ex} = 2G \left[\cos \lambda \cos \varphi \frac{1+v}{1-v}(f-\gamma) - \cos \varphi \frac{1-v/4}{4\pi(1-v)(h/b)} \left(\ln \frac{\eta h}{e^2 b} + 1 \right) \right]. \quad (2)$$

Here f is the lattice misfit, v is the layer's Poisson ratio, φ is the angle between surface normal and dislocation slip plane, η is the dislocation core parameter, b is the Burgers vector length and h is the film thickness ($h=vt$ where v is the growth rate). The relieved strain γ is given by $\gamma = \rho b \cos \lambda$.

The essential difference between the two models is that the DT model includes the MD multiplication [10]. This leads to (i) the inclusion of the instantaneous MD density beside the initial defect density in Eq. (1), (ii) a quadratic dependence of the relaxation rate on the excess stress and (iii) a possible deviation of the rate constants ($C_{M,DT}$).

The extension of these models to vicinal epitaxy is straightforward. We consider a substrate miscut towards [011]. An isotropic relaxation of the layer is assumed. The substrate miscut divides the eight slip systems to two groups [2,5]: one is more stressed over the other. Tilt is caused by those MDs whose slip plane normals are coplanar with the surface normal [5]. The substrate miscut β modifies the angular terms for these MDs as follows [4]: $\lambda_{A,B} \approx 60° \mp \beta$ and $\varphi_{A,B} \approx 35.3° \pm \beta$ [subscripts A and B refer to the two MD groups; upper signs are for group A (the more stressed ones)]. Equation (2) is then written for both groups; the relieved strain sums up from both groups as $\gamma = \gamma_A + \gamma_B$. Equations (1) are then integrated numerically to find the MD densities. The tilt angle $\Delta\beta$ is finally calculated from the MD densities by [3]:

$$\Delta\beta = \beta[f(1+v)-2\gamma v]/(1-v)-(\sqrt{3}/2)b \cos \varphi(\rho_A - \rho_B). \quad (3)$$

(Negative tilt means tilts away from surface normal.) We note that tilting on the basis of the Matthews model has been discussed by Ayers et al. [2]; however the full potentials of the model were not explored.

3. Results and Discussion

Tilts were calculated as a function of growth temperature, substrate miscut angle, initial defect density, and relieved strain. Figure 1 shows the equilibrium tilt values (that is, tilts upon total relaxation) as a function of growth temperature and initial defect density. The maximum tilt angle is also shown (that is, when all the MDs are from group A).

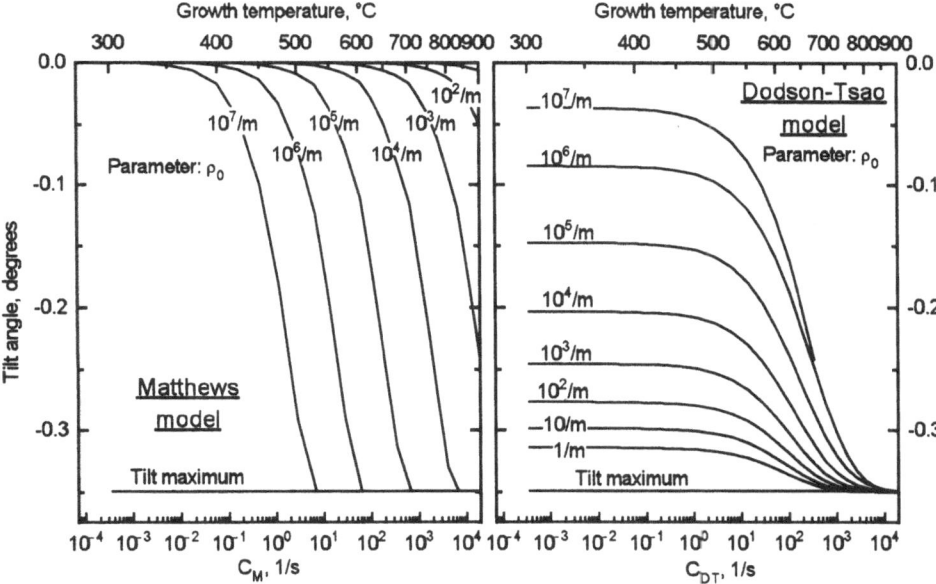

Figure 1. Equilibrium tilt angles as a function of growth temperature and initial defect density as calculated on the basis of the Matthews and the Dodson-Tsao relaxation models. The following parameters were assumed: f=0.005, η=4, β=4°, C_{0M}=C_{0DT}=5× 10^{11} 1/s, E_a=1.7 eV, v=1 μm/h, b=4×10^{-10} m, υ=0.33.

Both models yield increasing tilts with growth temperature. This is because higher temperature means growth more close to equilibrium, therefore group-A MDs can relax a major amount of strain (and induce a significant tilt) before group-B MDs appear. The major difference between the two models is the effect of initial defect densities on tilt. In the Matthews model, $\partial\rho/\partial t$ is proportional to the product of the initial defect density and the rate constant [see Eq. (1)], therefore increasing defect density is equivalent to a shift of the temperature scale, as shown in Fig. 1. On the other hand, in the DT model, the presence of the multiplication involves that $\partial\rho/\partial t$ is proportional to the sum of the initial and the MD density that has already been formed [see again Eq. (1)]. That is, as the strain relaxation proceeds, the relaxation is increasingly dominated by the newly generated dislocations. Initial defect density has only a "symmetrizing" effect. Note that at a fixed growth temperature, the effect of initial defect densities on the tilt magnitude is *opposing* in the two models.

Additionally, calculations not detailed here yield for both models *(i)* an approximately linear dependence of tilt on relaxation in partially relaxed layers, *(ii)* a linear dependence of tilt on substrate miscut angle for small tilts and saturation at higher tilts and *(iii)* a dependence of tilt on the asymmetry of the initial defect densities.

142

Although a large amount of experimental data has been accumulated in the literature, a systematic work that allows comparison to these models is still lacking. Nevertheless, the increase of tilt with growth temperature has been observed by several authors [5,6]. A review of the early [2] and more recent works [3] show that most experimental tilt angles lie below one half of the maximum value. This is in agreement with the common feature of strain relaxation to be far from equilibrium under usual heteroepitaxial growth conditions [7]. Our results show the decisive role of dislocation multiplication in tilt formation. Thus, using carefully designed growth parameters—growth temperature and varying initial defect density through the use of different substrates or engineered buffer layers—the measured tilts could provide important additional information on the strain relaxation mechanism.

Acknowledgements

This work was supported, in part, by the (Hungarian) National Scientific Research Fund (OTKA) through Grant F 016278.

References

1. Pesek, A., Hingerl, K., Riesz, F., and Lischka, K. (1991) Lattice misfit and relative tilt of lattice planes in semiconductor heterostructures, *Semicond. Sci. Technol.* **6**, 705-708.
2. Ayers, J.E., Ghandhi, S.K., and Schowalter L.J. (1991) Crystallographic tilting of heteroepitaxial layers, *J. Cryst. Growth* **113**, 430-440.
3. Riesz, F. (1995) A Dodson-Tsao relaxation approach to the crystallographic tilting in (100) heteroepitaxial systems, *Mat. Res. Soc. Symp. Proc.* **379**, in press.
4. Riesz, F. (1994) The surface and interface nucleation of misfit dislocations as a possible source of asymmetric strain relaxation in vicinal heterostructures, *Czech J. Phys.* **44**, 131-137 (1994); Erratum, **44**, 799.
5. Goldman, R.S., Wieder, H.H., and Kavanagh, K.L. (1995) Correlation of anisotropic strain relaxation with substrate misorientation direction at InGaAs/GaAs(100) interfaces, *Appl. Phys. Lett.* **67**, 344-346.
6. Suzuki, T., Mori, M., Jiang, Z.K., Soga, T., Jimbo, T., and Umeno, M. (1992) Tilt deformation of metalorganic chemical vapor deposition grown GaP on Si substrate, *Jpn. J. Appl. Phys.* **31**, 2079-2084.
7. Fitzgerald, E.A. (1991) Dislocations in strained-layer epitaxy: theory, experiment, and applications, *Mat. Sci. Rep.* **7**, 87-142.
8. Matthews, J.W., Mader, S., and Light, T.B. (1970) Accommodation of misfit across the interface between crystals of semiconducting elements or compounds, *J. Appl. Phys.* **41**, 3800-3804.
9. Dodson, B.W., and Tsao, J.Y. (1988) Relaxation of strained-layer heterostructures via plastic flow, *Appl. Phys. Lett.* **51**, 1325-1327; Erratum, **52**, 852.
10. Fox, B.A., and Jesser, W.A. (1990) The effect of frictional stress on the calculation of critical thickness in epitaxy, *J. Appl. Phys.* **66**, 2801-2808.

OPTIMIZATION OF MOVPE GROWTH FOR INGAAS ON (001)Si

H.-H. WEHMANN, G.-P. TANG, H. IBER, S. MO, and
A. SCHLACHETZKI
Institut für Halbleitertechnik, Technische Universität Braunschweig
Postfach 3329, D-38023 Braunschweig, Germany, IHT@tu-bs.de
L. MALACKY
Institute of Electrical Engineering, Slovak Academy of Sciences
Dubravska 9, SK-84239 Bratislava, Slovakia, elekmala@savba.savba.sk

1. Introduction

The highest ever reported velocity of an integrated circuit of 50 GHz is achieved with Si [1]. However, with increasing speed and complexity the connection of electronic circuits to the outer world becomes more difficult. Optical in- and outputs would solve this problem. Since Si has an indirect bandgap it cannot be used for the fabrication of optical emitters. Apart from costly hybrid solutions the monolithic integration of direct III/V-compound semiconductors on Si will be the optimum solution. However, the differences in lattice constant (4% for GaAs and 8% for InP and $In_{0.53}Ga_{0.47}As$ relative to Si), in crystal symmetry (diamond – zincblende) and in thermal expansion coefficient (164%, 82%, and 118%, respectively) between layer and substrate represent formidable hindrances.

In this paper an optimized growth process for $In_{0.53}Ga_{0.47}As$ (InGaAs for short) on Si is described. The layer properties most important for practical applications are presented.

2. Growth Process

At least two prerequisites for lattice-mismatched growth on Si are mandatory for the realization of an optical extension of the well-established Si-electronics. These are the compatibility of the III/V-processes to the Si-technology and the use of a growth technology suitable for industrial applications. Both requirements are fulfilled by the use of metalorganic vapourphase epitaxy (MOVPE) on exactly (001) orientated Si substrates which are standard in Si-MOS-technology. The growth procedure has been proven to be compatible with Si-MOS-technology [2].

Since exactly orientated (001) substrates are used, special care has to be taken to avoid the formation of antiphase domains [3]. As the direct growth of the ternary $In_{0.53}Ga_{0.47}As$ on Si is not possible by MOVPE, it is obvious to use InP as intermediate

143

J. Novák and A. Schlachetzki (eds.), Heterostructure Epitaxy and Devices, 143–146.
© 1996 *Kluwer Academic Publishers.*

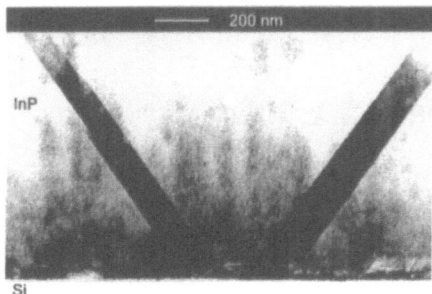

Figure 1. InP-on-Si XTEM-photograph with two microtwins originating from the interface.

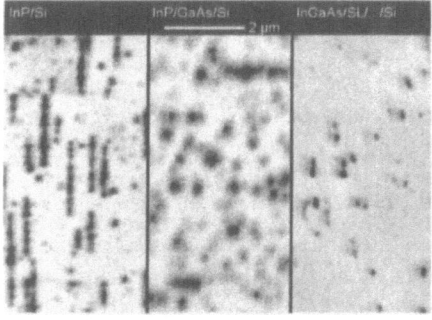

Figure 2. Surface-microphotographs of defect-etched InP on Si without (left) and with (centre) GaAs intermediate layer, as well as InGaAs grown on the optimzed layerstack (right).

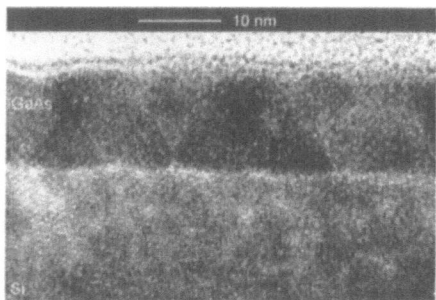

Figure 3. XTEM photograph of the optimized GaAs low-temperature buffer on Si.

layer. Apart from inevitable misfit and threading dislocations the main defects in InP-on-Si are microtwins. These are clearly visible as dark bands in Figure 1 taken by cross-sectional transmission electron microscopy (XTEM). On the one hand, these twinlamellae hinder the motion of threading dislocations and thus collect them in rows. These are clearly visible in the left part of Figure 2 after defect etching of the surface. On the other hand, at the intersecting lines of the microtwins with the layer surface the morphology is disturbed, causing growth defects in the following ternary layer. However, in the case of GaAs-on-Si the density of microtwins is strongly reduced if three-dimensional growth is suppressed [4]. This was achieved by a reduced V/III-ratio of 4 during the growth of a low-temperature (400 °C) buffer layer. Its XTEM photograph which as shown in Figure 3 proves the suppression of island growth since already the 14 nm thick buffer exhibits a blanked coverage of very homogeneous thickness. This buffer layer is followed by a 2.5 μm thick GaAs main layer grown at 700 °C to improve the crystallographic properties.

The growth at the lattice-mismatched interface to InP again has to be carried out at 400 °C to avoid island growth and thereby the formation of microtwins [4]. On the 80 nm thick InP low-temperature buffer-layer the InP main layer is grown at 640 °C to a thickness of 2 μm. By this procedure the density of microtwins could substantially be reduced leading to a more homogeneous distribution of the dislocation-related etch pits (Figure 2, centre). However, the subsequent InGaAs is heavily deteriorated if directly grown on this layer. We attribute this to the high anisotropy of the InGaAs growth [5]. Nevertheless, by inserting an InP/InGaAs superlattice (SL), consisting of 29 pairs of InP and InGaAs, each 6.6 nm thick the surface is smoothed [6] and thus the InGaAs grows as a closed layer. Since our SL is not intentionally strained, its influence on the threading dislocations is small. The respective etch-pit densities in Figure 2 decrease from 6×10^8 cm^{-2} to 3×10^8 cm^{-2} and 1.5×10^8 cm^{-2} (left to right).

3. Layer Properties

The surface morphology of the resulting optimized InGaAs layers on Si is similar to that of lattice-matched layers on InP (Figure 4). The two segments of straight lines visible on the lattice-mismatched layer (a) represent two remaining microtwins.

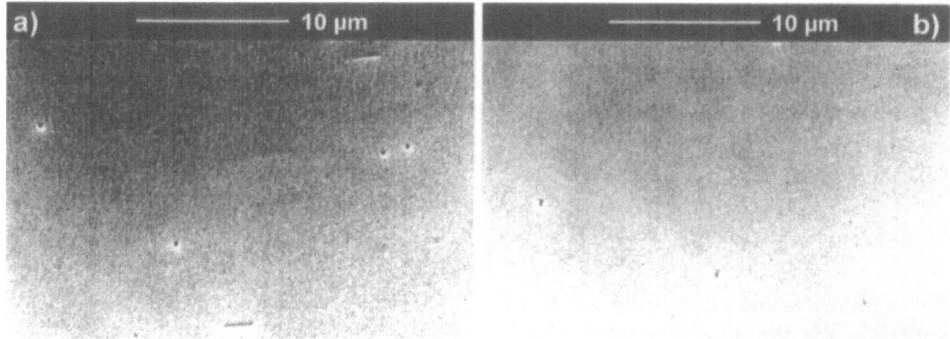

Figure 4. Surface morphology of simultaneously grown In$_{0.53}$Ga$_{0.47}$As on Si (a) and InP (b).

The full widths at half maximum (FWHM) of the (004) X-ray rocking-curves of the lattice mismatched GaAs and InP layers are 119 and 155 arcsec, respectively. The FWHM of the central SL peak is 200 arcsec, which is a superposition of the InP and InGaAs reflexes as well as the 0th order SL-peak. These values show the good crystallographic quality of the layers and thus support the findings with the surface morphology.

The electronic layer properties are examined by electrochemical capacitance/voltage profiles. We found an average electron concentration in the unintentionally doped In-containing layers of 3 to 5×10^{16} cm^{-3} and a factor of 10 less in the GaAs layer. Si from the substrate which is incorporated through the gasphase is the main contribution.

Photoluminescence spectra taken at 20 K show a maximum emission at 0.795 eV which is shifted by 16 meV to lower energies with respect to the peak emission of lattice-matched InGaAs on InP. This is due to strain related with the differences in thermal expansion coefficients of the substrates. The FWHM of the main peaks are 7 and 2 meV, respectively.

The favourable properties of the lattice-mismatched InGaAs on Si encouraged us to fabricate metal-semiconductor-metal (MSM) photodetectors with these layers. The resulting low dark-current densities, which are only a factor of 5 higher than those measured with simultaneously fabricated lattice-matched devices [7], confirm that the optimized ternary layers on Si are comparable to those on InP.

146

4. Conclusion

$In_{0.53}Ga_{0.47}As$ is grown by MOVPE on (001) Si for lattice-mismatched monolithic integration of III/V optoelectronic devices with the well-established Si electronics. To reduce the density of two-dimensional lattice defects special process steps are developed. Antiphase domain boundaries are avoided by an AsH_3 preflow after the initial high-temperature cleaning step of the Si substrate. By introducing low-temperature GaAs and InP buffer layers at the respective lattice-mismatched heterointerfaces the tendency to initial island-growth is suppressed and thus the density of microtwins in the InP layer is largely reduced. The residual surface roughness is smoothed by an unstrained InP/InGaAs superlattice which serves as a basis for the InGaAs device layer.

Its crystallographic, electronic, and optical properties are compared to lattice matched layers. First lattice mismatched MSM detectors were fabricated. Their dark-current densities are comparable to that of lattice matched devices demonstrating their suitability for application in optical systems.

5. Acknowledgements

The authors acknowledge the work of K. Dettmer for X-ray diffractometry, M. Hollfelder for photoluminescence, and M. Seibt for transmission electron microscopy. The work was partially funded by the German Ministry of Education and Research (BMBF) under contract X263.1.

6. References

1. Rein, H.-M. (1995) Very-high-speed Si and SiGe bipolar ICs, in de Graaff, H.C., van Kraneneburg, H. (eds.) *ESSDERC'95*, Editions Frontieres, Gif sur Yvette Cedex, France, 45-56.

2. Lubnow, A., Tang, G.-P., Wehmann, H.-H., and Schlachetzki, A. (1994) The effect of III/V-compound epitaxy on Si-metal-oxide-semiconductor circuits, *Japan. J. Appl. Phys.* **33**, 3628-3634.

3. Tang, G.-P., Lubnow, A., Wehmann, H.-H., Zwinge, G., and Schlachetzki A. (1992) Antiphase-domain-free InP on (100) Si, *Japan. J. Appl. Phys.* **31**, L1126-L1128.

4. Wehmann, H.-H., Tang, G.-P., Koch, A., Seibt, M., and Schlachetzki, A. (1995) Twin formation during epitaxial growth of InP on Si, *Solid State Phenomena* **47-48**, 547-552.

5. Zwinge, G., Wehmann, H.-H., Schlachetzki, A., and Hsu, C.C. (1993) Orientation-dependent growth of InGaAs/InP for applications in laser-diode arrays, *J. Appl. Phys.* **74**, 5516-5519.

6. Xu, X. (1994) Smoothing effect of $GaAs/Al_xGa_{1-x}As$ superlattices grown by metalorganic vapro phase epitaxy, *Appl. Phys. Lett.* **64**, 2949-2951.

7. Wehmann, H.-H., Tang, G.-P., Klockenbrink, R., Mo, S., and Schlachetzki, A. (1995) Dark-current analysis of InGaAs-MSM-photodetectors on Silicon substrates, in de Graaff, H.C., van Kraneneburg, H. (eds.) *ESSDERC'95*, Editions Frontieres, Gif sur Yvette Cedex, France, 447-450.

SEM-BASED CHARACTERIZATION TECHNIQUES FOR STRONGLY MISMATCHED HETEROEPITAXY

E. PEINER, S. MO, H. IBER, G.-P. TANG, AND A. SCHLACHETZKI

Institut für Halbleitertechnik, Technische Universität Braunschweig
Postfach 3329, D-38023 Braunschweig, Germany

1. Introduction

Heteroepitaxy of largely mismatched structures is of considerable interest, e. g., for the fabrication of blue-violet emitting nitride-based devices on sapphire or the monolithic integration of III/V-compound-based optoelectronics on Si [1]. Conventionally, growth proceeds in two steps whereby the crystal quality of the main layer is strongly affected by the structural characteristics of a thin nucleation layer deposited initially. The layers exhibit mechanical strain and lattice defects, which conventionally are characterized by transmission electron microscopy (TEM), X-ray diffraction (XRD), wet-chemical defect etching or optical techniques, e.g., photoluminescence spectroscopy (PL). These techniques suffer from drawbacks: they are either indirect (PL), destructive (defect etching), have relatively low lateral resolution (XRD) or require elaborate and destructive sample preparation thus being not suitable for routine inspections of entire wafers (TEM).

In this contribution we utilize the electron-channeling pattern (ECP) technique for crystal-quality assessment [2] and the determination of mechanical strain and energy-dispersive X-ray spectrometry (EDX) for the determination of the layer thickness [3]. Both techniques are performed in a scanning-electron microscope (SEM). We concentrate on GaAs and InP on Si (GaAs/Si and InP/Si for short) which can be considered as a model system for strongly mismatched heteroepitaxy.

2. Experimental

Growth of GaAs and InP layers on exactly (001)-oriented Si substrates was performed by metalorganic chemical-vapour epitaxy. Prior to growth the substrates were annealed for 15 min at 950°C under H_2 flow. For stabilization of the surface reconstruction, AsH_3 was added at 850°C during the cool-down phase. Thin buffer layers were deposited at 400°C with a ratio of the group V and III components (V/III) in the reactor of 4 or 41 for GaAs and 2700 for InP. Subsequently, several-μm-thick GaAs and InP main layers were grown at 700°C and V/III = 80, and 640°C and V/III =135, respectively. The total

J. Novák and A. Schlachetzki (eds.), Heterostructure Epitaxy and Devices, 147–151.
© 1996 *Kluwer Academic Publishers.*

pressure in the reactor during growth was 50 and 20 mbar leading to growth rates of 2 and 1 μm/h, respectively. The thicknesses of the buffer layers were determined by EDX using a novel technique [3] and by area-selective removal of the layers and subsequent measurement of the resulting step heights by mechanical surface tracing employing a Dektak 3030. For the EDX analyses we used the intensity ratio R of characteristic X-ray fluorescence generated in the layer and the substrate, respectively. For a calibration bevels were fabricated from thick layers on Si substrates by mechanical polishing. We measured R in dependence on the effective layer thickness d which was precisely defined by the position of the electron beam on the bevel [3].

ECPs are generated by rocking the electron beam by 2 to 10° about the sample's normal and collecting the electrons backscattered in the top few tens of nm of the sample. Depending on the necessary depth resolution the acceleration voltage was selected within 10 to 20 keV. The beam current was 1.6 nA and the working distance was below 10 nm. Further details on the principle of ECP can be found in Ref. [4]. For an ECP analysis at different vertical positions in an InP layer on Si we prepare layers of stepwise decreasing thickness by consecutive area-selective anodic oxidation, oxide stripping [5,6] and reduction of the masked area.

ECP and EDX analyses were performed in a SEM (Leica Stereoscan 360). Spectroscopic ellipsometry (SE) was performed with the buffer layers in the optical wavelength range around the band edge, using an automated nulling setup on the basis of a 436 ellipsometer by Rudolph Research. A three-layer model consisting of Si substrate, epitaxial buffer layer (GaAs or InP), and a native-oxide layer (3 nm) was used to extract the optical constants. The structural quality of the InP and GaAs main layers was characterized by X-ray diffraction (XRD) and in the case of InP/Si by defect etching using a $2H_3PO_4$:1HBr solution at 5°C [7].

3. Results and discussion

The ECP of a bulk single crystal is composed of bright bands of width $2\Theta_B$ with Θ_B as the Bragg angle. ECPs taken under identical conditions with strongly mismatched epitaxial layers reveal strain and distortion with respect to the bulk reference. The latter is already visible by inspection of the ECPs with the naked eye. Although the accuracy for the determination of absolute values of lattice constants by ECP is inherently limited to around 10^{-2} [4], it can be much better for the determination of ε which corresponds to a relative change of the lattice constant. For this purpose the points of minumum intensity in the ECP are marked by digital image processing. We obtain a pattern of sharp bright lines whose location is determined by the Bragg condition. Even small variations of Θ_B corresponding to ε in the range of a few 10^{-4} can be detected. In Fig. 1 the strain profile within InP/Si determined by measuring the variation of Θ_B of the (440) band is displayed. Good agreement was found with ε obtained by the strain dependence of the bandgap energy as measured by PL. For the latter calculation we used elastic stiffness coefficients C_{11} and C_{12} of 101.9 GPa and 57.3 GPa, respectively. The hydrostatic and shear deformation potentials were $a = -6.35$ eV and $b = -2.0$ eV [8]. By

PL with homoepitaxial n-type InP the bandgap energy of the unstrained reference was determined to be 1.346 eV. The solid line in Fig. 1 was calculated using a model which describes stress relaxation in mismatched layers by a kinetic process governed by the motion and multiplication of misfit dislocations [9]. In this calculation we employ the values of growth rate and temperature of the present study. By a fit to the ECP and PL results we obtained values

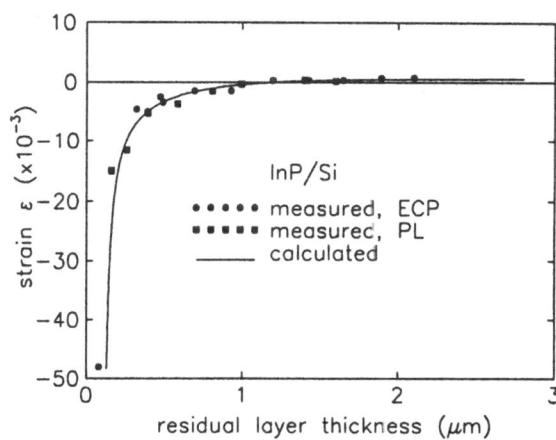

Figure 1. Depth profile of the mechanical strain of an InP layer on Si substrate; solid curve: theory (see Ref. [9]).

of the proportionality constants K and v_{cl} which are reduced with respect to the original data by factors of 1/2 and 1/4, respectively. K and v_{cl} enter the formulae describing the increase of the dislocation density and the process of dislocation climbing during growth, respectively [9].

For evaluation of the crystal quality of the epitaxial layers we investigate the intensity distribution of its ECP after subtraction of the ECP of a bulk reference [2]. In the ideal case the epitaxial layer has the same crystal quality as the bulk reference and the variance σ^2 of the intensity distribution vanishes. On the other hand inferior crystal quality causes a loss of structure of the ECP corresponding to an increase of σ^2. Before subtraction the ECPs of layer and bulk reference must be translated or rotated until they are precisely aligned to each other. For this purpose we employ the binary line patterns obtained by digital image processing discussed above. We investigated several InP/Si samples whose dislocation density N_{dis} was determined by defect etching and XRD and found a linear increase of σ^2 with $\ln(N_{dis})$. Using this calibration we examined the area-selectively thinned InP/Si sample of Fig. 2 by ECP. Figure 2 shows the obtained dislocation density vs. the residual layer thickness. For comparison a theoretical curve is displayed which was calculated using the stress-relaxation model already employed in the context of Fig. 1 (for details see Ref. [9]). No additional fitting was performed. This curve was extended toward the heterointerface where the density of inclined dislocations was estimated from the avarage distance of 5 nm between the misfit dislocations [1]. We find good agreement between the experimental points and the theoretical curve. The scatter of the experimental data points can be attributed to a major degree to the inferior surface appearance caused by the sample preparation which adversely affected the ECP analysis.

Finally, we used ECP and EDX to investigate the buffer layers and their effect on the main-layer characteristics. Buffer-layer thicknesses were measured by the intensity ratio R of the characteristic fluorescence radiation emerging from substrate and layer,

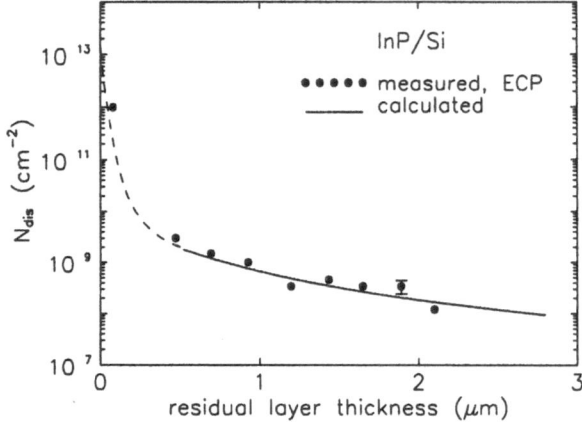

Figure 2. Depth profile of the dislocation density of an InP layer on Si substrate; solid curve: theory (see Ref. [9]).

respectively. By preliminary measurements we found that especially for small thicknesses d in the range typical for epitaxial buffer layers the thickness dependence of R can be very well described by $R = (a/d)\exp(-bd)$ where a and b are parameters determined by a fit. Mechanical surface tracing with area-selectively removed GaAs layers of 50 to 120 nm thickness exhibited agreement with the thickness values by EDX analysis within ± 8%. In the range of 10 to 30 nm the maximum deviation of EDX from mechanical surface tracing as well as cross-sectional TEM was 20%. Figure 3 shows σ^2 in dependence on the buffer-layer thickness for several GaAs/Si and InP/Si samples. In both cases we find that the crystal quality deteriorates for large thicknesses as well as below 10 nm for GaAs and 20 nm for InP where an irregular coverage of the Si substrate was observed. The quality of the GaAs buffer layer was slightly affected by the ratio of the group V and III components in the MOVPE reactor. These findings could be confirmed by SE as indicated by the absorption coefficient $\alpha_{0.9E_g}$ at a photon energy of $0.9E_g$. The large values of α in the transparent wavelength range below the band edge, which are not expected for direct semiconductors, can be related to the high dislocation density within the buffer [10,11].

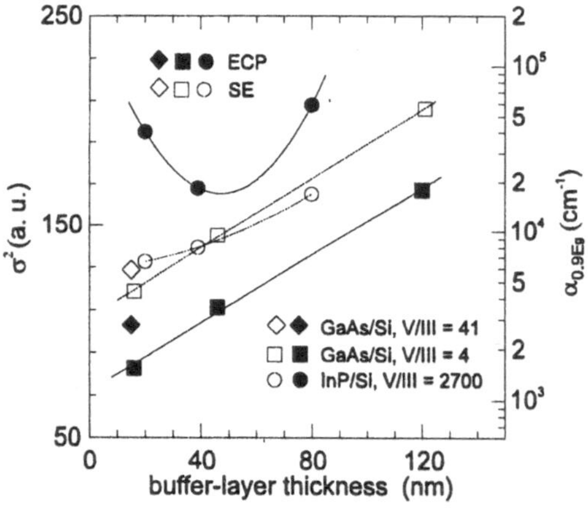

The effect of buffer-layer quality on mismatched heteroepitaxy was evaluated by the morphology and crystal quality of the main layers grown on top. By SEM and mechanical surface tracing we found a mirror-like surface finish for the GaAs main layer grown on the 16-nm-thick buffer layer. A considerably increased rms roughness of 20 nm and 30 nm was observed with main layers grown under identical conditions as above on the 46-nm-thick and 120-nm-thick buffer layers, respectively. These results are in agreement with reported findings [12,13].

Figure 3. Crystal quality of buffers for heteroepitaxy vs. layer thickness.

For InP/Si, the number of microtwins identified by cross-sectional TEM [14] is strongly reduced for the main layer grown on the 39-nm-thick buffer layer. Thus the morphology and the defect density of the main layers clearly reflected the crystal quality of the buffer.

These results show that structural characterization of epitaxial films including thin buffer layers which play a key role in mismatched heteroepitaxy was accomplished by EDX and ECP. Both techniques were performed in a SEM, which is a standard tool in research and development as well as in industrial laboratories.

We wish to thank J. Graßmann and L. Kunze for active technical support.

References

1. Schlachetzki, A. III/V-compound semiconductors on silicon, *These proceedings*.
2. Ophir-Arad, E., Fastow, R., and Kalish, R. (1990) Quantitative electron channeling measurements for high sensitivity surface analysis, *Appl. Phys. Lett.* 57, 2098-2100.
3. Peiner, E., Hansen, K., and Schlachetzki, A. (1995) Thickness control of InP and $In_{0.53}Ga_{0.47}As$ thin films by energy-dispersive X-ray spectrometry, *Thin Solid Films* 256, 143-147.
4. Joy, D. C., Newbury, D. E., and Davidson, D. L. (1982) Electron channeling patterns in the scanning electron microscope, *J. Appl. Phys.* 53, R81-R122.
5. Hollinger, G., Joseph, J., Robach, Y., Bergignat, E., Commère, B., Viktorovitch, P., and Froment, M. (1987) On the chemistry of passivated oxide-InP interfaces, *J. Vac. Sci. Technol. B* 5, 1108-1112.
6. Bartels, A., Peiner, E., and Schlachetzki, A. (1995) A procedure for temperature-dependent, differential van der Pauw measurements, *Rev. Sci. Instrum.* 66, 4271-4276.
7. Peiner, E. and Schlachetzki, A. (1992) Automatic Counting of Etch Pits in InP, *J. Electron Mater.* 21 887-892.
8. Swaminathan, V. (1992) Properties of InP and Related Materials, in A. Katz (ed.), *Indium Phosphide and Related Materials: Processing, Technology, and Devices*, Artech House, Boston, pp. 1-43.
9. Wehmann, H.-H., Tang, G.-P., and Schlachetzki, A (1993) Strain relaxation and threading dislocation density in lattice-mismatched semiconductor systems, *Solid State Phenomena* 32-33, 445-450.
10. Vignaud, D. and Farvacque, J. L. (1989) Charged dislocation induced optical absorption in GaAs, *J. Appl. Phys.* 65, 1261-1264.
11. Zwinge, G., Ziegenmeyer, I., Wehmann, H.-H., Tang, G.-P., and Schlachetzki, A. (1993) InP on Si substrates characterized by spectroscopic ellipsometry, *J. Appl. Phys.* 74, 5889-5891.
12. Lum, R. M., Klingert, J. K., Davidson, B. A., and Lamont, M. G. (1987) Improvements in the heteroepitaxy of GaAs on Si, *Appl. Phys. Lett.* 51, 36-38.
13. Itoh, Y., Sugou, M., and Mori, H. (1992) The effect of III/V ratio on the initial layer of GaAs on Si, *J. Appl. Phys.* 71 3050-3052.
14. Wehmann, H.-H., Tang, G.-P., Koch, A., Seibt, M., and Schlachetzki, A. (1995) Twin formation during epitaxial growth of InP on Si, 6th Intern. Autumn Meeting on Gettering and Defect Engineering in Semiconductor Technology, (GADEST '95), Berlin, Germany.

DEFECT CHARACTERIZATION OF STRAINED InGaAs STRUCTURES PREPARED ON InP AND GaAs

R. SRNANEK, J. KOVAC, I. NOVOTNY,
J. SKRINIAROVA, S. NEMETH
Microelectronics Dept., STU Bratislava, Slovak Republic
B. OPITZ, A. KOHL
RWTH Aachen, Inst. Halbleitertechnik, Aachen, Germany

1. Introduction

Recently the strained InGaAs material system attracts increasingly more attention thanks to its promising device potential for several applications, such as lasers, electro-optical modulators and field effect transistors.

Wannier-Stark localization in $In_xGa_{1-x}As/In_yGa_{1-y}As$ superlattices can result in strong modulation of the absorption edge which can be used to realize an electro-optical modulator operating near 1550 nm [1]. One major restriction of strained lattice mismatched materials is the requirement upon the layer thickness to be below the critical value. In this case the mismatch is accomodated by elastic deformation and can exhibit good crystallographic, optical and electrical properties. On thick relaxed layers the surface shows usually a cross-hatch pattern due to misfit dislocations (MDs) [2].

In this work the chemical etching and photoetching were used for defect analysis on MOVPE grown PIN diodes containing $In_xGa_{1-x}As/In_yGa_{1-y}As$ strained superlattices and MBE grown $In_xGa_{1-x}As/GaAs$ relaxed structures.

2. Experimental

The growth of PIN structures was performed by LP-MOVPE on an InP (001) sulphur doped substrate. A PIN structure consisting of different superlattices with ten periods of $In_xGa_{1-x}As$ wells and $In_yGa_{1-y}As$ barriers (x > y) was investigated. Each superlattice was embedded in the intrinsic region of a quaternary $In_{0.69}Ga_{0.31}As_{0.67}P_{0.33}$ PIN diode [1].

J. Novák and A. Schlachetzki (eds.), Heterostructure Epitaxy and Devices, 153–156.
© *1996 Kluwer Academic Publishers.*

The thick $In_xGa_{1-x}As/GaAs$ strained layers (x = 0.06, 0.09 and 0.225 gradually with steps) were grown by MBE on silicon doped GaAs (001) substrates.

For defect etching of InGaAs/InP structures the SN 11 [3] and Lourenco [4] etchants consisting of $K_3Fe(CN)_6$, KOH, H_2O were used. Examination of the defects were performed by optical interference microscopy.

3. Results and discussion

We examined superlattices in which the well material was lattice matched to InP : 8 nm well of $In_{0.53}Ga_{0.47}As$, 7 nm barrier of $In_{0.40}Ga_{0.60}As$. X-ray diffraction pattern these samples together with a calculated difractogram show that the simulated diffraction pattern exhibits a smaller width of a satellite peaks and a more pronounced thickness of the superlattice which might be caused by MDs [1]. The MDs were visible on the sample surface after MOVPE growth. After chemical etching they were clearly visible in both {110} directions as shown in fig.1a. The density of MDs in the <110> direction is lower than in the perpendicular direction. Various types of MDs in <110> direction are visible in fig.1b. MD1 is created from a threading dislocation which is prolongated from the substrate and then is bending and directed paralell to the (001) surface. MD2 has no endtails with round etch pits. This indicates that this MD was created in the superlattice and then bent to the surface.

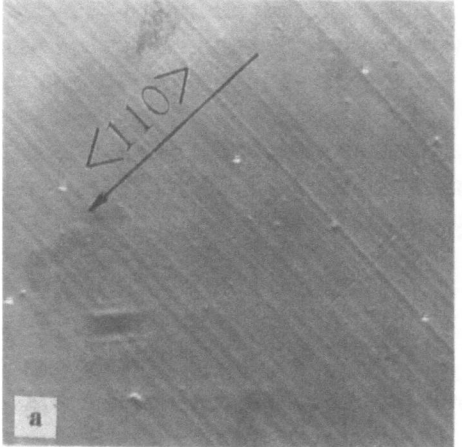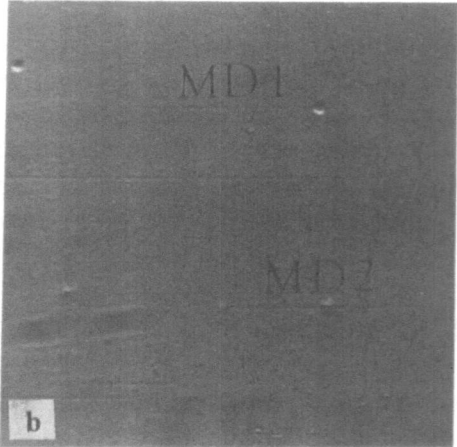

Fig.1. Surface morphology of PIN structure after etching in SN 11. a) MDs in the middle of the sample. b) Various types of MDs in <110 > direction.

An approach to improve the crystal quality might be to compensate the compressive strain of the barrier by tensilely strained wells ("strain balance"). For this purpose the structures of $In_{0.64}Ga_{0.36}As$ were grown instead of the lattice matched material. A reduced width of the satellite peaks is observed as well as an enhanced intensity of thickness fringes on the X-ray rocking curve. This indicates that the layer quality was improved by strain balance [1] which was also confirmed by microscopy observations (fig.2). MDs were observed only at the edge of the sample. In the middle part of the sample no MDs were observed or revealed by chemical etching.

Fig.2.   The surface of the structure with superlattice $In_{0.64}Ga_{0.36}As/In_{0.40}Ga_{0.60}As$ after etching in SN11. MDs are only on the edge of the sample.

Crystallographic defects in InGaAsP layers are revealed by photoetching using Lourenco etchant [4]. We applied this etchant also to reveal defects in the InGaAs strained superlattice. Defects, such as dislocations or microdefects, were not visible after etching (due to high roughness of the surface), only macroscopic defects (e.g. precipitates, oval defects, mechanical damages) were clearly seen surrounded by "halo" as shown fig.3. "Halo" indicates the range of defect free material around certain macroscopic defects.

To compare the MDs in relaxed structures, $In_xGa_{1-x}As/GaAs$ layers with an increasing value of x were investigated. In this case different types of surface morphologies (cross-hatch patterns) due to the increasing lattice misfit become more apparent. As an example, in fig.4 the MDs on as grown surface of an $In_{0.09}Ga_{0.91}As/GaAs$, 1040 nm thick structure are shown. Besides MDs, also a lot of V shaped defects were observed.

156

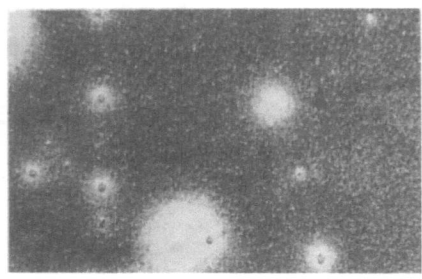

Fig.3. The surface of the top of the strained superlattice after etching in Lourenco etchant. "Halo" features are around precipitates.

Fig.4. The surface morphology of as grown $In_{0.09}Ga_{0.91}As$/GaAs structure.

Conclusion

Crystallographic defects in strained and relaxed layers were investigated by different chemical etching. "Halo" features around precipitates, oval defects and mechanical damages were observed in InGaAs superlattices.

References

1. Opitz,B., Kohl,A. Kovac,J., Brittner,S., Grunberg,F.,Heime,K. and Woitok,J., (1994) Wannier-Stark Effect in $In_xGa_{1-x}As/In_yGa_{1-y}As$ Superlattices of Different Compositions on InP , Proc.6 th Int.Conf.on InP and Rel.Comp., Santa Barbara, 1994, 459-462
2. Ito,H. and Ishibashi,T. (1986) GaAs/$In_{0.08}Ga_{0.92}As$ Double Heterojunction Bipolar Transistors with Lattice-Mismatched Base, Jap.J.Appl.Phys. **25**,L421-L424
3. Smanek,R., Nandraska,S. and Kovac,J. (1987) Observation of dislocation etch pits on (001)InGaAsP, J.Mater.Sci.Lett.**6**, 185-187
4. Lourenco,J.A. (1984) A Defect Etchant for (100) InGaAsP, J.Electrochem.Soc.**131**, 1914-1916

INFLUENCE OF THE TEMPERATURE ON THE MORPHOLOGY AND CRYSTAL QUALITY OF MBE GROWN InAs/GaAs HETEROSTRUCTURES

L. FRANCESIO, P. FRANZOSI, S. GENNARI, L. NASI, AND
G. SALVIATI,
MASPEC-C.N.R. Institute, Via Chiavari 18/A, Parma, Italy

M. R. BRUNI, G. PADELETTI, AND M.G. SIMEONE
ICMAT-C.N.R. Institute, Via Salaria Km 29, Monterotondo, Roma, Italy

2 µm thick InAs single layers have been grown by Molecular Beam Epitaxy on (001) oriented GaAs substrates at growth temperatures, T_g, ranging from 350 to 600°C. The correlation between crystal defects, surface morphology and growth temperature is discussed by comparing high resolution X-ray diffraction, Transmission Electron Microscopy and Atomic Force Microscopy investigations. Surface hexagonally shaped holes have been observed at the lowest growth temperatures to be correlated to spiral growth around threading dislocations with b=a/2 [110] on the interface plane. The presence of a InGaAs layer due to interdiffusion of Ga from the substrate has also been found at higher growth temperatures.

1. Introduction

In the frame of the increasing interest devoted to the growth and physical characterization of semiconductor lattice-mismatched heterostructures, the InAs/GaAs system is largely studied for its potential application in both microwave and optoelectronic device fabrication. Moreover, because of its large misfit ($\approx$7.2%), the system is attractive for the comprehension of the strain release mechanisms [1]. It is known that in highly mismatched systems, the very beginning of the growth follows the Stransky-Krastanov model with at first the deposition of a continuous film 1 or 2 monolayer thick and then with the formation of islands of different size according to the growth conditions [1,2]. The onset of 3D growth can involve a damage for the previously deposited film and even for the substrate, as observed by Zhang et al [2].

This study reports the correlation between the substrate temperature and the layer morphology at high layer thickness. Atomic Force Microscopy (AFM) and Transmission Electron Microscopy (TEM) have been used to study the sample surfaces and crystal defect nature and distribution respectively, whereas High Resolution X-ray Diffraction (HRXRD) has been used to test the overall crystal quality of the layers.

2. Experimental

InAs heterolayers have been grown by MBE on (001) oriented GaAs substrates,

J. Novák and A. Schlachetzki (eds.), Heterostructure Epitaxy and Devices, 157–160.
© 1996 *Kluwer Academic Publishers.*

158

following a procedure described elsewhere [3]. The films were deposited at seven different T_g values (350, 400, 450, 500, 520, 550 and 600°C). The growth rate was 0.68 µm/h and the layer thickness (expressed in equivalent layer by layer coverage) was 2 µm. For temperatures higher than 520°C it was very difficult to grow thick layers: previous theoretical calculations [4] predicted a dramatic decrease in the growth rate starting at about 580°C. A more complete study including also the influence of layer thickness on surface morphology and defect distribution is reported elsewhere [3].

HRXRD experiments have been performed by a high resolution diffractometer equipped with a two crystal-four reflections monochromator (Ge, 220). The Cu $K\alpha_1$ line and the symmetric 004 reflection have been used. TEM investigation has been carried out with a JEOL 2000FX instrument, working at 200 KV. The (001) oriented plan view and the (110) oriented cross sectional samples have been prepared by standard mechanochemical procedures followed by Ar ion milling in a 600 DUOMILL GATAN system. Contact mode AFM analyses have been performed in a Nanoscope IIIA Digital Microscope.

3. Results and discussion

The HRXRD measurements showed an almost complete relaxation of the InAs lattice in the 2 µm thick layers. The residual elastic strain, probably due to the thermal mismatch, was nearly constant in all the samples investigated and on the order of a few 10^{-4}, close to the resolution limit of the present measurements.

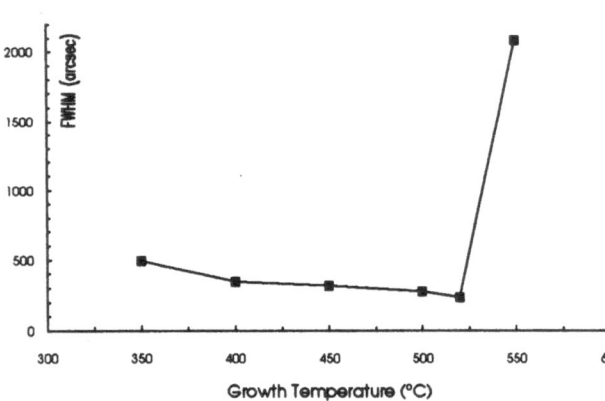

The Full Width at Half Maximum (FWHM), β, of the epilayers 004 Bragg peaks of are reported in Fig.1 as a function of the growth temperature T_g. Not shown in the picture is the β of the sample with T_g=600°: in this case no detectable XRD response was found. It is known that β depends on the presence of defects in the layer [5]: since it is apparent that β decreases by increasing T_g, an improvement in the epilayer crystal quality is evident up to a treshold temperature above which the growth degrades.

Figure 1. Width β of the 004 Bragg peak of the InAs heterolayers as a function of the growth temperature.

As a matter of fact, the peak of the sample with T_g=550° is 2078 arcsecs broad and is shifted toward higher angles, indicating a diffusion of the group III elements from the substrate to the layer and vice versa, with the consequent formation of an InGaAs layer with non-uniform composition. This effect is amplified in the sample grown at 600°C: the lack of XRD response in this case suggests the growth of a polycrystalline layer. This hypotesis is supported by the very irregular mophology (mean roughness of about 150÷200 nm) found during AFM investigations.

Different morphological features were found on the other samples according to the different growth temperature. For T_g=450°C the film is nearly flat (mean roughness 3nm), whereas for T_g=500 and 520°C, 40+100nm tall stepped rectangular pyramids oriented along one <001> direction appear. We assume that in this case the diffusion lenght of the atoms on the surface is ideal and the growth, after the coalescence of the islands detected by AFM in the first stages of growth, goes on step by step, with a very low defect density (about 1.3×10^8 cm^{-2} as detected by plan view TEM and as indicated by the narrow β). For lower T_g (350° and 400°C) the InAs surface appears smooth, with hexagonally shaped holes aligned along one of the two [110] directions. For T_g=400°C the density of the holes is lower with respect to the sample at T_g=350°C, but their size is bigger (Fig.2). It has been verified that for a thickness of the order of ten nanometers these features are not yet present, whereas in thicker samples their density reduces by a factor of five.

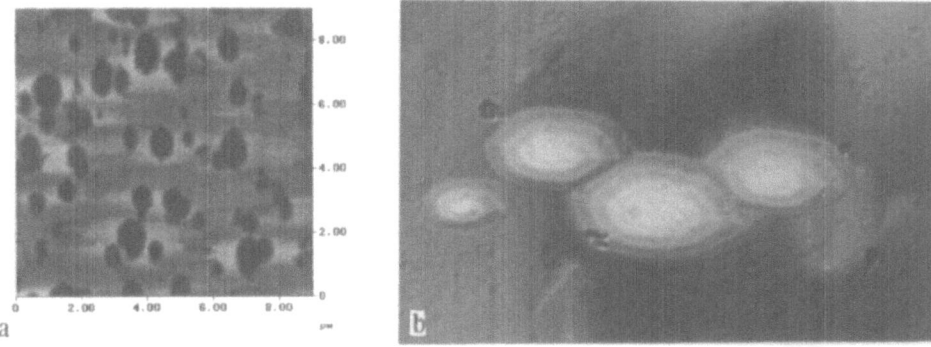

Figure 2: a) AFM image of the sample grown at T_g=400°C; b) details of the spiral growth as seen by plan view TEM

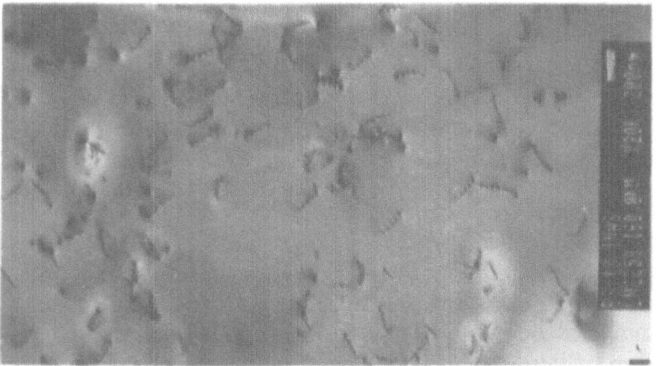

Figure 3. (001) oriented plan view TEM micrograph of an InAs layer grown at 350°C.

As shown in Fig.3, two types of dislocations have been found by plan view TEM investigations. One type (b=a/2 [110] on the interface plane) was strictly correlated with spiral growth. Further, dislocations with b= a/2 [110] at 45° with respect to the (001) plane were also found. Plan view TEM also revealed a threading dislocation density of about 2.5×10^8 cm^{-2} in the sample grown at T_g=350°C. Finally, (110) oriented XTEM investigations (Fig.4) evidenced that threading dislocations were mainly confined at the heterointerface, so confirming the good crystal quality predicted by HRXRD studies.

4. Conclusions

The crystal quality and the morphology of MBE (001) InAs/GaAs samples have been studied by HRXRD, TEM and AFM. The layers have been grown at different temperatures (from 350 up tp 600°C) keeping constant the nominal thickness (2 μm). According to the FWHM study of the (004) Bragg peak and TEM defect investigations it has been found that the threading dislocation density is smaller in the layers grown at higher T_g. This is true for $T_g \leq 520°C$; further, in the samples grown at 550 and 600°C, the formation of a ternary compound is suspected, due to the interdiffusion of the group III elements.

Finally, hexagonally shaped holes have been observed to be correlated to spiral growth around threading dislocations with b=a/2 [110] on the interface plane, at the lowest growth temperatures.

The correlation between growth temperature, origin and size of those holes is not yet clear and deserves further investigations.

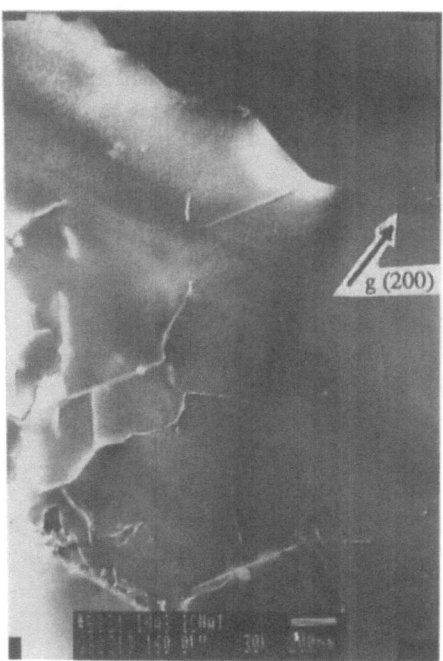

Figure 4. Weak beam XTEM image of the 400°C sample. The threading dislocation density decreases by approaching the top surface.

5. References

1. Chen., P., Xie, Q., Madhukar, A., Chen, Li, Konkar, A., (1994), Mechanisms of strained island formation in molecular-beam epitaxy of InAs on GaAs(100), *J.Vac.Sci.Technol.B*, **12**, 2568-2573
2. Zhang, X., Pashley, D.W., Neave, J.H., Zhang, J., Joyce, B.A., (1992), A trasmission electron microscopy study of a wedge-shaped InAs epitaxial layer on GaAs (001) grown by molecular beam epitaxy, *J.Cryst.Growth*, **121**, 381-393
3. Francesio L., Franzosi P., Gennari S., Nasi L., Salviati G., Bruni M.R., Padeletti G., Simeone M.G., Viticoli S., *Investigation of the structural properties of InAs/GaAs MBE heterolayers*, sent for publication to J.Appl.Phys.
4. Bruni, M.R., Lapicciarella, A., Scavia, G., Simeone, M.G., Viticoli, S., Tomassini, N., (1992), Thermodynamic study of molecular beam epitaxial growth of InGaAs/GaAs strained layer superlattices, *Thermochimica Acta*, **210**, 49-65
5. Ayers, J.E, (1994), The measurement of threading dislocation densities in semiconductur crystals by X-ray diffraction, *J. Crystal Growth*, **135**, 71-77

STUDY OF FUNDAMENTAL GROWTH MECHANISM BY ATOMIC FORCE MICROSCOPY

C.C. Hsu, J.B. Xu, and I.H. Wilson
Department of Electronic Engineering
The Chinese University of Hong Kong
Shatin, NT, Hong Kong

Abstract: We have studied the growth mechanism of GaAs, InP by MOVPE, and GaAs by MBE with atomic force microscope(AFM). Monolayer steps and 2D islands can be resolved by AFM. We can distinguish three different growth modes: 2D nucleation especially in the case of MBE, *step flow* on a vicinal surface and spiral growth near a screw dislocation. Concentric ring patterns were also discovered, which may be caused by edge dislocations or stacking faults.

1. Introduction

The study of epitaxial growth mechanism can be conducted using *in situ* and *ex situ* observations. A notable example of in situ observation is the reflection high energy electron diffraction (RHEED) intensity oscillations during molecular beam epitaxial (MBE) growth, where the two-dimensional (2D) nucleation mechanism has been observed. For *ex situ* observations, the Nomarski differential interference contrast (NDIC) optical microscope is normally used to examine the as-grown epitaxial layer surface. Due to the high vertical resolution of the NDIC microscope, monolayer growth steps can be observed under favorable conditions. Atomic force microscopy (AFM) is a more recently developed technique and is capable of providing atomic scale image

J. Novák and A. Schlachetzki (eds.), Heterostructure Epitaxy and Devices, 161–171.
© *1996 Kluwer Academic Publishers.*

of surfaces. With AFM or optical microscope, we can study the surface morphology of the as-grown materials. According to Frank, *we cannot undrstand the kinetics of growth without paying close attention to the surface morphology. When we fully understand the surface morphology, we know practically all about the kinetics*[1]. In this paper, we report our study of the surface morphology and growth mechanism of GaAs, InP epitaxial layers by AFM.

During growth, the nutrient atoms (or molecules) impinge upon the substrate surface as surface adatoms. These adatoms diffusue around the surface and seek nucleation sites. Most of the substrate surfaces are vicinal surfaces with misorientation from the nominally oriented directions. Monolayer steps are formed during growth on these vicinal surfaces. The steps provide the nucleation sites for the surface diffusing adatoms. If the surface diffusion length is less than 1/2 of the terrace width of monolayer steps, 2D islands will occur on the terrace. The growth mode is 2D nucleation. Otherwise, the step edge will act as sink for the surface diffusing species, and the steps will move forward as the adtoms are incorporated into the steps. This growth mode is called *step flow*. Besides the growth modes mentioned, 2D nucleation and *step flow*, there is spiral growth which is caused by the screw dislocations on the surface. The existence of screw dislocations on the surface makes the whole crystal a continuous non-Euclidean surface. Crystal growth is a continuing extension of this surface. Unlike typical steps, a screw dislocation has a step less than a monolayer height near the emergence point and it makes the screw dislocation a self-perpetuating step source during growth. The steps around the screw dislocations wind themselves in spirals and successive turns of steps are sent out by screw dislocations.

2. Experiment

The MOVPE growth was performed in a commercial (Aixtron) low pressure reactor. Trimethylindium, trimethylgallium, tertiarybutylarsine and tertiarybutylphosphine were used as the precursors. The growth temperature was $600^{\circ}C$ to $650^{\circ}C$ at a reactor pressure of 200 mBar. The group V to group III (V/III) ratio in the vapor phase was in the range of 10 to 30 during growth. The substrates were commercially available "epi-ready" GaAs and InP wafers nominally oriented in the [100] direction. After growth, the samples were examined using AFM. The measurements were made in air at room temperature with Digital Instruments Nanascope III. Images were acquired in the constant force mode using standard AFM cantilevers.

The MBE GaAs growth was performed at the Chalmers University of Technology in a Varian GEN II modular system. One sample was grown by starting at 550°C with a (2x4) reconstruction while the *step flow* growth condition was obtained by raising the temperature to 700°C in an increased As$_4$ flux. After growth, the substrate temperature was immediately ramped down, and the samples were examined using AFM.

3. Results and Discussions

Figure 1 is the surface morphology of a (100) GaAs substrate annealed at 750°C under an arsenic overpressure. Monolayer terraces and steps have developed as a result of evaporation and regrowth. During annealing, the surface atoms rearrange themselves or regrow through surface diffusion, while minimizing the total surface free energy. The density of steps and their direction reflect the misorientation of the substrate. A 280 nm terrace width is equivalent to a 0.06° misorientation from the [100] direction. Assuming this to be the type of surface that we start with our growth, then after thousands of monolayer growth (film thickness 600 nm) at 650°C, we obtained the surface shown in Figure 2. Relative regular monolayer terraces and steps can be seen along the [010] direction. The surface diffusion length must have been larger than 1/2 of the terrace width, otherwise we would see 2D nucleation islands on the terraces. We have observed terrace widths of 430 nm on different samples grown under the same growth conditions. The difference in terrace width originates from the initial unintended misorientation. Thus, we may argue that we may have a larger terrace width and a longer surface diffusion length if there were no misorientations from the [100] direction.

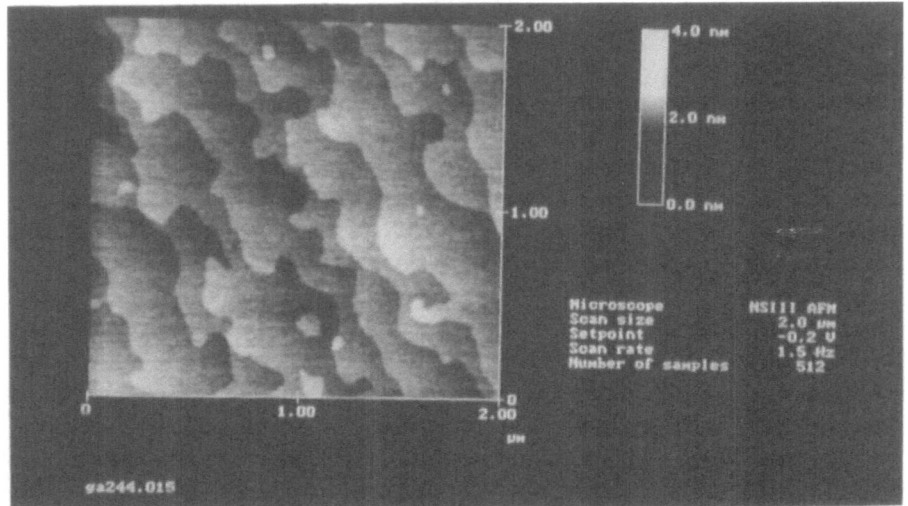

FIGURE 1. AFM image of a (100) GaAs substrate surface after annealing at 750°C in a TBA/hydrogen ambient.

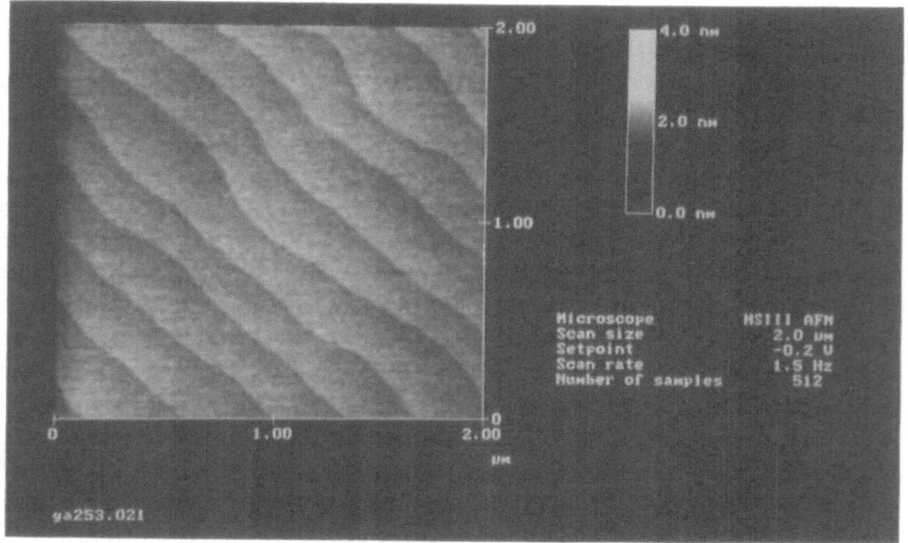

FIGURE 2. AFM image of a 600 nm thick MOVPE grown GaAs epilayer on a vicinal (100) substrate. Eleven atomic terraces with a mean width of 280 nm are seen in the [010] direction.

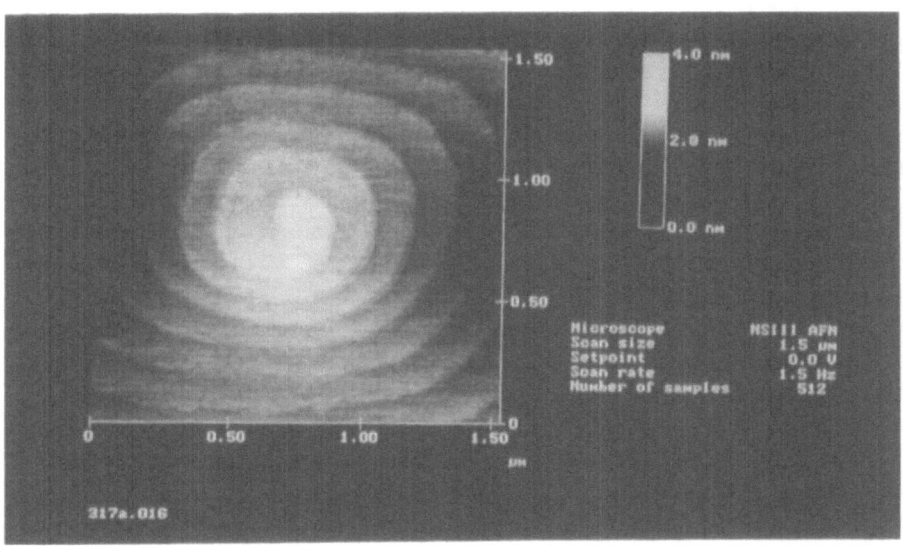

FIGURE 3. AFM image of GaAs growth spiral pattern. Terrace width is around 150 nm.

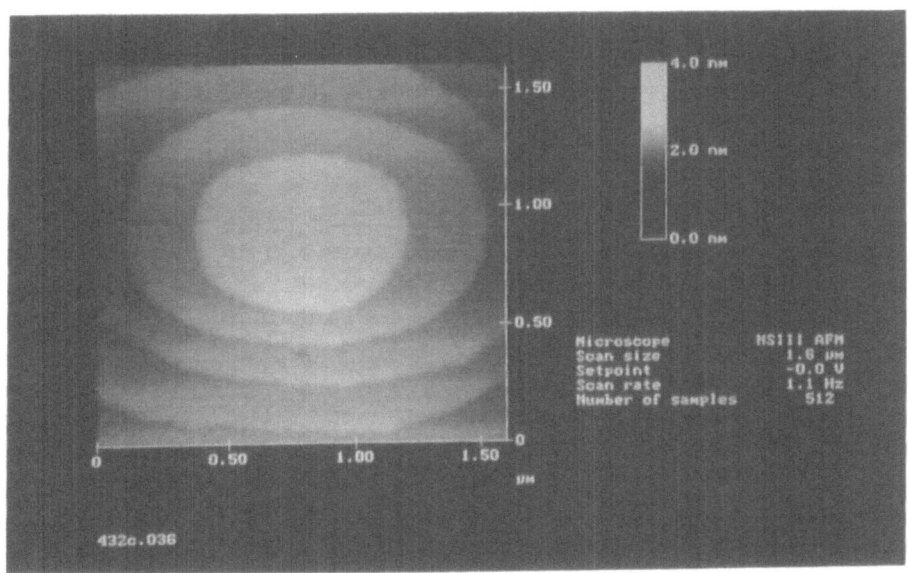

FIGURE 4. A concentric ring pattern on a (100) GaAs surface. Successive rings are sent out by the center ring.

Besides these regular steps, we also found spirals as shown in Figure 3. This spiral winds clockwise and the picture clearly demonstrates that the growth mechanism near the spiral is the screw dislocation controlled spiral growth. Near the screw dislocation, the interstep distance is around 150 nm. The spiral pattern is anisotropic and it also reflects the four-fold symetry of the (100) surface. The distance between the spiral steps is less than that of vicinal steps on the same substrate. This indicates that spiral growth is a very efficient growth process when competing with the *step flow* growth. Dislocation controlled spiral growth has a lower potential barrier and is an energetically more favorable mechanism. Besides the screw dislocations being very efficient step sources, edge dislocations and stacking faults are all possible step sources. These sources have submonolayer steps which can act as persistent step sources the same as the screw dislocations. We found a concentric ring pattern of steps as shown in Figure 4. Successive rings are sent out by the center ring as if the center ring is a persistent step source. Similar patterns were also observed on LPE grown GaAs surfaces by an optical microscope with Nomarski interference contrast[2]. If the concentric ring pattern is generated by a pair of screw dislocations of the opposite sign, the distance between two dislocations must be larger than the critical radius and there must exist a monolayer step between them. The critical radius is of the order of 100 nm. Within the noise level of our AFM, we cannot observe anything of the order of a monolayer height. If there is any morphological pattern in the center ring, it must be less than a monolayer in height. Frank suggested that it may be due to edge dislocation[3].

The same growth mechanisms were also found in InP grown by MOVPE at 600°C. Figure 5 is a left-hand InP spiral. The spiral pattern is also anisotropic, and different from that of GaAs. It is due to that the edge free energy of the steps on the (100) surface is different in different directions. According to Herring, the shape of the pattern is a reciprocal Wulff polar diagram of the edge free energy[4]. Figure 6 is a right hand spiral. The center of the spiral is the highest point in the area. To the left side of the spiral, steps look as though they are generated by the spiral or due

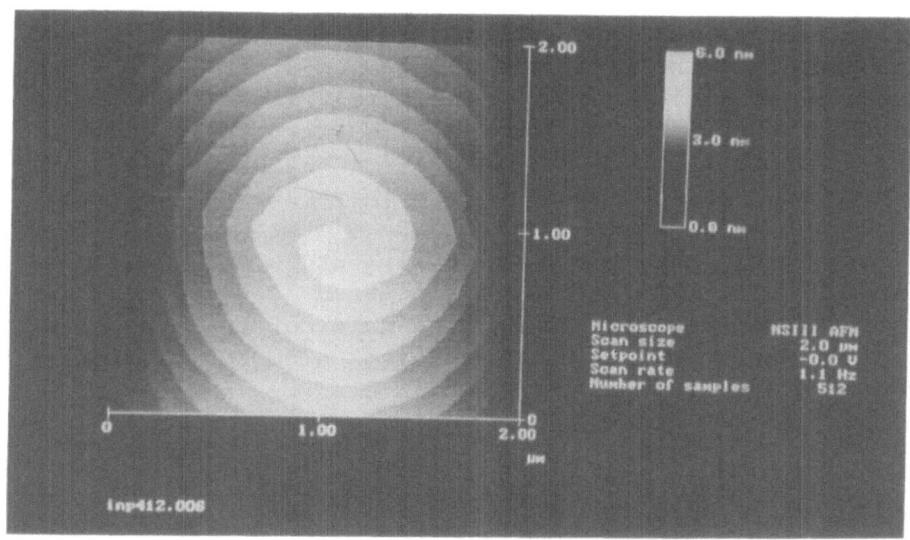

FIGURE 5. A left-hand InP growth spiral pattern.

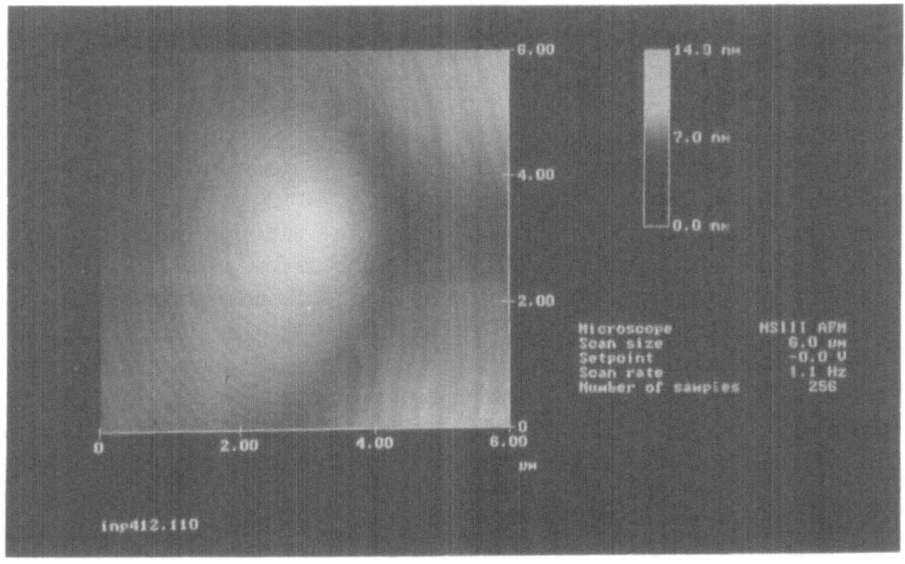

FIGURE 6. A right-hand spiral in the vicinity of misorientation steps. Spiral growth may become the dominant mechanism if there is no surface misorientation.

to the step train coming from the right side of the spiral. Steps meet and annihilate on the right side, and propagate to the right side merged together. The same surface morphology will repeat itself if we keep going to the right on the surface. In the extreme right end of the substrate, the original vicinal steps must be exhausted. In this area, new steps can be generated either by 2D nucleation or due to the spiral steps coming from the left side. Spiral steps are the most likely ones. The screw dislocation density on the surface does not have to be high in order to have spiral growth. Each one is a very efficient step source. Bauser reported a spiral generated surface across a few hundred microns on LPE grown Si[5].

There are two typical growth modes in MBE: 2D nucleation and *step flow*. During the MBE growth, RHEED intensity oscillations have been observed. Figure 7 shows the surface morphology of GaAs grown at 580°C under the 2D nucleation mode. The surface is rough, even though monolayer steps are observable in Figure 7. The main reason for the surface being rough is that the surface diffusion length is samll during growth. If we increase the growth temperature, which effectively lowers the supersaturation, the surface diffusion length becomes larger. When the surface diffusion length is larger than 1/2 of the vicinal step width, the growth mode will become the *step flow* mode. Figure 8 is the surface morphology of GaAs grown under the *step flow* mode at a growth temperature of 700°C where no RHEED intensity oscillations were observed except the streaky RHEED pattern. Steps and terraces can be seen, with some of the terrace width near 1000 nm. The misorientation is titlted toward the [011] direction. On the larger steps, 2D islands can also be observed. Arsenic dimer rows form on the surface to saturate the surface dangling bonds, even during the MBE growth. These dimer rows are aligned in the [011] direction on the (100) surface. This kind of surface reconstruction makes the surface properties anisotropic. During growth, the surface diffusing species experience different potential barrier along the two different [011], [011] directions. If steps are formed, they tend to be straight in the [011] direction and rough in the [011] direction. We shall call the steps on the (100) surface in the [011] direction "A steps", and "B steps" in the [011] direction.

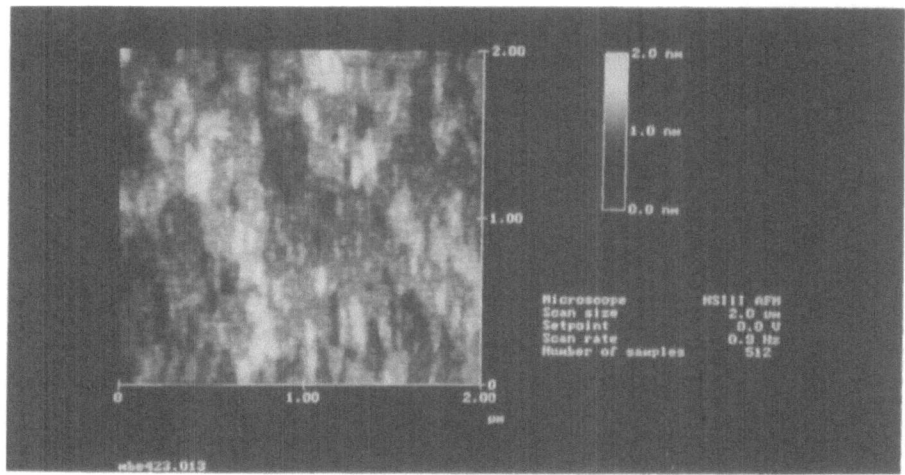

FIGURE 7. AFM image of MBE GaAs surface morphology grown at 580°C under the 2D nucleation mode. The surface is rough, even though monolayer steps are observable.

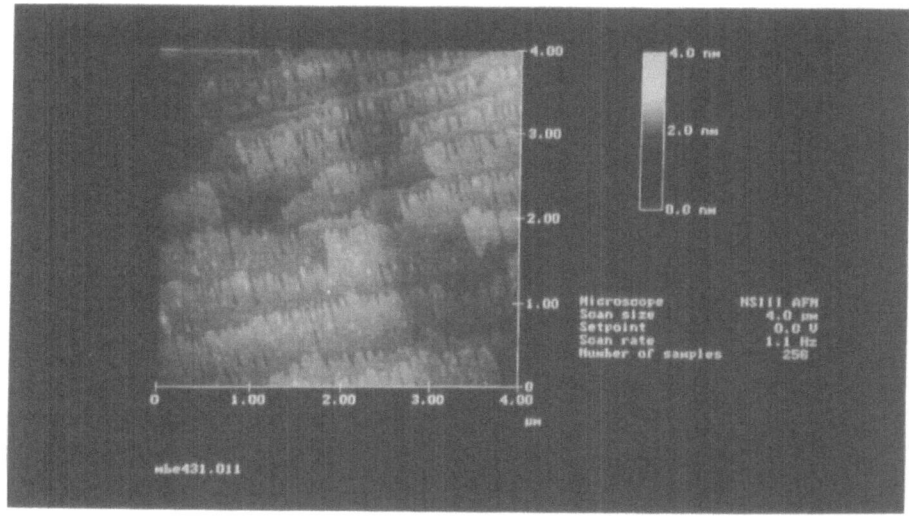

FIGHUR 8. AFM image of MBE GaAs surface morphology grown at 700°C under the *step flow mode*. Steps and terraces can be seen clearly, with some of the terrace widths near 1,000 nm. The surface misorientation is titlted toward the [011] direction.

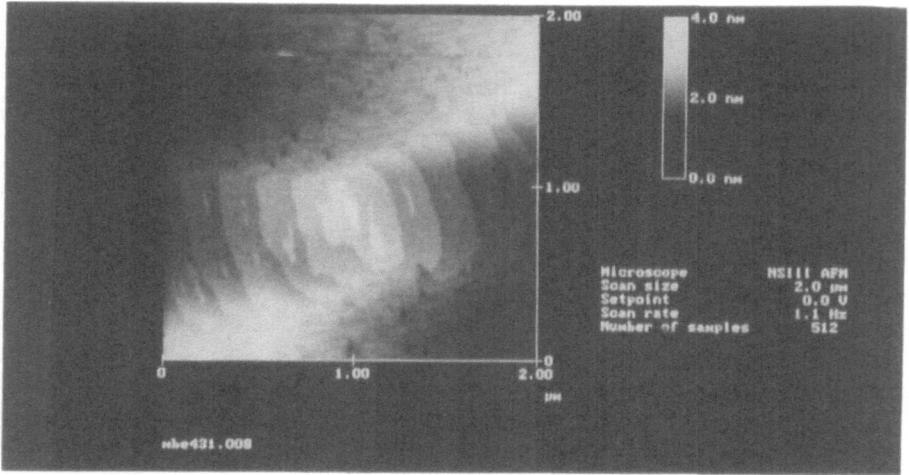

FIGURE 9. A spiral pattern due to a screw dislocation. *A* steps on the spiral are straight and *B* steps are rough due to surface reconstruction during growth. 2D islands can also be seen on the spiral steps.

We have also observed growth spirals on MBE grown GaAs. A typical spiral pattern is shown in Figure 9. On the spiral, the *A* steps are larger than the *B* steps. The surface reconstruction during growth makes the *A* steps straight and *B* steps rough. A few large 2D islands were also observed on the spiral steps.

Acknowledgement

We want to thank T.G. Andersson and J.V. Thordson of the Chalmers Univeeristy of Technology for providing the MBE GaAs samples.

4. References

1. F.C. Frank, F.C. (1958) Introductory Lecture, in R.H. Doremus, B.W. Roberts and D. Turnbull (eds.), *Growth and Perfection of Crystals*, Wiley, New York, pp. 3-10.
2. Bauser, E. and Strunk, H. (1981) Analysis of dislocation creating monomolecular growth steps, *J. Crystal Growth* **51**, 362-366.
3. Frank, F.C. (1981) "Edge" dislocations as crystal growth sources, *J. Crystal Growth* **51**, 367-368.
4. Herring, C. (1951), Some theorems on the free energies of crystal surfaces, *Phys. Rev.* **82**, 87-93.
5. Bauser, E. (1987) The preparation of modulated semiconductor structures by liquis phase epitaxy, in R.F.C. Farrow (ed.), *NATO ASI series Thin Film Growth Technique for Low Dimensional Structures*, Plenum, New York, pp. 171-194.

CHARACTERISATION OF THE EPITAXIAL LAYERS USING THE LIFT-OFF TECHNIQUE

J. NOVÁK
Institute of Electrical Engineering, Slovak Academy of Sciences,
SK-842 39 Bratislava, Slovak Republic

1. Introduction

Hybrid integration of electronic devices based on various semiconductor materials is limited by problems such as parasitics due to bond interconnnections, alignment of components, and the large size of hybrids. The application of thin film technology allows most of these problems to be overcome, but in this case one has to solve the additional problem of transportation of very small devices. A promising approach lies in the use of the epitaxial lift-off technique (ELO). In this case, devices are separately homoepitaxially grown on appropriate substrates, lifted off using wet chemical etching, and the they are finally bonded by Van der Waals forces onto host substrate. The ELO technique is very flexible and it can be simply modified for different applications such as the integration of devices prepared in noncompatible semiconductor materials, or the transport of devices to substrates with better thermal conductivity. It can also be applied to characterise epitaxial layers.

2. Principles of the ELO technique

The first version of the ELO technique was reported by Konagai et al in 1978 [1]. they used the ELO technique to transport GaAs solar cells onto the aluminium supporting plate. A new approach to epitaxial lift-off was proposed by Yablonovitch et al [2], who applied the black wax technique in the ELO process. The first step is the epitaxial growth of a device heterostructure on GaAs substrate including an intermediate AlAs lift-off layer. This layer's thickness is typically between 10 and 100 nm. The ELO layer separation is based on a high selectivity of etching of this lift-off layer by diluted HF. Its selectivity to AlAs compared to GaAs is higher than 10^8 (10% HF at temperature of 10 °C). This selectivity drastically drops with increasing aluminium content in the AlGaAs epitaxial layer. Therefore, special precautions to protect the aluminium contained layers in the heterostructure must be used [3]. It is very important to use a very thin intermediate AlAs layer and the HF acid must be diluted. Otherwise, the hydrogen produced during the etching process can form gas

J. Novák and A. Schlachetzki (eds.), Heterostructure Epitaxy and Devices, 173–181.

174

bubbles and, consequently, the etching process is stopped. Therefore, it is necessary to bend up the corners of the structure lifted to allow the hydrogen to diffuse out.

As it was described in [2], the ELO process begins with covering the surface of an epitaxial heterostructure with black wax (usually with Apiezon W), which is diluted in trichlorethylene and melted at 180 °C . By this way we obtain a uniform wax film which plays two important roles in the lift-off technique. At first, this wax protects the device structure in the following etch process. Secondly, the stress induced by wax is important to bend up the corners of the epitaxial structure, providing a channel for the outdiffusion of the hydrogen, which is produced during the etching (see Fig. 1). After wax deposition the substrate is dried up and baked (typically 15 min at room temperature and 30 min at 100 °C) and while the substrate cools down, it makes the wax contract, which involves the stress required. Another technique developed makes use of a heated syringe and a needle by means of which the pure molten wax is directly applied in the form of lines to the structure surface. An advantage this method is the possibility to selectively deposit wax on mesas etched through the epitaxial structure. There exist two basic possibilities to combine the processing of an

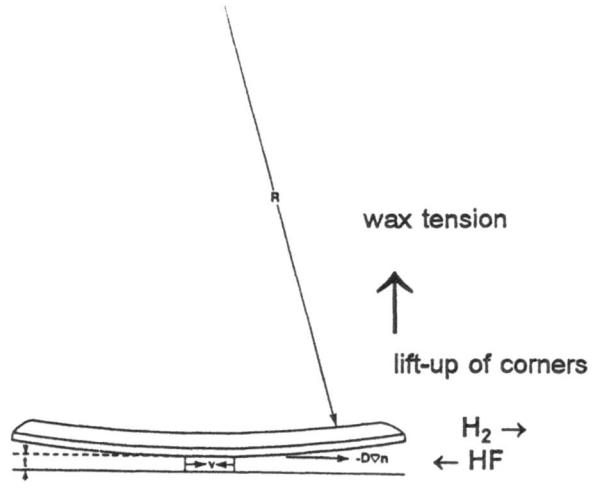

Fig. 1 Basic principle of the epitaxial lift-off technique [2]

epitaxial structure with the ELO technique. The processing can be done using standard technologies (pre-processing approach), then the complete structure is covered by wax and lifted-off. In the post-processing approach, the epitaxial structure is first lifted-off , transferred to the host substrate, and then it is processed by modified processing technology. The main advantage of the pre-processing method is that devices on both the substrates (growth and host) can be processed by standard technology. On the other hand, the finished devices must be transported with incorporated strain, which is caused by metallisation, dielectric passivation and antireflection coatings. Therefore, these devices must be handled very carefully.

One of the major problems of the ELO technique is the mechanical adhesion of the transferred epitaxial heterostructure to the host substrate. The main requirements for the bonding process are low defect density, low temperature processing, minimal degradation of device parameters, low electrical and thermal contact resistance between the lifted structure and host substrate. Generally, the most frequently applied bonding method uses bonding van der Waals forces. In this case, a structure is transported in deionised water, then it is dried up under small pressure (about 20 gmm^{-2}) for typically 24 hours, which yields very good mechanical contact. Recently, to improve electrical contact resistivity, Yablonovich et al [4] introduced the use of a palladium contact interlayer (this method is called topotaxy). They found that palladium is the only metal which reacts with both compound and elemental semiconductors at relatively very low temperature (< 200 C^{o}). Demesteer et al improved this method by using e-beam evaporated 30nm Pd on 50 nm polycrystalline silicon on silicon substrate. A short bake-out of this sandwich structure at 250 C^{o} for 30 min. significantly improved the quality of the bonding [5].

Recently, the application of ELO for the transportation of various optoelectronic devices from growth to host substrates has been reported. Justice et al [6] studied dark currents in PIN InGaAs/InP photodiodes fabricated by the preprocessing and postprocesing alternative. They showed the advantage of preprocessed photodiodes over the post-processed ones with respect to the dark current. As it follows from published results, preprocessed photodiodes have dark currents lower more than one order of magnitude. Very useful transportation of InGaAs:Fe/InP metal-semiconductor-metal photodetectors for the long wavelength region from InP to the silicon substrates was reported in [7]. The transferred detectors, with a finger spacing and width of 1.5

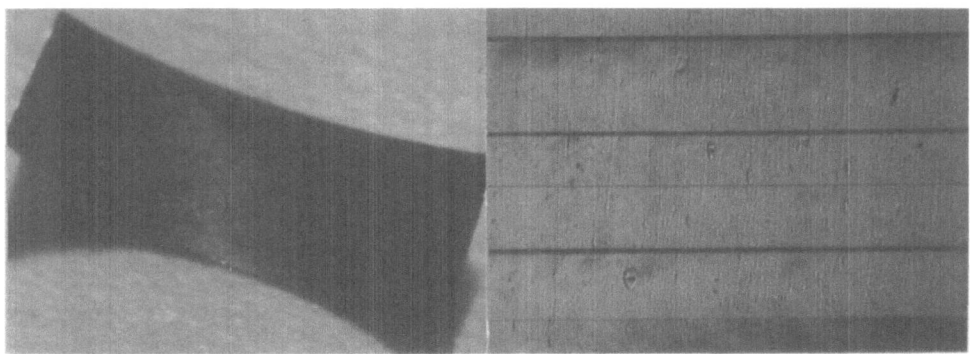

Fig. 2 Deformation and demaging of the epitaxial layer by incorporated strain. after substrate etching -off (sample of LT-GaAs grown at temperature of 250 ^{o}C.)

µm and 1.0 µm respectively, showed no deterioration of the device performance. A fast impulse response with an FWHM of 23 ps and an external quantum efficiency of 48 % was measured at 7V bias and 1300 nm wavelength. A low leakage current level of 250 nA at 7V bias was obtained. Similar good results were obtained when vertical resonant cavity ligth emiting diodes and photodetectors were formed[8]. In this case, an InGaAs/InP diode structure was grown on InP substrate and then transported using ELO into a double gold mirror cavity prepared on GaAs substrate. The total thickness of tranferred structure was 2.76 µm and mesas with surface area of 2x3 mm were transferred. Electroluminescence from the vertical cavity was measured CW at room temperature and the spectral width of 9 meV in contrast to the spectral width of 51

176

between 200 and 600 °C. An As$_4$/Ga beam-equivalent pressure ratio of 19 and a growth rate of 1 μm/h were used to grow 2 and 6 μm thick epitaxial layers. An 0.5 μm thick AlAs or AlGaAs etch-stop interlayer was grown between the substrate and the LT-GaAs.

In general, etching proceeds by an oxidation-reduction reaction at the semiconductor surface, followed by dissolution of the oxidized material. The typical etchant for this purpose consists of two separate components, one of which is an oxidizing agent (usually H$_2$O$_2$), the other dissolves the resulting oxide (for example by NH$_4$OH or an acid). Selective etching of GaAs (step I.) in relation to AlGaAs and AlAs was investigated in the past with respect to the etching selectivity of different materials in the epitaxial multilayer structures [13,14].The removal procedure of GaAs substrates consists of the following steps:

- lapping of the sample from the substrate side
- epoxy the sample, layer down, to a piece of microscopic glass,
- polishing etch of the GaAs substrate,
- etching down the rest of the substrate,
- etching of the interlayer, and
- rinsing of the separated layer in DI water.

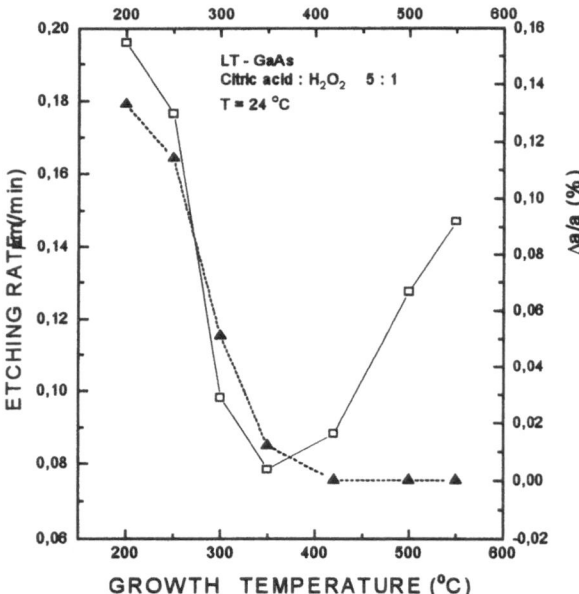

Fig 3 Citric acid etching rate ( ) and lattice mismatch ( ) of the LT-GaAs as a function of the growth temperature.

We found that the second step - epoxy the sample- is very important to be successful at the separation procedure. It is necessary to use highly firm and tempe-rature stable epoxy resin which can substitute for the role of the substrate instead of GaAs removed. The application of commonly used butylphtalate resin or Apiezon wax has been found unsui-table because the strain in-corporated in the epitaxial layer can cause total distor-tion or deformation of the epitaxial layer, as it is de-monstrated in Fig. 2. The GaAs semiinsulating sub-strate was etched away in a selective etchant based on citric acid. This etching process stops at the AlAs (or AlGaAs) etch-stop interlayer automatically. The separation procedure was finished by removing the interlayer with diluted hydrofluoric acid.

meV for the nontransferred devices was obtained. All these applications were made by a modified ELO technique. Its modification is based on the change of the etching direction. In the classical ELO version an AlAs interlayer is etched paralell to the layer plane from the sides to the centre, i.e. on epitaxial structure is covered with wax and undercut by etching-off the interlayer. In the modified method, the surface of the epitaxial structure is also covered with wax, but the substrate is etched off perpendicularly to the surface plane. In this case, the interlayer serves as an etch-stop layer. This method is very effective, but it may be used only for epitaxial structures without incorporated strain. To characterise epitaxial structures with incorporated strain, it is necessary to compensate for the supporting role of the substrate after it has been removed. Otherwise, the strain incorporated in the epitaxial layer can cause its total distortion or deformation, as it is demonstrated in Fig. 2. A typical example of such behaviour is low temperature GaAs (LT-GaAs) grown by molecular beam epitaxy (MBE). In the case of LT-GaAs, the epitaxial layer must be supported by another holder with better mechanical properties as those of wax. This problem can be solved very simply - usually a microscopic glass plate and very firm epoxy resin are used [9,10]. This modified ELO method is frequently used for characterisation of epitaxial layers in such cases when the presence of a substrate could influence characterisation results.

3. Characterisation of epitaxial layers

Electrical properties of LT-GaAs layers are strongly dependent on the growth and annealing conditions, and therefore, detailed knowledge of this behavior is very important for the design and preparation of advanced devices with LT-GaAs .

The room temperature resistivity of as-grown LT-GaAs layers increases strongly with the growth temperature T_g = 200-450 ^oC and reaches values up to 10^7 Ω cm [10,11]. Many applications require annealing of device structures at temperatures higher than T_g. The annealing results in an increase of the LT-GaAs resistivity, e.g. from about 10 Ω cm in 200 °C as-grown layers up to 10^5-10^7 Ω cm after their annealing at temperatures higher than 500 °C [10,12]. The sheet resistivity of the annealed LT-GaAs layer (usually about 1 μm thick) becomes comparable with the resistivity of the substrate in above mentioned cases. On the other hand, in the as grown LT-GaAs samples, the hoping conductivity dominates at room temperature due to extremely high concentration of native defects and the apparent Hall mobility is very low, less than 0.15 cm^2/V s in 200-250 °C as-grown layers [10]. From this it follows that it is practically impossible to obtain correct conductivity and Hall-effect parameters of LT-GaAs layers without separating the layers from their substrates. This was not considered in earlier studies of LT-GaAs, and later, on the basis of measurements on separated layers, corrections in results presented before were necessary . However, details about the layer separation have not been reported up to now.

GaAs layers were grown in a Varian Mod GEN II MBE system on indium-free mounted 2 inch (100) semi-insulating GaAs substrates with the resistivity of 1.5×10^8 Ω cm. The growth temperature was adjusted by a calibrated thermocouple in the range

The main advantage of the AlAs etch-stop layer and the citric acid etching system is that after etching away the remaining parts of the GaAs substrate and removing the AlAs etch stop layer by diluted hydrofluoric acid, the surface of the LT-GaAs layer is unaffected and clean. All AlGaAs etchants have a lower selectivity to GaAs and etching is influenced by imperfections in the epitaxial layer. Consequently, the total and planar removal of the AlGaAs layer without damaging the LT-GaAs surface is very difficult. In addition, the colour of AlAs differs from that of GaAs, while colours of GaAs and AlGaAs are similar. This colour difference is very important because it allows continual control and direct observation of the etching process, if it is necessary.

The main advantage of the citric acid etching system is the sufficient selectivity and the planar etching of GaAs unaffected by crystal imperfections. It is true that the selectivity of this etching system to GaAs/AlAs is low (only about 5) in comparison with that to GaAs/$Al_{0.3}Ga_{0.7}As$ (higher than 110) [13], but considering all advantages and problems mentioned above, we have decided for the combination of an AlAs etch-stop layer and the citric acid / H_2O_2 etchant (5 : 1).

The etching rate of the citric acid/H_2O_2 etchant composition (5:1) to the semiinsulating GaAs substrate is near 0.3 μm/min. As all other properties of LT-GaAs, the etching rate of this material is also influenced by its nonstoichiometry, which is a function of the growth temperature T_g. The etching rate vs T_g is illustrated in Fig 2. At first the etching rate decreases from 0.196 μm/min for $T_g = 200$ °C to the minimum value of 0.78 μm/min for $T_g = 300 - 400$ °C. Then, the etching rate increases up to 0.2 mm/min for $T_g = 550$ °C. To exclude possible mistakes in the determination of the etching rate, all samples were etched at the same time in the same etching bath at the temperature of 25 °C. It is generally believed that special properties of LT-GaAs are connected with a high amount of As incorporated into the GaAs lattice, [15, 16] which causes remarkable nonstoichiometry. This nonstoichiometry leads to the lattice mismatch in layers grown at T_g lower as 350 °C, which is demonstrated in Fig. 4. The lattice mismatch depends on the growth temperature and it decreases from $\Delta a/a = 1.3 \times 10^{-3}$ for $T_g = 200$ °C to zero for the layer grown at $T_g = 400$ °C or higher. The slope of this T_g is similar to that of the etching rate vs. T_g for the temperature interval

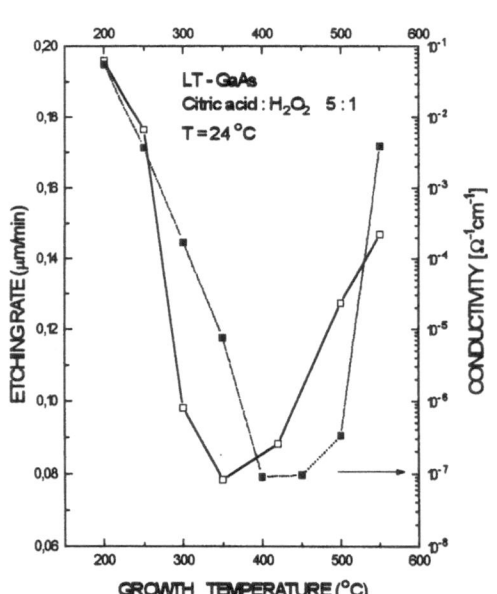

Fig. 4 Citric acid etching rate and electrical conductivty of LT-GaAs as a function of the growth temperature

between 200 and 400 C. But for a higher T_g, the lattice mismatch $\Delta a/a$ goes to zero while the etching rate increases. Therefore, we can deduce the etching along the misfit dislocations is not the main initializing mechanism of the LT - GaAs etching.

On the other hand, it is interesting to make comparison between the plots of the etching rate vs. T_g and those of the electrical conductivity of the epitaxial layers vs. T_g. As it follows from Fig. 4, both dependences have very similar characteristics. Therefore, it may be expected that both the etching rate and the electrical conductivity

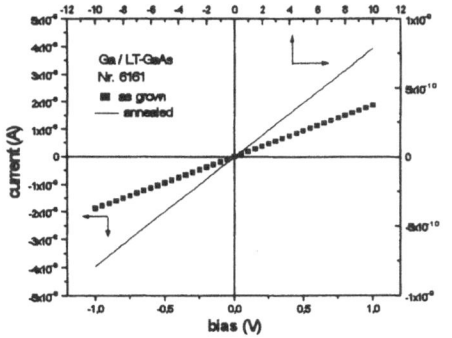

Fig 5. I-V characteristics of Ga/LT-GaAs ohmic contacts

are inicialized by the same mechanism connected with the material structure of the epitaxial layer. The arsenic excess (about 1 atomic %) is uniformly distributed in the crystal lattice and leads to the creation of deep levels, which are responsible for high resistivity of GaAs layers grown by MBE at low and intermediate temperatures rather than arsenic precipitates [10], which can be observed in layers grown at 200-300 °C after annealing at temperature 600 °C or higher [15]. So, we suppose that also the etching rate is related to the arsenic excess.

The method of wet chemical separation presented was applied to processing different MBE epitaxial structures. Separated LT-GaAs epitaxial layers were characterised by using a high impedance system and the van der Pauw method. Ohmic contacts were made by rubbing pure gallium into the samples surface. The main advantage of the gallium contacts is that they are liquid in the whole temperature range in which the electrical parameters were measured. It is a very important property because epitaxial layers without substrate support are very frail and a minimal additional stress may lead to the fatal distortion of a structure measured. As it follows from Fig.5, the Ga/LT-GaAs contacts are ohmic having a linear I-V characteristics in the whole range measured and are suitable for both high resistive as well as low resistive epitaxial layers.

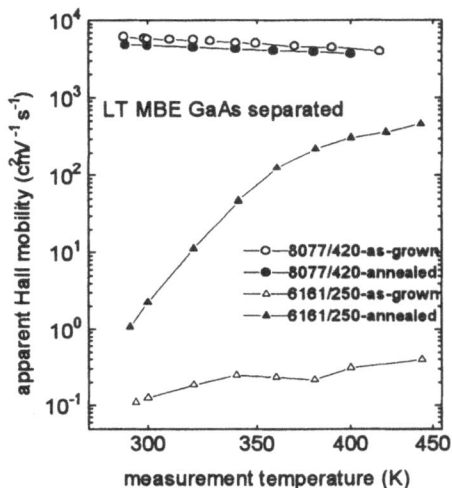

Fig 6. Apparent Hall mobility of two different LT-GaAs samples measured before and after annealing

For illustration, temperature dependencies of the Hall mobility measured

180

on four LT-GaAs layers are in Fig. 6. The layers were grown at T_g =250 and 420 °C, both separated before (as grown) and after annealing (590 °C). The separation made it possible to obtain reasonable results even in the case of extremely low Hall mobility values connected with a hopping mechanism of conductivity. In all samples investigated the Hall coefficient was negative (according to the convention: the Hall coefficient is negative for electrons and positive for holes).

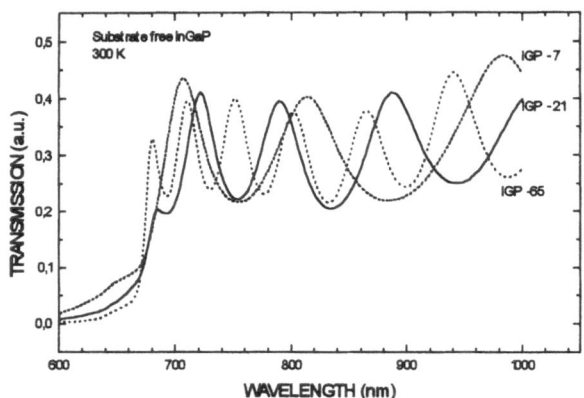

Fig 7 Transmission of thre different InGaAs /GaAs layers separated from the substrate. Signal is modulated by layer thickness.

It is very interesting to apply the ELO technique to study optical properties of various epitaxial materials. Fig 7 shows results of the transmission measurement of three different InGaP epitaxial layers lattice matched to GaAs substrate. These layers were separated from the substrate by etching the substrate off. As this semiconductor material has sufficiently different chemical properties in comparison with GaAs, no special etch-stop layer must be grown between the substrate and the epitaxial layer under study. The substrate was removed using an NH_4 OH etchant and a selectivity higher as 10 [4] was observed.

This figure shows a modulation of the transmission signal by thickness of the epitaxial layer (the layer thickness increases from 0.7 μm - sample IGP 7 to 1.6 μm at sample IGP 65). In special cases , in which an etchant selectivity reliable low, the removal of the substrate must be controlled to prevent the etchant`s influence on the results of characterisation. The GaAs substrate removal from an InGaAs layer is controlled my means of room temperature photoreflectance measurements, which is shown in Fig. 8. In this figure, three signals measured near the GaAs band gap energy are compared - full GaAs , islandlike remaining part of the substrate and only noise signal after substrate total removing.

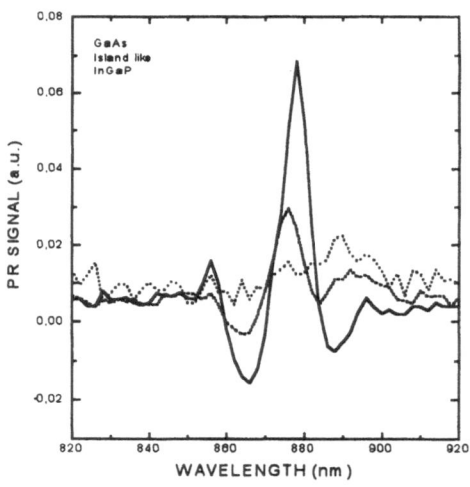

Fig.8 Control of substrate removing using room temperature photoreflectance measurement

4. Conclusion

As we have shown, the epitaxial lift-off technique is a very useful method which can be used in the cases when the presence of the substrate can influence results of the epitaxial layer characterisation or make this characterisation impossible.

The experimental part of this work was done in the close collaboration between Institute of Electrical Engineering , Bratislava and Institute of Ion Technology, Juelich. Also, P. Kordoš, A. Förster, J. Betko, M. Morvic and M. Kučera have substantially contributed to this work which is kindly acknowledged.

5. References

1. Konagai, M., Sugimoto, M. and Takahashi, T. (1978), *J. Crystal Growth*, **45**, 277- 279
2. Yablonovitsch, E, Gmitter, T, Harbison, J.P. and Bhat, R (1987) Extreme selectivity in the lift-off of epitaxial GaAs films, *Appl. Phys. Lett.*,**51**, 2222-2224
3. Pollentier, I., Buydens, L., Van Daele,and Demeester, P. (1991), *IEEE Photonics Technol. Lett.***3**, 115-118
4. Yablonovitch, E, Sands,T., Hwang, D.M., Schnitzer, I., Gmitter,T.J., Shastry,S.K. and Fan J.C. (1991), *Appl.. Phys. Lett.* **59**, 3159-3160
5. Demeester P., Pollentier,I., De Dobbelaere, P., Brys, C. and Van Daele, P. (1993) Epitaxial lift-off and its applications, *Semicond. Sci.& Technol.* **12**, 1124-1135
6. Justice,J., Corbett, S., Walsh, S, Considine, L and Kelly W.M. (1995), Dark currents in PIN photodetectors fabricated by preprocessing and postprocessing techniques of epitaxial lift-off, *Electronics Lett.*,**31**, 1382-1383
7. Herscher,M, Grundmann, M., Droge, E., Kollakowski, St., Bötcher, E.H. and Bimberg, D. (1995) Epitaxial lift-off InGaAs/InP MSM photodetectors on Si, *Electronics Lett..*, *31*,1383-1384
8. Corbett,B, Considine,L., Walsh, S. and Kelly W.M., (1993) Resonant Cavity Light Emitting Diode and Detector Using Epitaxial lift-off, *IEEE Photonics Techn. Letters*, **5**, 1041-1043
9. Look, D.C., Walters, D.C., Robinson, G.D., Sizelove, J.R., Mier, M.G. and Stutz, C.E. (1993) Annealing dynamics of molecular-beam epitaxial GaAs grown at 200 °C, *Journal of Applied Physics* **74**, 306-310.
10. Kordoš, P.,Förster, A., Betko, J., Morvic, M. and Novák, J. (1995) Semi-insulating GaAs layers grown by molecular beam epitaxy, *Applied Physics Letters* **67**, 983-985.
11. Kordoš, P., Betko, J., Förster, A., Kuklovský, S.,Dieker, Ch. and Ruders, F. (1993) Electrical and structural characterisation of MBE GaAs grown at temperatures between 200 and 600 °C, in H. Goronkin and U. Mishra (eds.), *Compound Semiconductors 1994*, (Institute of Physics Conference Series Number 141), IOP Bristol, p. 295-300.
12. Betko, J., Kordoš, P., Kuklovský, S., Förster, A., Gregušová, D. and Lüth, H. (1994) Electrical properties of molecular beam epitaxial GaAs layers grown at low temperature, *Material Science Engineering* B **28**, 147-150.
13. DeSalvo,G, Tseng, W.F. and Comas, J, (1992) Etch rates and selectivities of citric acid/Hydrogen peroxide on GaAs, InGaAs, InAlAs and InP,*J. Electrochem. Soc.* **139**, 831- 835
14. Fink, T. and Osgood, R.M.Jr. (1993) Light-induced Selective etching of GaAs in GaAlAs/GaAs Heterostructures,*J. Electrochem. Soc.* **140**, L73-L74
15. Lilienthal-Weber,Z., Lin, X.W., Washburn,J. and Schaff, W. (1995) Rapid thermal annealing of low temperature GaAs layers,*Appl. Phys. Lett..*, **66**, 2086-2087

MANY CRYSTAL X-RAY DIFFRACTOMETRY ON SUPERLATTICES

V. HOLÝ,
Department of Solid State Physics, Masaryk University, Kotlářská 2, 61137 Brno, Czech Republic.

1. Introduction

X-ray diffractometry is a frequently used method for monitoring the epitaxial technology. In a row of routine studies, from the x-ray diffraction curves the basic structural parameters of a layered sample are determined, as the layer thicknesses and the lattice constants. In this paper we show the capability of x-ray diffraction of a more detailed investigation of the sample structure.

In the simplest arrangement of an x-ray diffraction experiment (a double-axis diffractometer), a nearly parallel and nearly monochromatic x-ray beam produced by the monochromating crystal irradiates the sample and the detector measures the intensity of the diffracted beam regardless of its direction of propagation. The dependence of the diffracted intensity on the direction of the primary beam (the reflection curve) is influenced both by the geometrical parameters of the sample (thicknesses, lattice constants) and by the structure defects, the interpretation of the measured curves, however, is not straightforward.

In a triple-axis diffractometer, an analyzing crystal is placed into the diffracted beam [1,2]. Then, the device makes it possible to measure the directional distribution of the diffracted beam that can be expressed as a distribution of the diffracted intensity in *reciprocal* plane (reciprocal space map). This experimental arrangement, however, is not sensitive to the distribution of the diffracted intensity in *real* space. The reciprocal space map can be measured both around the origin of reciprocal space (reciprocal lattice point 000) and around a non-zero reciprocal lattice point. In the former case (reflection mode), the scattering process can be described by optical reflection of x-rays from the interfaces in the sample. The angles of the primary and exit beams with the sample surface are small and the method is called x-ray reflectometry. The latter case (diffraction mode) represents the x-ray diffraction in usual sense, the difference of the wave vectors of the diffracted and primary beams is close to a reciprocal lattice vector (diffraction vector).

It will be shown in this paper that the reciprocal space map of the diffracted intensity is closely connected with the Fourier transform of the correlation function of the deformation field, thus this map yields a direct information of the structure

J. Novák and A. Schlachetzki (eds.), Heterostructure Epitaxy and Devices, 183–192.
© 1996 *Kluwer Academic Publishers.*

defects in the sample.

In the first part of the paper, the basic principle of an ideal high resolution x-ray diffractometer is formulated. Then, the distribution of the intensity scattered by an ideal layered sample is calculated and the possibility is shown for determining of basic structural parameters (thicknesses and lattice constants) of the sample from the measured reciprocal space map. In the last section, the method is shown for simulating the reciprocal space maps of both in the reflection and the diffraction modes on the basis of simple defect models. Some examples are shown illustrating the capabilities of the method.

2. High resolution x-ray diffractometer

We introduce the ideal x-ray diffractometer in the following way. The sample (assumed very large in lateral dimensions) is irradiated by a plane, perfectly monochromatic wave with the wave vector $\mathbf{K}_0$. In a general case, the sample produces a divergent scattered wave with a set of the wave vectors $\mathbf{K}_s$. An angularly sensitive detector is placed into the scattered beam so that it measures the directional distribution of the scattered intensity. This detector is realized by means of an ideal crystal with a very narrow reflection curve, and a radiation detector. If we denote xz-plane the plane of incidence of the primary beam and if the analyzing crystal can rotate with respect to an axis perpendicular to that plane, the signal measured by the detector is [3]

$$J(K_{sx}, K_{sz}) = \int dK_{sy} \Phi(\mathbf{K}_0, \mathbf{K}_s) \tag{1}$$

where $\Phi(\mathbf{K}_0, \mathbf{K}_s)$ is the intensity of the component of the scattered beam having the wave vector $\mathbf{K}_s$. Thus, our ideal diffractometer resolves the distribution of the diffracted intensity in xz-plane and it has no resolution perpendicular to it.

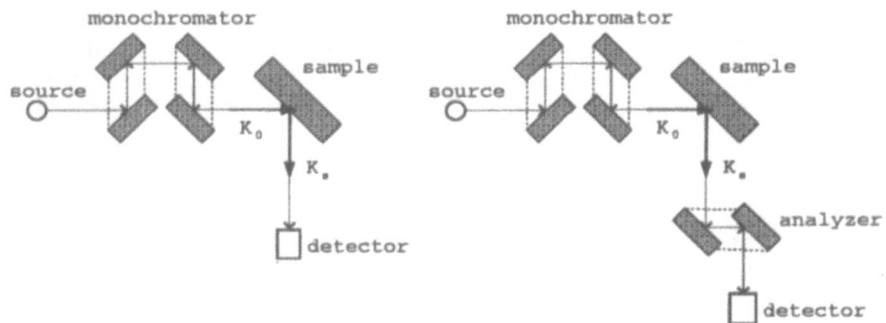

Figure 1. The setups of the double crystal diffractometer (left) and the triple axis diffractometer (right).

In most cases, the intensity Φ of the scattered beam is a function of the difference

$$\mathbf{Q} = \mathbf{K}_s - \mathbf{K}_0$$

of the wave vectors of the scattered and the primary beams, thus the diffractometer signal J can be expressed being the function of Q_x, Q_z (*reciprocal space map*).

If we do not use the analyzing crystal (double-crystal arrangement), the radiation detector collects all the scattered waves propagating with different $\mathbf{K}_s$'s. In this case, the measured signal is

$$I = \int_{ES} d\mathbf{K}_s J(K_{sx}, K_{sz}) \tag{2}$$

where the integration is performed over the Ewald sphere $\mathbf{K}_s = 2\pi/\lambda = \text{const.}$ The position of the Ewald sphere is determined by the direction of the primary beam $\mathbf{K}_0$, thus the double-crystal diffractometer signal I is a function of the angular deviation α of the primary beam from its diffraction position. This function is called *reflection curve*.

If the interaction of the x-ray beam with the sample is weak, multiple scattering processes can be neglected. This is the essence of the *kinematical approximation*. In the diffraction mode, this approximation is valid for samples much thinner than some critical thickness (extinction length). This length is proportional to the Fourier coefficient of the electron density and, usually, it is several microns for perfect layers. The extinction length of a disturbed layer depends on the disturbance and it is always greater than that in the ideal case.

In the reflection mode, the kinematical approximation might be applied for angles of incidence much greater than the critical angle of total external reflection (below 1 deg). In many cases, however, the kinematical approximation is *not applicable* at all in the reflection mode.

Within the kinematical approximation, the intensity Φ of the plane component of the scattered radiation can be expressed by the Fourier transform of the autocorrelation function of the electron density $\varrho(\mathbf{r})$ [3-5]

$$\Phi(\mathbf{Q}) = \text{const.} \int_V d\mathbf{r} \int_V d\mathbf{r}' \langle \varrho(\mathbf{r})\varrho(\mathbf{r}') \rangle e^{i\mathbf{Q}\cdot(\mathbf{r}-\mathbf{r}')} \tag{3}$$

where the statistical averaging $\langle\ \rangle$ is performed over the statistical ensemble of all defect configurations and the integral is calculated over the sample volume V.

In the reflection mode, the wave vector transfer $\mathbf{Q}$ is very small with respect to all reciprocal lattice vectors, thus, instead of $\varrho(\mathbf{r})$, we can put its value ϱ_0 averaged over the elementary cell, into Eq. (3), i.e. the 0-th coefficient of its Fourier series. Thus, in this mode, the crystallographic structure of the sample has no influence on the scattered intensity. In an ideal layered sample, ϱ_0 is a function of the coordinate z perpendicular to the sample surface. In a sample with randomly rough interfaces, ϱ_0 is a random function of $\mathbf{r}$.

In the diffraction mode with diffraction vector $\mathbf{h}$ the electron density in Eq. (3) can be replaced by

$$\varrho_h e^{-i\mathbf{h}\cdot\mathbf{r}}$$

where ϱ_h is the h-th Fourier coefficient of the electron density. Then, the diffractometer signal can be expressed as a function of the reduced wave vector transfer

$$\mathbf{q} = \mathbf{Q} - \mathbf{h}$$

that is measured with respect to the reciprocal lattice point $\mathbf{h}$. In an ideal layered sample, ϱ_h is a function of z, again. If the sample contains a deformation field $\mathbf{u}(\mathbf{r})$ (due to defects), the autocorrelation function of the electron density is then

$$\langle \varrho(\mathbf{r})\varrho(\mathbf{r}') \rangle = \langle \varrho(\mathbf{r})\varrho(\mathbf{r}') \rangle_{ideal} \cdot G(\mathbf{r},\mathbf{r}') \tag{4}$$

where

$$G(\mathbf{r},\mathbf{r}') = \langle e^{i\mathbf{h}\cdot[\mathbf{u}(\mathbf{r})-\mathbf{u}(\mathbf{r}')]} \rangle \tag{5}$$

is the correlation function of the deformation field in the sample [4].

The practical realization of a high-resolution diffractometer requires a conditioning of the primary beam in order to limit its divergence and the spectral width, and an application of an analyzing crystal with narrow reflection curve with reduced tails. The first item can be achieved by means of a crystal monochromator. The most versatile type of a monochromator is the Bartels monochromator using two channel-cut crystals with four successive diffractions. For instance, if we use four symmetrical diffraction 220 in two Ge crystals and the CuKα_1 line, the beam emitted by the last diffraction has the divergence about 12 sec of arc and its relative spectral width is about $\Delta\lambda/\lambda \approx 10^{-4}$. The second condition is realized by a channel-cut crystal with two or three successive diffractions. If we use 220 diffractions in Ge again, the angular resolution of the detector in xz-plane is 12 sec of arc again. In review papers [1,2], principles of a high resolution x-ray diffractometer are explained and examples of its application are reviewed.

3. Diffractometry of ideal layered structures

An ideal multilayer has perfectly flat interfaces and its crystal structure is ideal, i.e. it contains no structure defects. Thus, an ideal multilayer means an ideally pseudomorphic structure with no relaxation of internal stresses, since the stress relaxation is accompanied by structure defects (misfit dislocations).

In this ideal case, all the Fourier coefficients ϱ_0, ϱ_h depend only on z, and, therefore, the diffractometer signal $J(Q_x, Q_z)$ is concentrated in very thin stripes going through the reciprocal lattice points in Q_z direction - so called *truncation rods*. The intensity distribution along a truncation rod is determined by the functions $\varrho_0(z)$ (in the truncation rod going through the origin) and by $\varrho_h(z)$. For instance, if the sample contains only one homogeneous layer with thickness T, the main intensity maximum on the truncation rod has the width

$$\Delta Q_z = 2\pi/T.$$

In the case of a periodical superlattice, periodically distributed satellite maxima can be detected along the truncation rod, their distance is $2\pi/D$ again, where D is the superlattice period.

Within the kinematical approximation, the distributions of the diffracted intensity along all the truncation rods (i.e. both in the reflection and diffraction modes) are similar. If the kinematical approximation is not valid, a complete dynamical simulation of the diffracted (or reflected) intensity is necessary for determining the layer thicknesses.

As shown above, the double-crystal diffractometer signal $I(\alpha)$ is an integral of the reciprocal space map J over the Ewald sphere. Thus, if the sample is perfect, similar maxima can be obtained in the reflection curve as well. However, an attention must be paid to the conversion between the angular variable $\Delta\alpha$ and ΔQ_z.

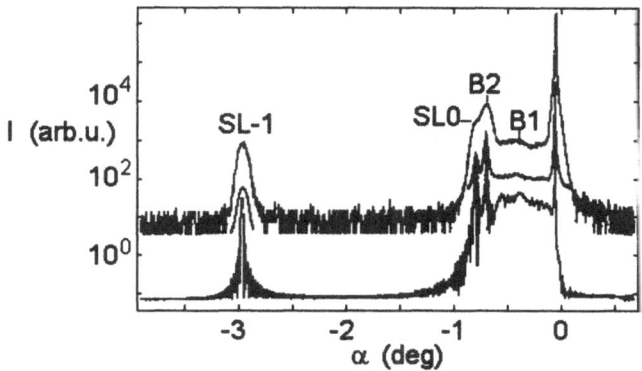

Figure 2. Reflection curve of a SiGe superlattice measured by a double crystal diffractometer (upper curve), measured distribution of the diffracted intensity along the truncation rod (middle curve), dynamical simulation (lower line) [6]. The distance between the satellite maxima SL0 and SL-1 is inversely proportional to the superlattice period. The peaks B1 and B2 correspond to the buffer layers. The difference between the upper and the middle curve is caused by the diffuse scattering.

In Fig. 2, the reflection curve of a SiGe superlattice measured by the double-crystal diffractometer and the intensity distribution along the truncation rod measured by the triple axis diffractometer are compared with the results of a dynamical simulation assuming a perfect structure [6].

From the positions of the intensity maxima of the non-zero truncation rods measured around several reciprocal lattice points, the components of the diffraction vectors can be determined, and, consequently, the lattice constants and the internal strains can be stated. In a fully pseudomorphic structure, the in-plane lattice constants of all layers are same as that of the substrate and the truncation rods belonging to the same reciprocal lattice point in different layers and in the substrate must lie along the same line perpendicular to the sample surface.

The internal strains in a fully relaxed structure are zero and the diffraction vectors in the layer and in the substrate belonging to the same reciprocal lattice point must be parallel. Then, the intensity maxima from the layer and from the substrate belonging to the same reciprocal lattice point must lie at the same line connecting the reciprocal lattice points of the substrate and the layer by the origin of reciprocal space.

In many cases, the structure can be classified as partially relaxed that means, the in-plane lattice constants of the layer and the substrate are not same and the layer is not strain-free. Then, from the position of the intensity maximum in the reciprocal space map both the in-plane and the vertical lattice constants of the layer can be determined, and, consequently, the degree of relaxation of the internal stresses in the layer can be stated. If the position of the intensity maximum belonging to the layer is given by the coordinates (q_x, q_z) with respect to the substrate reciprocal lattice point with the diffraction vector $\mathbf{h} = (h_x, h_z)$, the in-plane lattice constant $a_{\parallel}$ and the vertical lattice constant $a_{\perp}$ of the layer are (see [7], for instance)

$$\frac{a_{\parallel} - a_{\parallel}^{S}}{a_{\parallel}^{S}} = -\frac{q_x}{q_x + h_x} \tag{6}$$

$$\frac{a_{\perp} - a_{\perp}^{S}}{a_{\perp}^{S}} = -\frac{q_z}{q_z + h_z} \tag{7}$$

where $a_{\parallel,\perp}^{S}$ are the in-plane and vertical lattice parameters of the substrate, respectively.

The advantage of this method for determining the lattice parameter is that it enables us to distinguish the intensity maxima of individual layers, and, therefore, the relaxation degree of each layer in a stack can be determined from a single reciprocal space map. In the double-crystal diffractometry arrangement, this determination is not possible, since the intensity maxima on the reflection curve belonging to different layers overlap and an exact determination of their positions is not possible.

4. Investigation of structure defects

Structure defects affect the distribution of the scattered intensity and they can be studied by means of the reciprocal space map measurements. The structure defects diminish the intensity of the truncation rod and they give rise to waves scattered into another directions as those corresponding to the truncation rod. These waves are called *diffusely scattered waves*.

In the kinematical approximation of the **diffraction mode**, from the measured reciprocal space map the correlation function G of the deformation field can be obtained by means of the inverse Fourier transform following from the theory in Section 2. The shape of the autocorrelation function can be compared with theoretical predictions perform on the basis of simple defect models. The defect structure in a relaxed epitaxial layer can be modelled using the *mosaic block model*. Within this model, the layer structure is assumed consisting of randomly rotated mosaic blocks. The model contains two parameters – the mean block size R and the root mean square misorientation of the blocks Δ. Using the correlation function of the mosaic structure [4,11] and expressions (3,4) we can calculate the reciprocal space map of a mosaic layer. The diffuse intensity maximum has always an elliptical shape, being elongated perpendicular to the diffraction vector $\mathbf{h}$. This elongation is given by $\Delta|\mathbf{h}|$, thus it is proportional to the distance of the reciprocal lattice point from the origin of reciprocal space. The width of the intensity maximum in the direction parallel to $\mathbf{h}$ does not depend on $|\mathbf{h}|$ and it is inversely proportional to R.

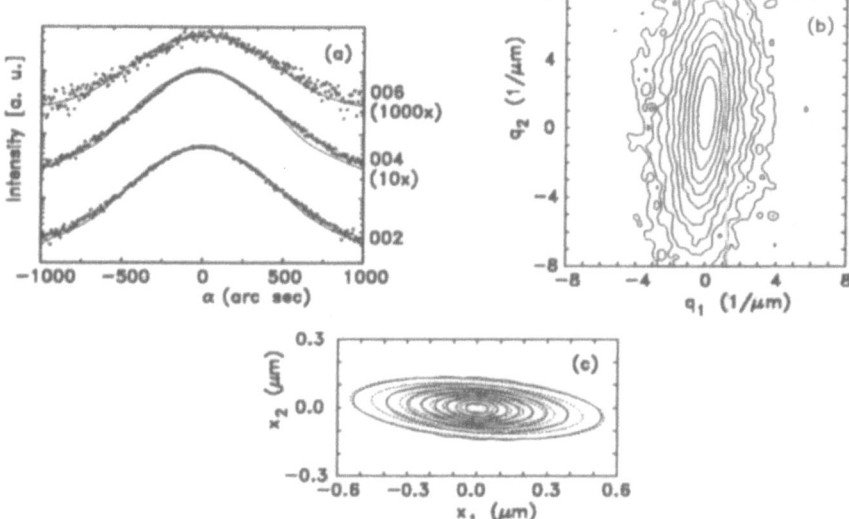

Figure 3. Part a: double crystal reflection curves measured from a ZnTe/GaAs epitaxial layer in three different diffractions (points) and their theoretical simulations (lines) after [3], part b: measured reciprocal space map, part c: correlation function of the deformation field obtained from the measured reciprocal space map (full) and its simulation on the basis of the mosaic block model (dashed) after [4]. The axes $x_1(q_1)$ and $x_2(q_2)$ are parallel and perpendicular to the diffraction vector, respectively.

In many cases, the diffuse intensity maxima in the reciprocal space map are elongated not exactly in direction perpendicular to $\mathbf{h}$, but they are slightly rotated [4]. This rotation might be caused by random strains in the mosaic blocks. The mosaic block model has been used for characterizing the structure of a ZnTe epitaxial layer grown by MOCVD on a GaAs substrate. Fig. 3 shows the double crystal reflection curves (from [3]),. the measured reciprocal space map and the correlation function of the deformation field is compared with that calculated from the mosaic block model [4].

The connection of the parameters of the mosaic blocks with the actual defect structure is only indirect. It has been demonstrated that the diameter of the block is comparable with the mean distance between the threading segments of the misfit dislocations, the misorientation angle Δ depends on the Burgers vector of these segments.

In another model [8], the correlation function G is calculated using the assumption of a random distribution of misfit dislocations in a relaxed epitaxial layer by means of the deformation field of a single dislocation. Comparing the simulated

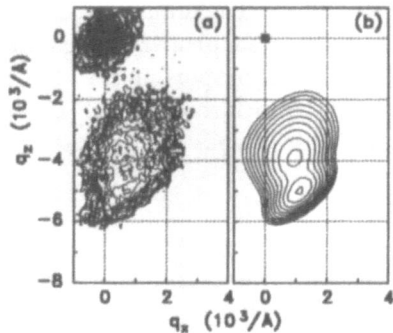

Figure 4. Measured (a) and calculated (b) reciprocal space map of a SiGe layer with graded Ge content (after [8]).

and measured reciprocal space maps, the density of the misfit dislocations can be estimated. It has been demonstrated that this density compares well with the value following from the lateral lattice misfit of the layer and simple geometrical considerations. This method is demonstrated in Fig. 4, where the reciprocal space map measured from a SiGe layer with linearly graded Ge content is compared with the calculated one.

The reciprocal space map measured in the diffraction mode might also be sensitive to the interface roughness in the multilayer. Calculating this effect we express the autocorrelation function of the electron density by means of the correlation function of the interface roughness [9,10]. Its form can be calculated, among others, using the formalism of fractals [9]. Each interface (say, the j-th) in the multilayer is characterized by the following parameters: σ_j is the root mean square (r.m.s.) roughness, Λ_j is the in-plane correlation length (i.e. some mean width of the "hills" and the "valleys") and D_j is the fractal dimension of the interface (if $D_j = 2$ the interface has a non-fractal nature, for $2 < D_j < 3$ the interface has a fractal character). In addition, further parameters characterize the correlation of the roughness profiles of different interfaces ("vertical" correlation) [10].

From the calculations performed within the kinematical theory it follows that the shape of the diffuse intensity maxima in reciprocal plane substantially depends on the vertical roughness correlation [11]. If the roughness profiles of different interfaces were not correlated, no distinct maxima (in addition to the truncation rods of the perfect structure) could be observed. In the case of maximum vertical correlation, in addition to the narrow vertical truncation rods, horizontally elongated diffuse intensity maxima can be observed. These horizontal sheets go through the reciprocal lattice points, and, in the contradiction to the mosaic structure, they are always parallel to the sample surface. The distribution of the intensity along the truncation rod in the diffraction mode is practically not affected by the interface roughness.

The diffuse intensity maxima are distinct especially in the case of periodical

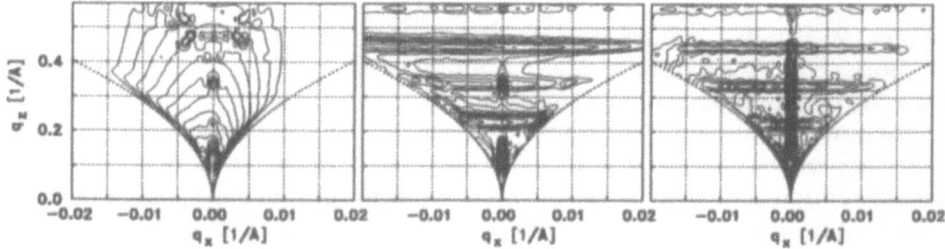

Figure 5. Reciprocal space map measured in the reflection mode from a Fe/Sc calculated without (a) and with (b) vertical roughness correlation, and the measured reciprocal space map (c) (after [13]).

multilayers, where they intersect the satellite maxima on the truncation rod (for detalis see [11]).

Comparing the intensity of the diffuse scattering from rough interfaces with that from mosaic structure or other volume defect we find that the former process is very weak and, in most cases, it cannot be distinguished from the influence of the volume defects. Only in samples with no volume defects (perfectly pseudomorphic multilayers) the diffuse scattering from the interface roughness can be observed in the diffraction mode using synchrotron radiation.

We have shown in Section 2 that the reciprocal space map in the **reflection mode** is not sensitive to the crystal structure of the sample. Thus, this map is only affected by the interface roughness and not by the volume structure defects. Calculating the distribution of the diffusely scattered intensity in reciprocal plane we have to take into account the dynamical phenomena, above all, refraction and total external reflection of x-rays.

The interface roughness diminishes the intensity distribution along the truncation rod, i.e. the specular reflectivity of the sample. Comparing the measured specular reflectivity with that calculated dynamically, the r.m.s. roughnesses σ_j of individual interfaces can be estimated, the specular reflectivity is nearly insensitive to other roughness parameters Λ_j and D_j. These parameters substantially affect the distribution of the diffusely scattered intensity in reciprocal plane [12]. Similarly to the diffraction mode, if the roughness is vertically correlated, the diffusely scattered intensity is concentrated in a set of sheets going through the satellite maxima on the origin truncation rod. Due to the refraction of x-rays, these sheets are slightly bent. In Fig. 5, the reciprocal space map measured from a Fe/Sc multilayer is compared with the map calculated with and without vertical roughness correlation [13].

5. Conclusions

We have demonstrated that reciprocal space maps of layered samples measured by a triple-axis diffractometer can bring not only basic structural parameters of the samples but also they can give an information concerning the defects in the samples. Measuring the intensity distribution along the truncation rods, the layer thicknesses can be determined and, in addition, in the reflection mode, the roughnesses of the interfaces can be estimated.

Analyzing the intensity distribution outside the truncation rods (diffuse x-ray scattering) we can characterize the structure defects. In the diffraction mode, the diffusely scattered intensity is mainly connected with the random deformation field of the volume defects (dislocations, precipitates etc.), while in the reflection mode the diffuse scattering originates from the interface roughness.

In some cases, the interpretation of the measured reciprocal space map might be ambiguous. Then, a choice of a suitable defect model is necessary and a comparison of the obtained defect parameters with those stated by other independent methods (TEM, for instance) is inevitable.

The work has been supported by the Grant Agency of Czech Republic (Project 202/94/1871). Stimulating discussions with J. Kuběna are greatly acknowledged.

1. Fewster, P.F. (1993) X-ray diffraction from low-dimensional structures, *Semiconductor Sci. Technol.* **8**, 1915–1934 and the references therein.
2. Tanner, B.K. and Bowen, D.K. (1993) Advanced x-ray scattering techniques for the characterization of semiconducting materials, *J. Cryst. Growth* **126**, 1-18 and the references therein.
3. Holý, V., Kuběna, J., Abramof, E., Lischka, K., Pesek, A., and Koppensteiner, E. (1993) X-ray double and triple crystal diffractometry of mosaic structure in heteroepitaxial layers, *J. Appl. Phys.* **74**, 1736-1743.
4. Holý, V., Wolf, K., Kastner, M., Stanzl, H., and Gebhardt, W. (1994) X-ray triple crystal diffractometry of defects in epitaxic layers, *J. Appl. Cryst.* **27**, 551-557.
5. Cowley, J.M. (1975) *Diffraction Physics*, North-Holland, Amsterdam.
6. Koppensteiner, E., Bauer, G., Holý, V., and Kasper, E. (1994) High resolution x-ray triple axis diffractometry of short period SiGe superlattices, *Jpn. J. Appl. Phys.* **33**, 2322-2328.
7. Heinke, H., Möller, M.O., Hommel, D., and Landwehr, G. (1994) Relaxation and mosaicity profiles in epitaxial layers studied by high resolution x-ray diffraction, *J. Cryst. Growth* **135**, 41-52.
8. Holý, V., Li, J.H., Bauer, G., Schäffler, F., and Herzog H.J. (1995) Diffuse x-ray scattering from misfit dislocations in SiGe epitaxial layers with graded Ge content, *J. Appl. Phys.*, the October number.
9. Sinha, S.K., Sirota, E.B., Garoff, S., and Stanley, H.B. (1988) X-ray and neutron scattering from rough surfaces, *Phys. Rev. B* **38**, 2297-2311.
10. Spiller, E., Stearns, D., and Krumrey, M. (1993) Multilayer x-ray mirrors: Interfacial roughness, scattering and image quality, *J. Appl. Phys.* **74**, 107-118.
11. Holý, V. (1994) Diffuse X-ray scattering from non-ideal periodical crystalline multilayers, *Appl. Phys. A* **58**, 173-180.
12. Holý, V., and Baumbach, T. (1994) Nonspecular x-ray reflection from rough multilayers, *Phys. Rev. B* **49**, 10668-10676.
13. Holý, V., Kuběna, J., van den Hoogenhof, W.W, and Vávra, I. (1995) Effect of interfacial roughness replication on the diffuse x-ray reflection from periodical multilayers, *Appl. Phys. A* **60**, 93-96.

QUANTUM MAGNETOTRANSPORT IN TWO-DIMENSIONAL ELECTRON GAS IN InGaAs/InP HETEROSTRUCTURES

B. PŐDÖR, G. GOMBOS
Research Institute for Technical Physics of the Hungarian Academy of Sciences
1325 Budapest, Hungary
GY. KOVÁCS
Department of Low Temperature Physics, Eötvös Loránd University
1088 Budapest, Hungary
I. G. SAVEL'EV, V. S. NOVIKOV
A. F. Ioffe Physico-Technical Institute of the Russian Academy of Sciences
194021 St. Petersburg, Russia
G. REMENYI
CNRS, CRTBT and Laboratoire des Champs Magnétiques Intenses
38042 Grenoble, France

Quantum magnetotransport measurements were performed on $In_{0.53}Ga_{0.47}As/InP$ heterostructures in magnetic fields up to 23 Tesla at temperatures from 4.2 K to 60 mK. Shubnikov-de Haas measurements in tilted field led to the resolution of the spin splitting of Landau levels up to N = 3. The values of effective Landé factor g^* were deduced separately for the spin-split Landau levels. A considerable, magnetic field dependent enhancement of g^* was found. In the mK range of temperatures non-linear current-voltage characteristics were seen for filling factors $0.3 \leq v \leq 0.4$ in the two-dimensional electron gas in heterostructures with strong disorder. These observations are interpreted in terms of magnetic field induced Wigner solidification.

1. Introduction

The transport phenomena of two-dimensional electron gas (2DEG) in high magnetic fields and at low temperatures are very rich in various quantum effects, like integer and fractional quantum Hall effect, Shubnikov-de Haas oscillations, weak-localization and electron-electron interaction effects, and the occurrence of magnetic field induced Wigner solidification of the electron system to mention but a few (e.g. see [1,2]). While in the past the bulk of experimental work in these fields was related to the GaAlAs/GaAs heterostructure, other material systems, like InGaAs/InP might present interesting opportunities too for the investigation of these phenomena. This and related heterostructures have recently received much attention for the use in optoelectronic and high speed electronic devices, so investigations of basic physical phenomena in this material system have an important bearing on practical applications too.

Here we report on new experimental results concerning the spin-splitting of the Landau levels and the anomalous enhancement of the effective Landé factor of the

J. Novák and A. Schlachetzki (eds.), Heterostructure Epitaxy and Devices, 193–196.

electrons in the $In_{0.53}Ga_{0.47}As/InP$ heterostructure as well as on possible Wigner solidification of the 2DEG system in this heterostructure. Detailed account of the experiments and their analysis has been and is being published elsewhere [3 to 5].

2. Experimental Details

The samples used were liquid phase epitaxial $In_{0.53}Ga_{0.47}As/InP$ heterostructures [6]. The layer sequence consisted of a 1 μm p-type InP buffer ($p=10^{15}$ cm^{-3}), a 0.2-0.5 μm n-type InP layer ($n=(2-10)x10^{16}$ cm^{-3}), and a 0.5-1.5 μm p-type $In_{0.53}Ga_{0.47}As$ layer ($p=10^{15}$ cm^{-3}). The electron densities and mobilities were in the range of $(1.2-4)x10^{11}$ cm^{-2} and $(1-3)x10^4$ cm^2/Vs respectively. Photolithographically defined Hall bar devices were used, with currents low enough to avoid any non-ohmic effects.

Magnetotransport measurements were carried out at 4.2 K temperature in a superconducting magnet up to 6 T, and down to 50 mK in a resistive magnet providing fields up to 23 T. Measurements in the resistive magnet were performed in tilted fields too. Both conventional dc and 8 Hz lock-in measuring techniques were used.

3. Results and Discussion

3.1. SPIN-SPLITTING AND LANDÉ FACTOR AT 4.2 K TEMPERATURE

Fig. 1 shows the resistance and its second derivative versus the magnetic field for a sample with a low 2DEG density. The second derivative curve eliminates the slowly varying magnetoresistance, and greatly enhances the resolution of the oscillatory component, which can be seen up to Landau index N = 5. It also clearly shows the large spin-splitting of the N = 1 level.

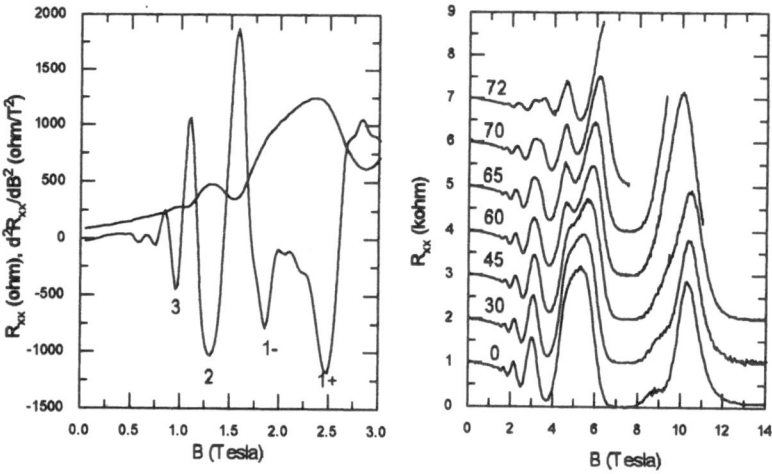

Figure 1. Longitudinal resistance and its second derivative with respect of the magnetic field versus the magnetic field for a sample with $n_s = 1.6x10^{11}$ cm^{-2}.

Figure 2. Longitudinal resistance versus the perpendicular magnetic field component for tilt angles 0°, 30°, 45°, 60°, 65°, 70°, and 72°.

Tilting the magnetic field with respect to the normal to the 2DEG plane enhances the resolution of the spin-splitting. The resistance and its second derivative versus the perpendicular component of the magnetic field for a sample with about two times larger 2DEG density ($n_s = 3.7 \times 10^{11}$ cm^{-2}) for various tilt angles are presented in Figs. 2 and 3. Here too, for zero tilt angle the second derivative curve exhibits a well-resolved double peak for the N = 1 Landau level, corresponding to the spin-splitting of this level. For the N = 2 peak a depression starts to develop at $\Theta = 60^0$ and the spin-split peaks are separated beginning from $\Theta = 65^0$ tilt angle. The N = 3 peak begins to show a doublet structure for $\Theta = 72^0$, and is clearly separated into the spin doublet for $\Theta = 76^0$ tilt angle. As far as we know this is the first direct observation of resolved spin-splitting for the relatively high index N = 3 level in In$_{0.53}$Ga$_{0.47}$As/InP.

The analysis of the above experimental results led to the conclusion that the g* factor depends on the Landau index (i.e. on the perpendicular magnetic field) and also on the parallel field. Its values are 13-24, 8-11, 5-8, 4-10, and about 6 for the levels 0$^-$, 1$^+$, 1$^-$, 2$^+$, and 2$^-$ respectively. These values are in a qualitative agreement with earlier results [7,8] and show the complicated behaviour due to exchange interaction.

3.2. POSSIBLE WIGNER SOLIDIFICATION

In the mK range (the lowest temperature was 50 mK) below ν=0.4 the resistance began to increase dramatically, approximately exponentially in the magnetic field. Increasing the magnetic field further led to a noisy resistance roll-off accompanied with the onset of a large out-of-phase component in the longitudinal resistance.

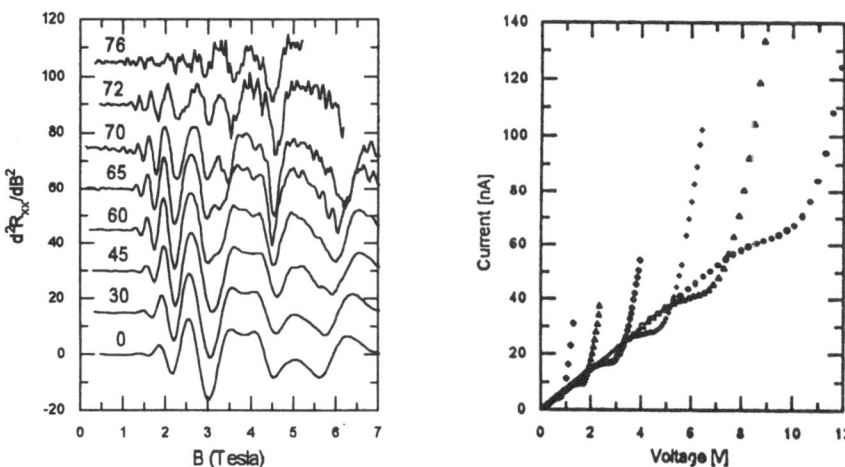

Figure 3. Second derivative of the longitudinal resistance versus the perpendicular magnetic field component for various tilt angles.

Figure 4. Two-terminal I-V characteristics at 70 mK in magetic fields increasing from 12 T to 17 T in.steps of 1 T from left to right. The corresponding filling factors are ν = 0.414, 0.382, 0.354, 0.331, 0.310, and 0.292 respectively.

Fig. 4 shows two-terminal I-V characteristics at 70 mK in various magnetic fields in a sample with n_s=1.2×10^{11} cm^{-2} and μ=3×10^4 cm^2/Vs. In the range of filling factors $0.3 \leq v \leq 0.4$ a characteristic non-linearity occurs, with a magnetic field dependent `plateau` and `threshold`. For higher filling factors these features are absent. The investigation of the temperature dependence of the I-V characteristics up to 150 mK revealed a saturation for higher temperatures ($T \geq 90$ mK) at lower `threshold` voltage, which is just the opposite to saturation expected for electron heating. This saturation behaviour is quite analogous to the one reported for GaAlAs/GaAs in [9,10], where it has been considered as a signature of Wigner solidification.

The values of the threshold electric field (a few tens of V/cm) are much larger than the value expected for an ordinary Wigner crystal. This is possibly explained by assuming the pinning of the crystalline electron solid by the large disorder. The strength of the disorder was estimated from the Landau level broadening to be about $e<V>=8$-10 meV, comparable in magnitude to the Coulomb energy U, $e<V>/U \cong 1$. This disorder is much stronger that usually encountered in GaAlAs/GaAs, where the value of this ratio can be smaller than 1/100. The incompressible fluid state of the 2DEG is fully destroyed and a glass-like condensate pinned by the disorder is formed [11].

4. Summary and Acknowledgements

To sum up, we have studied spin-splitting of Landau levels in $In_{0.53}Ga_{0.47}As/InP$ heterostructures and deduced effective g^* values for the low index spin-split levels.

Further on we have observed non-linear current voltage characteristics and determined the respective threshold fields in two-terminal longitudinal resistance of 2DEG in $In_{0.53}Ga_{0.47}As/InP$ heterostructures in the range of filling factors below 0.4. The formation of a disorder-pinned Wigner solid was suggested as an explanation.

The high field measurements were performed in the CNRS High Magnetic Field Laboratory under project No. SE3394. The work was also supported from grants No. 14094 by the Hungarian National Research Fund (OTKA). B. P. was also supported by the Copernicus Project CP93:8252, and Gy. K. by the Hungarian Soros Foundation.

5. References

1. Ando, T., Fowler, A.B., and Stern, F. (1982) *Rev. Mod. Phys.* **54**, 437-672.
2 Chakraborty, T. and Pietiläinen, P. (1995) *The Quantum Hall Effects: Fractional and Integral*, 2nd Edition, Springer-Verlag, Berlin, Heidelberg.
3. Pődör, B., Novikov, S.V., Savel`ev, I.G. and Gombos, G. (1994) *Acta Phys. Hungarica* **74**, 147-153.
4. Kovacs, Gy., Novikov, S.V., Gombos, G., Podor, B., and Remenyi, Gy. (1995) *Acta Phys. Polonica A* **87**, 473-476.
5. Kovacs, Gy., Remenyi, G., Gombos, G., Savel`ev, I.G., Kreshchuk, A.M., Hegman, N., and Pődör B. (in press) *Acta Phys. Polonica.*
6. Golubev, L.V., Kreshchuk, A.M., Novikov, S.V., Polyanskaya, T.A., Savel`ev, I.G., and Saidashev, I.I. (1988) *Fiz. Tekhn. Poluprov.* **22**, 1948-1954.
7. Nicholas, R.J., Brummel, M.A., Portal, J.C., Razeghi, M., and Poisson, M.A. (1982) *Solid State Comm.* **43**, 825-828.
8. Nicholas, R.J., Brummel, M.A., Portal, J.C., Cheng, K.Y, Cho, A.Y., and Pearsall, T.P., (1983) *Solid State Comm.* **45**, 911-914.
9. Andrei, E.Y., Deville, G., Galttil, D.C., Williams, F.I.B., Paris, E., and Etienne, B. (1988) *Phys. Rev. Lett.* **60**, 2765-2768.
10. Goldman, W.J., Santos, M., Shayegan, M., and Cunningham J.E. (1990) *Phys. Rev. Lett.* **65**, 2189-2192.
11. Aoki, A. (1979) *J. Phys. C* **12**, 633- 645.

TRANSPORT PROPERTIES OF MBE GaAs LAYERS GROWN AT 420 °C

M. MORVIC, J. BETKO and J. NOVÁK
Institute of Electrical Engineering, Slovak Academy of Sciences,
SK-842 39 Bratislava, Slovak Republic

A. FÖRSTER and P. KORDOŠ
Institute of Thin Films and Ion Technology, Research Centre Jülich,
D- 52425 Jülich; Germany

1. Introduction

Semi-insulating (SI) GaAs layers are often needed for the growth of device structures and various procedures of their preparation are under development [1]. In this connection, the so-called low temperature GaAs prepared by molecular beam epitaxy (LT MBE GaAs) is intensively studied because of its interesting properties. Also, LT MBE GaAs appers to be one alternative of epitaxial growth of high resistivity GaAs material. Typically the LT layers are grown at T_g= 200-250 °C and their high resistivity can be obtained after annealing at about 600 °C. MBE GaAs layers grown at intermediate temperatures (400-450 °C) have resistivities of 10^6-10^7 Ω cm even in the as-grown state [2-4], as it has recently been reported. However, the transport properties of such GaAs layers are not known in detail up to now. In this contribution we present the transport properties - resitivity, Hall mobility and Hall concentration of both as-grown and annealed GaAs layers prepared by MBE at 420 °C, and compare them with the properties of a typical LT GaAs layer and bulk SI GaAs.

It is difficult to exactly determine the layer resistivities of $\approx 10^7$ Ω cm in the layer with the substrate, because the layer sheet resistivity can be comparable or even higher than that of the substrate. Moreover, it is impossible to evaluate Hall effect properties of unseparated layers [5] and only boundaries, i.e. a minimal value for the Hall concentration and a maximal value for the Hall mobility can be obtained from measured data [6]. Therefore, the temperature dependent resistivity and Hall effect measurements were carried out on layers separated from the substrate.

2. Experimental

The GaAs layers were grown in a Varian MOD GEN II MBE system on indium-free-mounted 2-in. diam. (100) SI GaAs wafers of resistivity about 1.5×10^8 Ω cm. The growth temperature 420 °C was adjusted by the calibrated thermocouple. An As/Ga

J. Novák and A. Schlachetzki (eds.), Heterostructure Epitaxy and Devices, 197–200.
© 1996 *Kluwer Academic Publishers.*

198

beam equivalent pressure ratio of 19 and a growth rate of 1 μm/h were used to grow 6 μm thick layers and no post growth annealing was performed in the MBE chamber. An AlAs etch-stop interlayer was grown between the substrate and the layer [5].

On a part of wafers thermal processing under local As overpressure was used for annealing at 590 °C for 10 min. The layers were separated by lapping, polishing and etching down the substrate. Nonalloyed contacts [6] were prepared by rubbing-in an In+Ga alloy at room temperature to avoid the influence of uncontrolled heating before measurements were performed. The Ohmic behaviour of the contacts was found from the I-V characteristics in the range of currents used. Precise D.C. resistivity and Hall effect measurements in the temperature range 295-415 K were carried out on separated layers and SI GaAs substrate using a high impedance apparatus and the van der Pauw method.

The samples, whose size was approximately 5x5 mm were inserted in a special holder with high electric insulation and good shielding from ambient interference. Currents used at measurements were in the range 10^{-7}-10^{-10} A, the Hall effect was measured in the magnetic field typically 0.5 T. Relative experimental errors of the parameters determined were about 1%. The apparatus allowed to measure the samples with a resistance up to 10^{13} Ω.

3. Results and discussion

Temperature dependences of the resistivity of separated GaAs layers are shown in Fig. 1. The layers prepared at 420 °C exhibit semi-insulating properties with a room temperature resistivity of $7x10^6$ Ω cm and $1.5x10^7$ Ω cm for as-grown and annealed samples, respectively. Results on a separated annealed LT GaAs layer grown at 250 °C (2 μm thick) and an SI bulk GaAs sample are shown for comparison. For the layers grown at 420 °C, the thermal activation energy is found at 0.68 eV whereas the thermal activation energy in the annealed layer is a little higher, 0.73 eV and the resistivity of annealed layers is higher than that of the as-grown material in the whole temperature range. This change in thermal activation energy can be also seen in the temperature dependence of the electron Hall concentration, Fig.2.

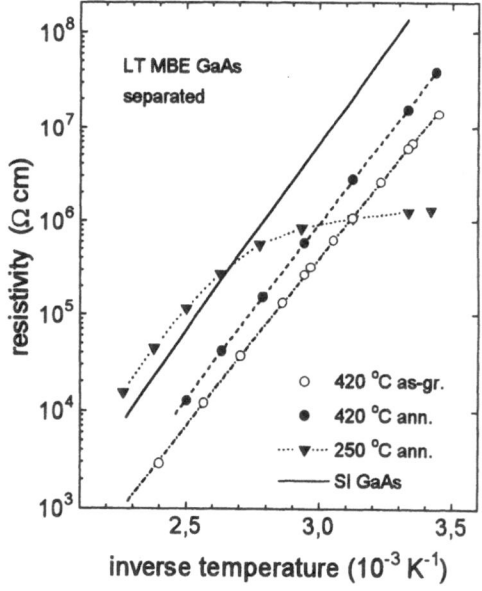

Figure 1. Resistivity vs inverse measurement temperature for GaAs MBE layers grown at 420 °C (as-grown - open dots, annealed - full dots), annealed layer grown at 250 °C (full triangels) and SI GaAs (solid line).

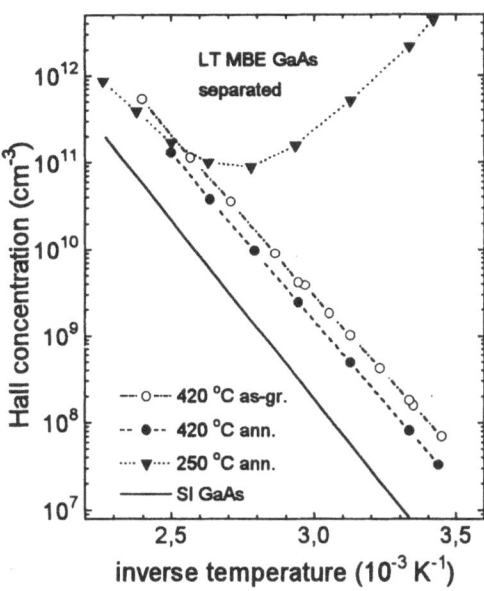

Figure 2. Hall concentration vs inverse measurement temperature for the same samples as in Fig. 1.

There is a difference of about four orders of magnitude between the room temperature Hall electron concentrations of the annealed layer grown at 250 °C and of both the 420 °C layers. As the hopping conductivity dominates in GaAs layers grown at low temperatures, the Hall concentration increases towards lower measurement temperatures. On the other hand, the layers grown at 420 °C have similar Hall concentration temperature dependence as the bulk semi-insulating GaAs material.

Electron Hall mobilities as a function of temperature in the MBE GaAs layers grown at 420 °C, a bulk SI GaAs and the annealed layer grown at 250 °C are shown in Fig. 3. Room temperature mobilities of ≈5900 cm^2 V^{-1} s^{-1} and ≈4800 cm^2/Vs in the as-grown and the

annealed layer, respectively have been evaluated for 420 °C MBE layers. It is comparable with the values measured on the bulk SI GaAs sample. These mobility values are markedly higher than those in as-grown LT GaAs [7], even for the annealed 250 °C layer, the difference at room temperature is ≈3 orders of magnitude, which can be seen in the Fig.3. The temperature dependencies of the Hall mobility of the 420 °C layers could be fitted as ~T$^{-1.1}$ and ~T$^{-0.83}$ for the as-grown and the annealed layer, respectively. Compensated bulk SI GaAs behaves in a similar way [8]. For typical LT layers, this dependence is influenced by a change from the hopping to band conductivity in the temperature range used in measurements.

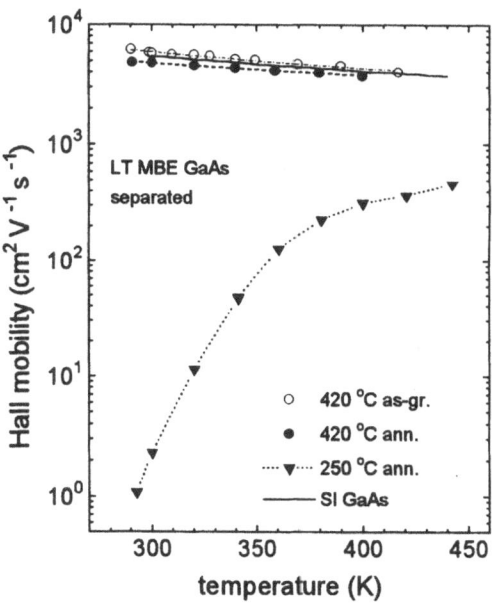

Figure 3. The temperature dependence of the Hall mobility of the same sample as in Fig. 1.

4. Conclusion

We have measured and analysed the temperature dependences of resistivity, Hall concentration and Hall mobility in as-grown and annealed MBE GaAs layers prepared at 420 °C and separated from their substrates. It has been found a distinct difference between the as-grown and the annealed layers - the thermal activation energies were 0.68 eV and 0.73 eV, the room temperature Hall mobilities were ≈ 5900 cm^2 V^{-1} s^{-1} and ≈ 4800 cm^2 V^{-1} s^{-1} with their temperature dependences fitted as $\sim T^{-1.1}$ and $\sim T^{-0.83}$, respectively. From the lower value and the weaker temperature dependence of the Hall mobility of the annealed layer it can be concluded that this layer is more compensated than the as-grown one. This is also supported by lowering the Hall concentration by annealing. Thus, a rise of a deep acceptor and/or a decrease of a deep donor concentration caused by annealing could be responsible for the observed changes due to higher compensation ratio. Moreover, during the annealing process a creation of additional deep donors with the higher thermal activation energy could be supposed.

In comparison with a typical annealed LT GaAs layer grown at 250 °C which exhibits a relative high resistivity at room temperature by hopping type conductivity, the 420 °C samples exhibit the band conductivity behaviour like bulk SI GaAs.

Acknowledgements. The authors thank J. Darmo for sample annealing. This work was partly supported by Slovak Grant Agency for Science (grant No. 2-1090/94).

1. Imaizumi, T., Okazaki, H., Yamamoto, T. and Oda, O. (1994) Undoped semi-insulating GaAs epitaxial layers and their characterisation, *Journal of Applied Physics* **76**, 7957-7965.
2. Kordoš, P.,Förster, A., Betko, J., Morvic, M. and Novák, J. (1995) Semi-insulating GaAs layers grown by molecular beam epitaxy, *Applied Physics Letters* **67**, 983-985.
3. Look, D.C., Fang, Z.-O., Sizelove, J.R. and Stutz, C.E. (1993) New As$_{Ga}$ related center in GaAs, *Physical Review Letters* **70**, 465-468.
4. Kordoš, P., Betko, J., Förster, A., Kuklovský, S.,Dieker, Ch. and Ruders, F. (1993) Electrical and structural characterisation of MBE GaAs grown at temperatures between 200 and 600 °C, in H. Goronkin and U. Mishra (eds.), *Compound Semiconductors 1994*, (Institute of Physics Conference Series Number 141), IOP Bristol, p. 295-300.
5. Look, D.C., Walters, D.C., Robinson, G.D., Sizelove, J.R., Mier, M.G. and Stutz, C.E. (1993) Annealing dynamics of molecular-beam epitaxial GaAs grown at 200 °C, *Journal of Applied Physics* **74**, 306-310.
6. Yamamoto, H., Fang, Z.-Q. and Look, D.C. (1990) Nonalloyed ohmic contacts on low-temperature molecular beam epitaxial GaAs: Influence of deep donor band, *Applied Physics Letters* **57**, 1537-1539.
7. Betko, J., Kordoš, P., Kuklovský, S., Förster, A., Gregušová, D. and Lüth, H. (1994) Electrical properties of molecular beam epitaxial GaAs layers grown at low temperature, *Material Science Engineering* B **28**, 147-150.
8. Betko, J. and Měřínský, K. (1979) Determination of the electrical properties of semi-insulating GaAs: A role of the magnetic field dependences of single carrier parameters, *Journal of Apllied Physics* **50**, 4212-4216.

DONOR NEUTRALISATION BY HYDROGEN IN S AND Se DOPED GaAs AND GaAlAs

K. SOMOGYI, B. THEYS*, L. CSONTOS, SZ. VARGA, and
J. CHEVALLIER*
*Research Institute for Technical Physics of the Hungarian Academy of
Sciences, H-1325 Budapest, P.O.B. 76., Hungary.
*Laboratoire de Physique des Solides of CNRS, Meudon-Bellevue,
1. Pl. A. Briand, 92125 Meudon Cedex, France.*

1. Introduction

Elements of the group VI and IV of the periodic table are two types of donors in III-V semiconductors, except of carbon. Elements of VI group occupy lattice sites of atoms of group V. The elements of group IV have by principle amphoteric character, nevertheless in vapour phase epitaxial layers they occupy preferentially group III atom sites, i.e. they also are convenient n-type dopants.

It is well known that the electrical activity of donors from group IV can be neutralised by atomic hydrogen [1]. The mechanism of the neutralisation is also well established: a donor atom is bonded by H atom (symmetrically to the As to donor bond) with the participation of the extra electron of the donor [2]. Much less is known about the neutralisation of donors of the VI group. Though some H neutralisation effects of group VI have been reported [3, 4], further study is necessary to know more about the interactions between hydrogen and group VI donors for a more complete understanding of the behaviour of H in GaAs and related compounds. Even less work is done in the case of GaAlAs with low Al content from the basic ternary alloys related to the GaAs [5].

In this work effect of the neutralisation in S doped GaAs and $Ga_{1-x}Al_xAs$, and Se doped GaAs epitaxial layers has been studied.

2. Experimental conditions

GaAs:S epitaxial layers have been grown by conventional chloride VPE at usual growth conditions [6]. GaAs:S, GaAs:Se and $Ga_{0.92}Al_{0.08}As$:S layers have been grown by liquid phase epitaxy. Layer thicknesses varied between 4 and 10 μm. All the samples were grown on semi-insulating GaAs substrates.

Galvanomagnetic measurements were carried out on the as grown samples. Then samples were exposed to radio frequency H or D plasma with different conditions. The plasma is excited between two plate electrodes. In one case the samples were placed

J. Novák and A. Schlachetzki (eds.), Heterostructure Epitaxy and Devices, 201-204.
© *1996 Kluwer Academic Publishers.*

202

directly between these electrodes, i.e. one of the electrodes served as the sample holder. This arrangement is called further as "direct" plasma. In the other arrangement the sample holder was placed outside of the plasma generation volume downstream at a distance of ~10 cm. This is called "indirect" plasma. In the latter case the surface damage is less important, since the samples are outside of the ambient of accelerated ions. The pressure in the plasma chamber was maintained at about 1 mbar. The details of the different plasma conditions are collected in TABLE 1.

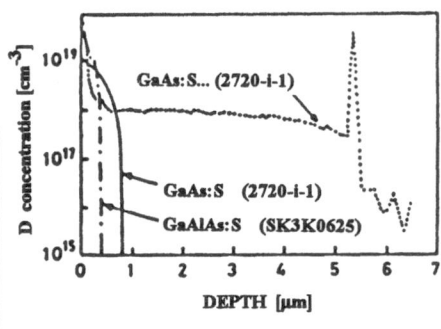

Figure 1. Deuterium SIMS profiles of deuterium diffused samples

TABLE 1. The applied plasma conditions

Nota-tion	Type	Temp [°C]	Time [h]	Gas	Power [W/cm²]
Type I	direct	220	6	D	0.26
II	direct	300	8	D	0.18
III	indirect	220	20	D	1.3
IV	indirect	220	13	H	1.3

After the plasma exposures the same electric measurements were carried out and the results have been compared with those obtained on as grown samples. On deuterated samples SIMS measurements were also performed.

TABLE 2. Variation of the parameters of the samples during different plasma treatments

Sample		Before plasma exposure				After plasma exposure				Plasma type
		n (x10⁻¹⁸) [cm⁻³]		μ [cm²/Vs]		n (x10⁻¹⁸) [cm⁻³]		μ [cm²/Vs]		
No.	Type	300K	77K	300K	77K	300K	77K	300K	77K	
2720	GA:S	4.7	-	1290	-	1.6	-	1550	-	IV
2720	GA:S	4.7	4.7	1300	1350	2.7	2.7	1590	1600	III
2720	GA:S	3.6	3.6	1470	1420	2.4	2.4	1770	1690	III
2922	GaAs	1	2.7	270	1600	0.2	0.06	300	3400	IV
SK3K	GAA:S	1.2	-	460	-	0.002	-	1630	-	I
SK3K	GAA:S	1.3	-	480	-	0.005	-	1480	-	II
SK3K	GAA:S	1.4	1.3	440	390	0.003	0.002	1630	1680	III
SWK0	GA:Se	3.9	-	1820	-	3.2	-	1970	-	IV
SWK0	GA:Se	4.1	-	1870	-	3.3	-	1940	-	III

Abbreviations: GA means GaAs and GAA means GaAlAs. Sample No. SK3K denotes SK3K06

3. Results and discussion

Figure 1. shows deuterium SIMS profiles of two GaAs:S and one GaAlAs:S samples. Two different plasma conditions were applied for GaAs. It is seen from the shape of the curves that neither of them can be described correctly by complementary error function. For the ternary alloy a surface accumulation range followed by a plateau is observed and then a drastic decrease of the deuterium concentration is seen. A similar solubility plateau appeared on the diffusion curve of the GaAs:S sample with high exposure time. Such effect was exhibited also by other alloy samples [4]. These results suggest the existence also of other types of interactions of hydrogen in GaAs, probably with some defects other than substitutional donors.

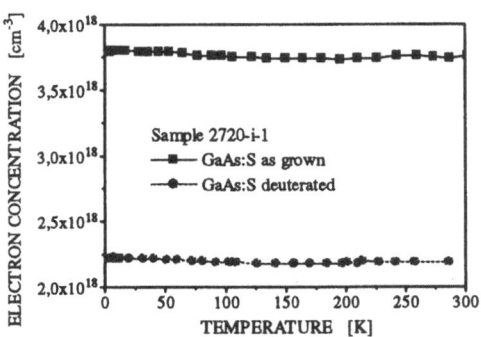

Figure 2. Temperature dependence of the concentration of a GaAs:S sample before and after the deuterium diffusion

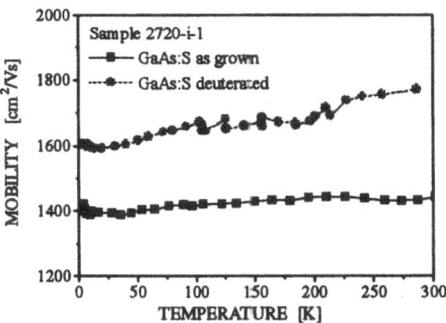

Figure 3. Temperature dependence of the mobility of a GaAs:S sample before and after the deuterium diffusion

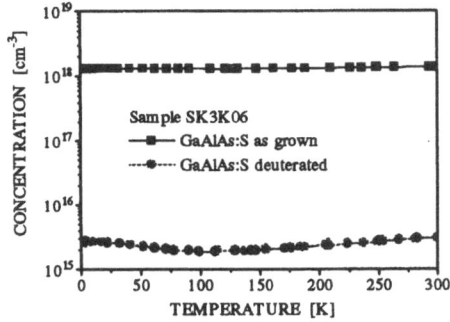

Figure 4. Temperature dependence of the concentration of a GaAlAs:S sample before and after the deuterium diffusion

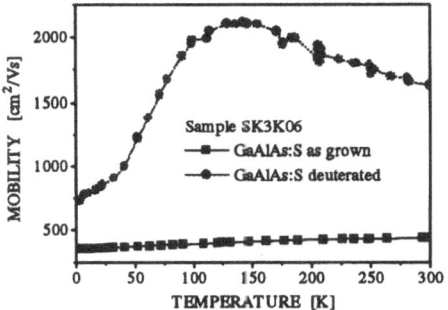

Figure 5. Temperature dependence of the mobility of a GaAlAs:S sample before and after the deuterium diffusion

Comparison of the D profiles of GaAs and GaAlAs samples exposed to the same plasma indicates a higher efficiency of the deuterium capture in the ternary alloy than in GaAs, though the aluminium content is low.

Table 2 and also the figures 2 and 4 show that the electron concentration decreased due to the plasma treatment. This is caused by the donor passivation process, similarly to that in Si doped samples [1]. (Heat treatment experiments [not shown here] were in agreement with this statement. A part of the samples underwent a 10 min annealing process at 400 °C and the initial as grown parameters were recovered in each case.) The

204

passivation efficiencies are different according to the different plasma conditions. More significant decrease of the electron concentration indicates a higher passivation efficiency for the ternary alloy in agreement with the SIMS measurements. This is consistent with the observations made on n-type Si and Te doped GaAlAs and GaAs [4, 7].

The mobility values (table 2) and their temperature dependence (figures 3 and 5) are in a full agreement with the electron concentration results. An increase in the mobility is observed, which corresponds to a decrease of the influence of the mobility determining scattering on ionised impurities. It is very well emphasised e. g. by the appearance of a maximum on the $\mu(T)$ curve of the most effectively passivated GaAlAs sample (fig. 5.). The weakest passivation effect was found with Se doped samples.

4. Conclusions

Decrease of the concentration of the conducting electrons and increase of their mobility as a result of the hydrogen/deuterium diffusion process is typical for the donor neutralisation effect by hydrogen. The mobility increase indicate mainly a donor neutralisation effect not a simple compensation. That is S and Se can be passivated in both GaAs and GaAlAs by such plasma treatment. The formation of neutral donor-hydrogen complexes is supposed. There are also infrared spectroscopic evidences of the existence of S-H and Se-H complexes in GaAs [3]. However the influence of H on the properties of GaAs and related compounds doped with atoms of group VI has to be further studied, since differences in the passivation efficiency of different dopants, the activation energies, the diffusion coefficients, the type of the bond of H in the lattice, etc. are not fully clarified yet.

5. Acknowledgements

This work was supported a part by the Hungarian National Research Foundation (OTKA, grants No. T4178, E12012, and T15619) and the intergovernmental French-Hungarian Balaton program (contract 35/95).

6. References

1. Chevallier, J., Clerjaud, B. and Pajot, B. (1991) in *Hydrogen in semiconductors*, eds. Pankove, J.I. and Johnson, N.M., Academic Press, San Diego, p. 447.
2. Pajot, B., Newman, R.C., Murray, R., Jalil, A., Chevallier, J., and Azoulay, R. (1988) *Physical Review B* 37, 4188.
3. Rahbi, R., Theys, B., Jones, R., Pajot, B., Öberg, S., Somogyi, K., Fille, M.L., and Chevallier, J. (1994) *Solid State Communications* 91, 187.
4. Theys, B., Machayekhi, B., Chevallier, J., Somogyi, K., Zahraman, K., Gibart, P., and Miloche, M. (1995) *J. Applied Physics* 77, 3186.
5. Chevallier, J., Machayekhi, B., Grattepain, C.M., Rahbi, R., and Theys, B. (1992) *Physical Review B* 45, 8803.
6. Görög, T., Gyúró, I., and Somogyi, K. (1983) *Acta Physica Hungarica* 57, 223.
7. Mostefaoui, R., Chevallier, J., Jalil, A., Pesant, J.C., Tu, C.W., and Kopf, R.F. (1988) *J. Applied Physics* 64, 207.

COMPARISON OF PHOTOLUMINESCENCE SPECTRA OF MOCVD AND VPE GROWN GaAs LAYERS

K. SOMOGYI
Research Institute for Technical Physics of the Hungarian Academy of
Sciences, Budapest, H-1325 Hungary, P.O.B. 76.

1. Introduction

Metal organic chemical vapour deposition (MOCVD) is the most widely used epitaxial
layer growth method of our days in the field of III-V compound semiconductors. At the
same time VPE remains one of the most effective, high yield technologies in GaAs, GaP
and InP wafer production. A comparison of the properties of epitaxial layers grown by
these two methods is of high interest [1].

It is assumed here that PL spectra can enlighten also some common properties of the
principally different layer growth methods not only the particular characteristics [2].
Undoped and heavily Ge doped GaAs epitaxial layers were studied. PL spectra of a
great number of VPE and MOCVD grown samples have been measured and compared.
Some characteristic differences and similarities have been demonstrated.

2. Experimental background

Layer thicknesses were between 0.1 μm and 40 μm. Background electron concentration
of the undoped layers varied between $<10^{14}$ cm^{-3} and $\approx 2 \times 10^{16}$ cm^{-3}. As a rule, VPE
samples have the smaller (and even very low) concentrations. PL measurements were
carried out at near liquid He temperatures.

VPE layers have been grown in our laboratory by chloride (Effer-Nozaki) type VPE
method [3]. The undoped samples were collected (about 40 layers) from wafers grown
in different periods during 12 years. Four pure VPE samples from other laboratories
were involved additionally. MOCVD samples have been prepared in different
laboratories by using TMG and AsH$_3$ sources (and GeH$_4$ for doping) [4]. Twenty
different samples were measured.

3. Results and discussion

Figure 1 shows two typical PL spectra of VPE GaAs layers. Figures 2, 3 and 4 show
similar spectra for three groups of MOCVD samples and Table 1 summarizes typical PL
peaks of different samples.

J. Novák and A. Schlachetzki (eds.), Heterostructure Epitaxy and Devices, 205–208.
© 1996 *Kluwer Academic Publishers.*

The intensity of the near band edge exciton peak of VPE samples is always significantly higher than that of the second peak, compared to MOCVD samples. These peaks are usually higher and narrower than in the case of MOCVD ones. (A direct comparison was possible in few cases.) Resolution of these main peaks was possible only for a few VPE samples with background concentrations remarkably below 10^{14} cm^{-3}. The difference can be explained probably by the difference of the near surface structure of the layers.

Table 1.
Characteristic photoluminescence peaks

No.	Techno-logy	PL peaks	
		Exciton [meV]	2nd peak [meV]
2793n3	VPE	1511.7	1474
2893n3	VPE	1512.7	1490
2814n3	VPE	1512.9	1494
PL942	MOCVD	1513.4	1490
200069	MOCVD	1513.4	1492
FL360	MOCVD	1514	1497
236	MOCVD	1514	1497
E-3-2/1	MOCVD	1515	1490
316	MOCVD	1515.8	1492
FL2	MOCVD	1515.8	1494

Figure 1. Typical photoluminescence peaks of VPE grown layers ⇒

The same peaks exhibit also difference in their position. Maxima of the peaks of VPE layers lay in the range between 1511.7 and 1513.7 meV, and that of MOCVD layers between 1513.2 and 1515.8 meV. It suggests that for VPE layers acceptor bound exciton peaks are characteristic, while for MOCVD ones (neutral and ionized) donor bound exciton peaks [5, 6].

The scatter of the position of the second peak and that of the appearance of a third and a fourth peak did not make us possible a grouping of the VPE samples on such basis. The most frequently detected maxima were at about 1480 meV, which is associated to Si on As sites electron-acceptor recombinations [5-7]. Then Zn related peaks at about 1488 meV were observed [5-8]. After some C doping experiments in the VPE reactor also C related (~ 1495 meV) peaks were detected [6, 8]. MOCVD samples show practically always one additional peak and they could be divided into three groups. One of them exhibits PL peaks at about 1490 meV, associated with Zn, and donor-acceptor transitions (figure 1) [9]. The second and the third ones show maxima (Figures 3 and 2, respectively) around 1492 and 1497 meV and both of these peaks can be related with carbon. (Though for the peaks at 1492 meV it is not very evident that it is related either with Mg, Be or C [8]).

The dominance of Si related peaks in VPE crystals can be connected directly with the fact that Si is supposed to be the main background impurity in hot wall chloride epitaxy. Since the V/III ratio can be varied in a narrow range only, a luminescence peak related

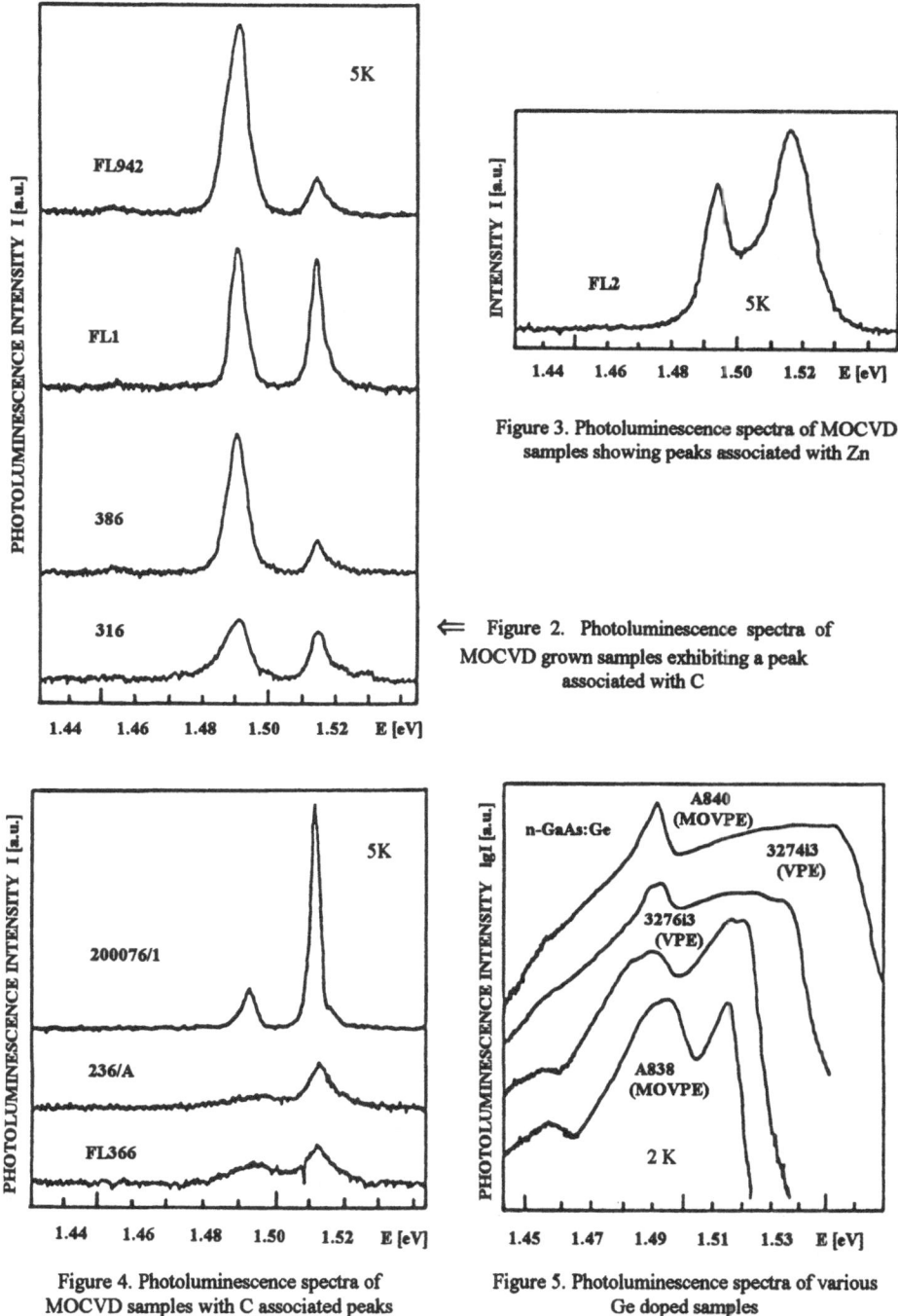

Figure 3. Photoluminescence spectra of MOCVD samples showing peaks associated with Zn

⇐ Figure 2. Photoluminescence spectra of MOCVD grown samples exhibiting a peak associated with C

Figure 4. Photoluminescence spectra of MOCVD samples with C associated peaks

Figure 5. Photoluminescence spectra of various Ge doped samples

also to stoichiometry defects is not surprising. On the other hand, one of the draw-backs of the MOCVD method is the incorporation of C from organometallics [9, 10]. This can explain the very frequent presence of C related PL peaks. The V/III ratio can be varied in a very wide range, therefore the probability of the stoichiometry defects can be controlled and they do not obviously appear.

Ge was chosen as a dopant because of its amphoteric character (related to III-V-s), its high solubility and of the ease of the doping. Figure 5 shows four PL spectra at increasing from $1x10^{17}$ up to $4x10^{18}$ cm^{-3} doping levels. The spectra show an evolution in the dependence on the increasing Ge concentration: the near band gap peak broadens in the direction of higher energies above the band gap. This is probably due to the increasing degeneracy of the GaAs above 10^{17} cm^{-3}, but lattice distortions can also play role. The homogeneous character of the change in the shape of the spectra indicates that the behaviour of the VPE and MOCVD layers are identical or, at least, very similar. No characteristic differences were observed.

4. Conclusions

Though on the basis of heavily doped GaAs layers a similarity in the PL properties of VPE and MOCVD layers can be stated, however undoped samples show characteristic differences. These differences are in agreement with other results on the characteristic background impurities of these technologies. It was also seen by PL peaks that the variety of the type of background impurities (complexes) is greater in the case of VPE layers than that for those grown by MOCVD. Differences in exciton peaks indicate, however, more advantageous properties of (very) pure VPE layers.

5. Acknowledgements

This work was supported in part by the Hungarian National Research Foundation (OTKA, grants No. T4178, E12012 and T15619) and from the contract with the National Committee for the Technical Development (OMFB, No. 91-97-07-0316). The author is very indebted to Yu. N. Sveshnikov, H. Hartdegen, J. Novak, Yu. V. Zhilyaev, M. Tlaczala, K. D. Yashin providing him samples, to L. Andor and J. F. Rommeluere for performing PL measurements, to B. Pődör for his advices and to Gy. Kiss for her assistance.

6. References

1. Somogyi, K. (1994) *Acta Physica Hungarica* **74**, 227-233.
2. Pődör, B., Andor, L., Nemcsics, Á., Somogyi, K., Gyúró, I. *Crystal Properties Preparation* **19/20**, 17-20.
3. Görög, T., Gyúró, I., and Somogyi, K. (1983) *Acta Physica Hungarica* **57**, 223-232.
4. Jain, B.P. and Purohit, P.K. (1984) *Prog. Crystal Growth Characterisation* **9**, 51-103.
5. Heim, U. and Hiesinger, P. (1974) *physica status solidi (b)* **66**, 461-470.
6. Gavrilenko, V.I., Grehov, A.M., Korbutyak, D.V., and Litovchenko, V.G. (1987) *Opticheskie svoystva poluprovodnikov* (in Russian), Naukova Dumka, Kiev.
7. Skromme, B.J., Low, T.S., Stillman, G.E., Kennedy, J.K., and Abrokwah, J.K. (1983) *J. Electronics Materials* **12**, 433-457.
8. Swaminathan, V., Van Hasen, D.L., Zilko, J.L., Lu, P.J., and Schumacher, N.E. (1985) *J. Applied Physics* **57**, 5349-5353.
9. Bhat, R., O'Connor, P., Temkin, T., Dingle, R., and Keramidas, V.G. (1982) Inst. Phys. Conf. Ser. **63**, 101-106.
10. Kuech,T.F. and Veuhoff, E. (1984) *J. Crystal Growth* **68**, 148-156.

X-RAY DIFFRACTION STUDY OF MOCVD GROWN InGaP

D. KORYTÁR, L. FRANCESIO[*], R. KÚDELA

Institute of Electrical Engineering, Slovak Academy of Sciences, Dúbravská 9, 842 39 Bratislava, Slovakia
[*]*Maspec, CNR, Via Chiavari 18/A, 43 100 Parma, Italy*

1. Introduction

Layers of $In_{1-x}Ga_xP$ epitaxially grown by MOCVD on (001) GaAs substrates are interesting from the point of view of optoelectronic and microelectronic devices. Their growth and properties are influenced by such parameters as substrate quality and orientation, growth temperature, V/III ratio, total gas flow, etc.

X-ray double crystal and multiple crystal high resolution diffractometry are used to characterise chemical composition, lattice mismatch, and lattice tilts of the epilayers routinely in various degrees of approximation [1,2]. Combination of a small beam size and precise X-Y translation stage allows to assess homogeneity of parameters across the wafer, too. Another way of evaluating the homogeneity and crystal perfection of the layers is provided by single and double crystal topography. These techniques are not directly compatible with specialised high resolution diffractometers and thus they are not so frequently used. However, a combination of the techniques can bring more information, especially when studying non homogeneous layers.

2. Experimental

$In_{1-x}Ga_xP$ epitaxial layers were prepared at various conditions in Aixtron MOCVD reactor on (001) GaAs substrates. Growth conditions are given in Table I.

Lattice mismatch and composition were measured by means of a double crystal diffractometer in (n,-n) arrangement with Ge(004) and Ge (115) monochromators and $CoK\alpha_1$ radiation and also with a Philips high resolution diffractometer with a two-crystal four-reflection monochromator (Ge,220, $CuK\alpha_1$). Symmetric (004), and two asymmetric (115) diffractions with reversed beampaths giving $\Delta\Theta_B^-$ and $\Delta\Theta_B^+$ Bragg angle shifts for the lower- and higher- incident angle, respectively, were used. Perpendicular and parallel relative lattice parameters can be determined [1-3] using the equations

$$\Delta\Theta_B = \left(\Delta\Theta_B^+ + \Delta\Theta_B^-\right)/2 \qquad \Delta\varphi = \left(\Delta\Theta_B^+ - \Delta\Theta_B^-\right)/2 \qquad (1,2)$$

$$\Delta a_\perp/a_s = \Delta\varphi . tg\varphi - \Delta\Theta_B . cot\,\Theta_B, \quad \Delta a_\parallel/a_s = -\Delta\varphi . cot\,\varphi - \Delta\Theta_B . cot\,\Theta_B \quad (3,4)$$

J. Novák and A. Schlachetzki (eds.), Heterostructure Epitaxy and Devices, 209–212.
© 1996 *Kluwer Academic Publishers.*

210

where a_s is the substrate lattice parameter, $a_\perp$ and $a_\parallel$ the perpendicular and parallel lattice parameters of tetragonally deformed epilayer lattice, respectively, $\Delta a_\perp = a_\perp - a_s$, φ and $\Delta\varphi$ the lattice planes angles, Θ_B the substrate and $\Delta\Theta_B$ the epilayer Bragg angle positions. Relaxed lattice parameter can be obtained from (3) using the equation

$$\Delta a_{rel}/a_s = \left(\Delta a_\perp/a_s - \Delta a_\parallel/a_s\right).\left(1-\nu\right)/\left(1+\nu\right)+\Delta a_\parallel/a_s \qquad (5)$$

where ν is Poisson's ratio, and transformed into composition (x) of the epilayer by means of Vegard's law. In analogy with [4] also a coherency factor can be calculated.

Single crystal (1/2 1/2 5/2) topographs in reflection Lang camera and also double crystal topographs with the asymmetric Ge (115) monochromator, both with CoKα_1 radiation and exposures up to 20 hours, were taken as well.

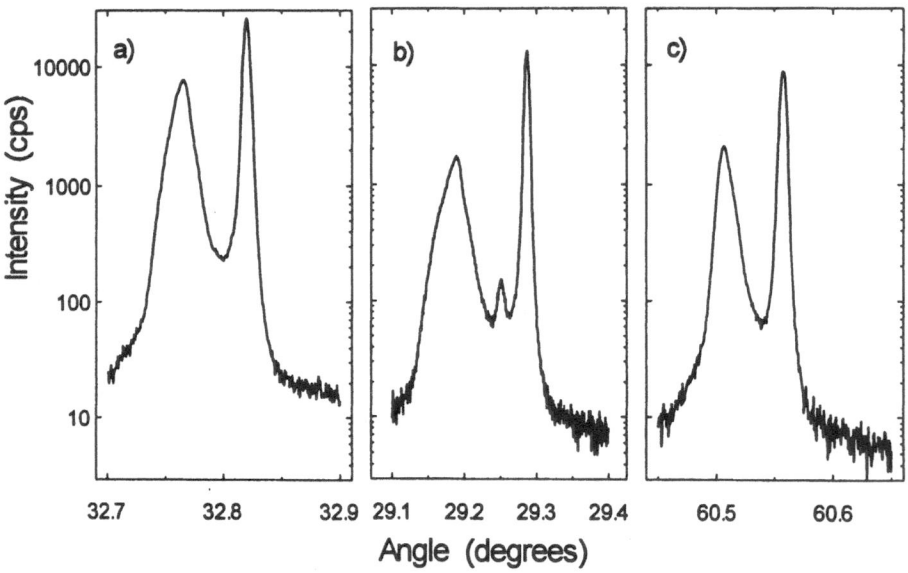

Fig.1. Rocking curves from IGP42 obtained with high resolution diffractometer in (004) reflection (a), and in reflection (115) at lower (b) and higher (c) angle of incidence.

3. Results and discussion

Table I shows growth and structural parameters of three InGaP epilayers chosen to show the variety of features encountered. IGP42 (see also Fig.1) represents a thick, structurally good and homogeneous epilayer with narrow single epilayer peaks, where the analysis of mismatch and composition is rather straightforward. IGP59 exhibits both lower- and higher- angle epilayer peaks similar to simulated graded composition epilayers [2], while IGP26 shows broad lower angle epilayer peaks.

211

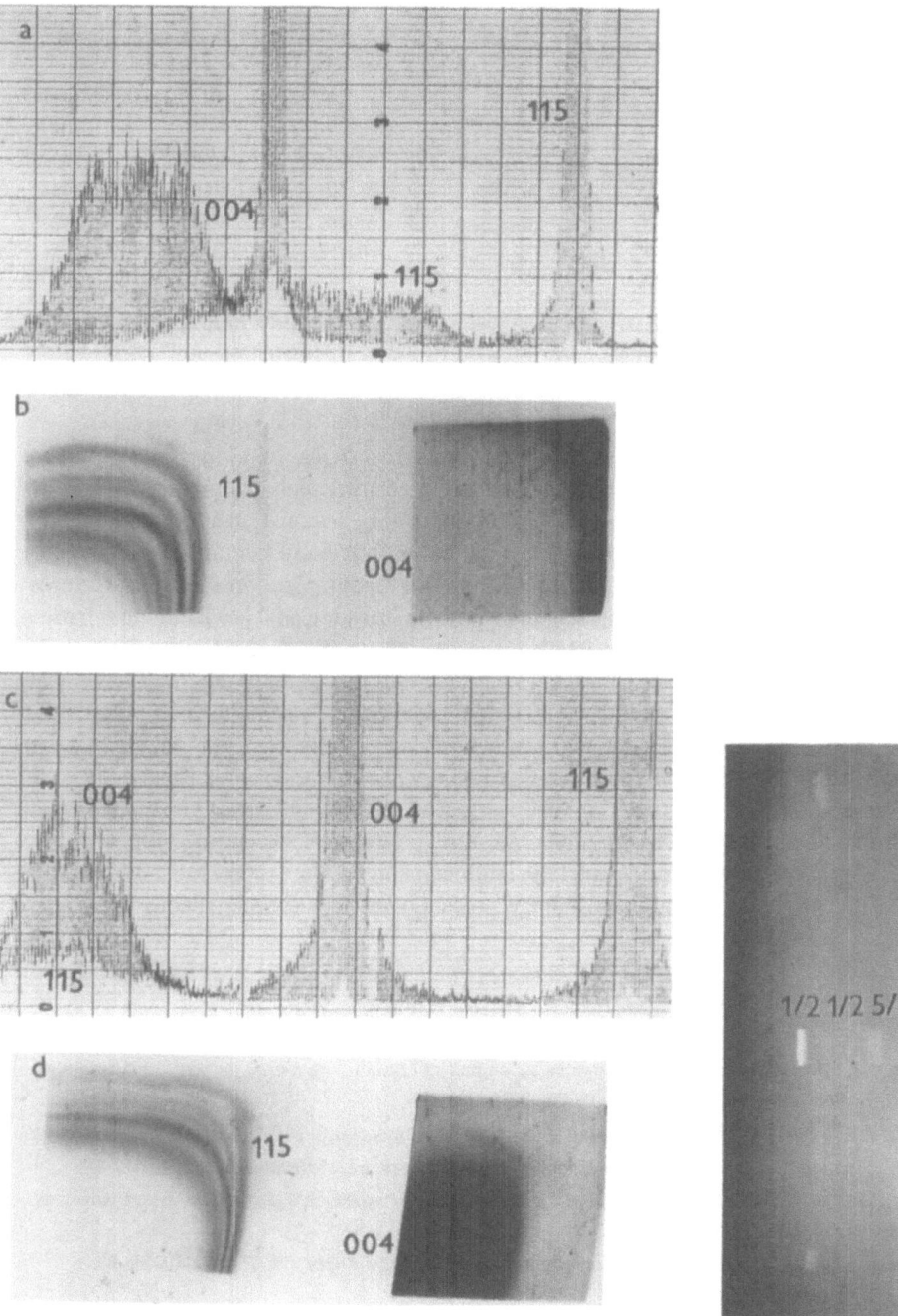

Fig.2. Simultaneous (004,115) CoKα₁ diffractions. Rocking curves taken at 5 mm (a) and 1 mm (c) from the right edge of the specimen. Vertical edge 10 mm. Topographs taken at -960 (b) and at -1320 (d) arcsecs from the substrate (115) peak.

Fig.3 Reflection Lang topograph in (1/2 1/2 5/2) diffraction. Weaker satellite diffractions present too.

TABLE 1. Growth and structural parameters of selected InGaP/GaAs specimens

Sample	h(μm)	In/Ga	III/V	T[°C]	$\Delta a_\perp/a_s$ [10^{-3}]	$\Delta a_\parallel/a_s$ [10^{-3}]	x
IGP42	1.57	1.42	396	720	1.460	-0.169	0.506
IGP26	0.7	1.14	160	640	1.656	0.028	0.504
					3.140	-0.037	0.494
IGP59	1.2	1.22	216	560	1.084	0.076	0.508
					-0.937	-0.113	0.523

Although the simple analysis of the strongest peaks is not quite appropriate for non homogeneous heterolayers, it may give rough estimation of mismatch and composition (Table I). All values of $\Delta a_\parallel/a_s$ in Table I are zero within 12 percent of $\Delta a_\perp/a_s$ values which means non-coherent interface.

When taking double crystal topographs, in some cases (e.g. in IGP 26), two topographs could be recorded in one angular setting. Surprisingly, they were caused by multiple 004 (stronger) and 115 (weaker) diffractions. Fig.2 shows such topographs (a,c) in two angular settings and also the rocking curves (b,d) taken in two different positions by a technique based on this observation [5]. On the topographs, more distinct features are observed on (115) topographs sensitive both to parallel and perpendicular lattice parameters and tilts, which are related to the growth itself. The rocking curves (b,d) show the overlap of the (004) and (115) peaks thus explaining the rough contrast behaviour. The origin of fine strips in (115) topographs was not as far identified. Inhomogeneous cationic ordering [6,7] was observed in this sample, too[8], but the Lang topograph in (1/2 1/2 5/2) shows rather homogeneous contrast (see Fig.3).

In conclusion, in addition to diffractometric study of lattice mismatch and composition the X-ray topography was used to study structural properties of InGaP/GaAs specimens. Rough and fine inhomogeneities were observed and tried to explain. The cationic ordering itself, though present, was rather homogeneous across the specimen.

1. Bartels W.J. (1983) Characterization of thin layers on perfect crystals with a multipurpose high resolution X-ray diffractometer, *J.Vac.Sci.Technol.* **B1**(2), 338-345

2. Tanner B.K. (1991) Double crystal X-ray diffraction and topography, in M. Grasserbauer and H.W. Werner (eds.), *Analysis of Microelectronic Materials and Devices*, John Wiley and Sons, pp. 609-635

3. Wang X.R., et al. (1987) X-ray double crystal diffraction studies of GaInAsP/InP heterostructures, *J.Vac.Sci.Technol.* **B6**(1), 34-36

4. Nanchang Zhu, et al. (1993) Determination of interface coherency, *J.Appl.Phys.* **75**(6), 2805-2808

5. Korytár D., to be published

6. Liu Q., et al. (1992) Analysis of ordering in GaInP, *J.Appl.Phys.* **73**(6), 2770-2774

7. Francesio L., et al. (1994) Investigation of the cationic ordering in InGaP/GaAs epilayers grown by low-pressure, metal organic vapour phase epitaxy, *Materials Science and Engineering* **B28**, 219-223

8. Dobročka E., et al. (1995) Ordering in InGaP prepared by MOCVD, presented at this conference

STRUCTURAL AND OPTICAL PROPERTIES OF ORDERD DOMAINS IN InGaP$_2$ ALLOYS

L. NASI, F. FERMI°, M. MAZZER*, C. ZANOTTI-FREGONARA*,
L. VITALI•, F. VIDIMARI• and G. SALVIATI

C.N.R.-MASPEC Institute, Via Chiavari 18/A, 43100 Parma, Italy
°INFM, Physics Department, University of Parma, Viale delle Scienze,
43100 Parma, Italy
*Materials Department, Imperial College of Science Technology and
Medicine, Prince Consort Road SW7 2BP, London, U.K.
•ALCATEL-TELETTRA, Via Trento 30, 20059 Vimercate(MI), Italy

Structural and optical studies on CuPt ordered Ga$_{0.51}$In$_{0.49}$P layers, grown by organometallic chemical vapour deposition on differently misoriented (001) GaAs substrates, show that the CuPt-doimains distribution strongly influences the homogeneity of ordering degree. For the first time, a correlation between cathodoluminescence emission from ordered regions and changes in surface steps distribution during growth is also shown.

1. Introduction

Ga$_{0.51}$In$_{0.49}$P alloys, grown lattice matched on (001) GaAs substrates, are excellent materials for optoelectronic devices. They recently attracted considerable attention because their optical and electrical properties are affected by CuPt-type ordering that spontaneously occurs during organometallic chemical vapor deposition (MOCVD). Atomic ordering formation and evolution are driven by surface mechanisms including surface reconstruction and step motion [1,2]. However, the role of the initial distribution of substrate surface steps and their subsequent arrangement during growth on the morphology and optical properties of ordered domains are still an open problem. In this paper, the correlation between structural and optical properties of ordered domains as a function of substrate misorientation is discussed. A correlation between the spatial distribution of ordered domains and the sample surface morphology is also reported.

2. Experimental

The lattice matched GaInP layers were grown in ALCATEL by low pressure MOCVD on (001) GaAs substrates at exact orientation and 2° and 6° towards [100] at temperature of 660 °C. The V/III ratio and the growth rate were 120 and 2.5 µm/h respectively.
The spectrally-resolved Cathodoluminescence (SCL) measurements were performed at 35K in a JEOL 840 Scanning Electron Microscope at the Imperial College. Atomic

J. Novák and A. Schlachetzki (eds.), Heterostructure Epitaxy and Devices, 213–216.
© 1996 Kluwer Academic Publishers.

Force Microscopy (AFM) contact mode images were taken in a Nanoscope III Digital Microscope. As for Transmission Elctron Microscopy (TEM) and Photoluminescence (PL) investigations, the experimental details have already been published elsewhere [3].

3. Results and discussion

TEM diffraction patterns showed the layers spontaneusly ordered in CuPt-like structures with alternating Ga- and In-rich planes along the (-111) and (1-11) directions [3].

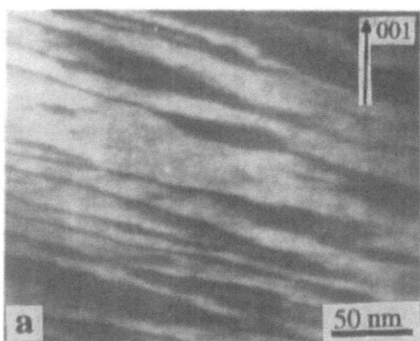
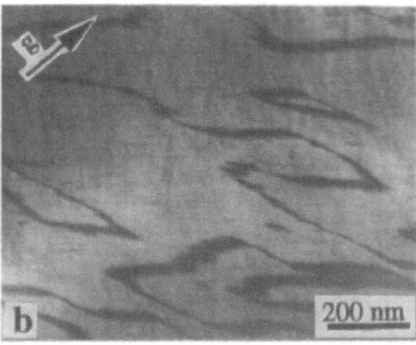

Figure 1 a and b. TEM Cross section dark field images of the 2° and 6° off sample respectively.

The spatial distribution of the two (-111) and (1-11) ordered variants was investigated by dark field (DF) TEM analyses of both [110] cross sections and [001] top views, as a function of substrate misorientation. In the 0° off sample the two variants are equally distributed in plate-like domains inclined at a shallow angle but in an opposite sense relative to the interfacial plane. The tilt of the substrate breaks down the symmetry of the system and allows only the (1-11)-type regions to be enhanced. The influence of the substrate misorientation on the enlargment of this strongest ordered variant is shown in figure 1 by comparing the (1/2,-1/2,1/2) DF images of the 2° and 6° off samples. The (1-11) CuPt structure is distributed over the entire epilayer in the 6° off sample, the dark bands being antiphase boundaries (APBs) between adjacent ordered regions. It can be seen that the wavy APBs density decreases by approaching the top surface, indicating a change in surface steps distribution during the growth. The shape of the samples ordered

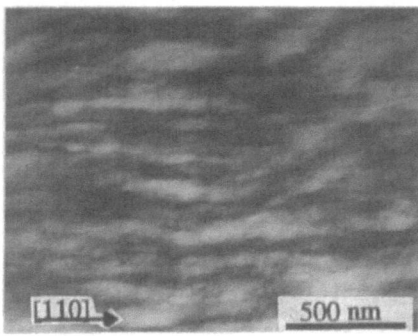

Figure 2. Plan view 1/2(3-31) dark field image of the 2° off sample.

domains in the [001] plan view dark field images appears very different. As for the boundaries between the dominant (1-11)-type domains in the 2° off sample, figure 2 shows that they appear as wavy dark lines nearly parallel to the [110] direction.

Further, in agreement with other authors [4], our previous HREM investigations [3] revealed the presence of a complex structure of (-111) and (1-11) ordered microdomains in the 0° and 2° off samples, while the 6° off specimen appeared as a more homogeneous structure, due to the absence of one of the two variants.

PL spectra (figure 3) evidenced two emission bands in all the samples. The high energy peaks are strictly related to the excitonic transitions detected in the PLE mode, while those at lower energy, whose nature is actually unknown, come from localized areas ($\phi \approx$ 5 μm) as confirmed by monochromatic CL images [5]. The excitonic PL bands, that show a substantial red shift only for the 0° (15 meV) and 2° off samples (10 meV), have large full width at half maximum (FWHM) and change their peak energy and FWHM by increasing the temperature. These results are consistent with a statistical distribution of order degree whose mean value is higher the lower is the optical band gap. Accordingly to TEM results, the large PL FWHM value in these samples is ascribed to structural inhomogeneities produced by the presence of partially ordered domains of the two variants, numerous APBs and alloy fluctuations. On the contrary, the highly homogeneous 6° off sample shows a very narrow excitonic PL band (FWHM=6 meV) that overlaps the excitonic PLE peak and is insensitive to temperature increase at least up to T=45 K, indicating a sharp distribution of ordering degree in the CuPt domains.

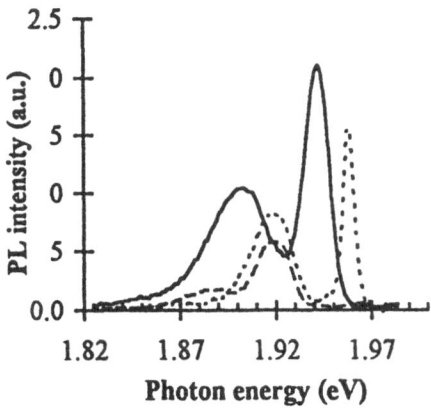 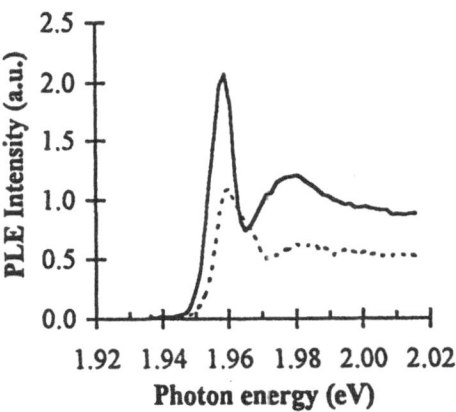

Figure 3. PL spectra observed at T=18 K of the 0°, 2° and 6° off samples. Excitation: λ=620 nm (2eV), Δλ=3 nm.

Figure 4. Linear polarized PLE spectra observed at T=18 K of the 6° off sample. Emission: λ=648 nm (1.9 eV), Δλ=2 nm.

Conclusions about the order parameter value, on the basis of the energy onset of the PLE, must be carefully drawn because it can be significantly affected by contributions from localized states, quantum dimensional effects, compositional variation and local elastic energy in the 0° and 2° off samples. A direct correlation between ordering degree and energy gap is instead clearly evident in the polarized PLE spectra of 6° off sample (figure 4). The onset of the band-to-band transitions is quite evident for light polarized along the [110] direction, while a second peak emerging in the [1-10] polarized PLE can be attributed to the exciton associated to the crystal field split-off band ($\Delta E \cong 9$ meV) due to CuPt ordering, according with the optical selection rules for this structure [6].

PL and SCL investigations gave identical emission spectra. Taking advantage from the possibility of getting monochromatic CL images, optical emissions from sample areas with different degree of ordering have been studied by scanning the excitonic peaks. Due to the CL resolution, only a slight difference between monochromatic images taken at different excitonic peak positions has been found. As an example, figure 5 shows SCL monochromatic picture of the 2° off sample taken at a wavelength coresponding to the excitonic peak maximum. The white areas are due to ordered domains, while the black ones are possibly due to APBs. This hypothesis is supported by the fact that the APBs direction on the (001) plane is the same as found by plan view TEM investigations (figure 2). It is worthmentioning that, to the best of our knowledge, it is the first time

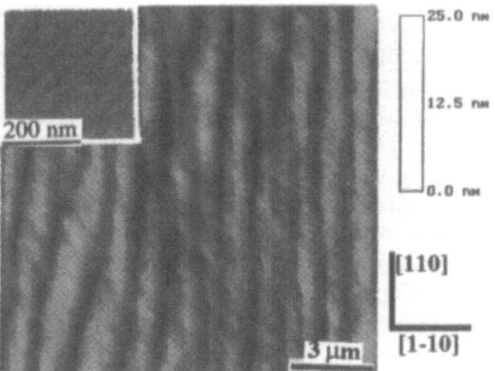

Figure 5. SCL picture of the 2° off sample taken at a wavelenght corespondig to the excitonic peak maximum.

Figure 6. AFM image of surface morphology of the 2° off sample. The presence of fine modulation is shown in the inset.

that non radiative CL recombination has been seen from APBs between ordered domains. A possible contribution to the SCL contrast variations from surface roughness has been ruled out by AFM micrographs (figure 6) showing ondulations only 6-8 nm high. A careful analysis of the AFM pictures revealed the presence of a fine modulation (0.4-0.6 nm tall) related to [100] oriented step bunching [7] and a coarse one consisting of supersteps aligned along [110] direction. The evident correspondence between SCL bright areas and AFM large-scale modulation can be ascribed to the loss of coherence of ordered regions at the superstep edges. Infact, APBs preferentially grow in correspondence of undimerized atoms at step edges sites that arise when they are not exactly aligned along the <110>-type directions [1]. At the same time, a possible surface step rearrangement during the growth can induce a different morphology of ordered domains and APBs.

This point deserves further investigations since so far it is not clear if the step edge rearrangement is simply due to growth kinetics or if it can be influenced by the interaction with inducing ordering processes [5].

References

1. Zunger, A., and Mahajan, S. (1994) Atomic ordering and phase separation in epitaxial III-V alloys, Handbook of Semiconductors 3(19), 1399.
2. Su, L.C., Stringfellow, G.B., Christen, J., Selber, H., and Bimberg, D. (1994) Control of ordering in GaInP and effect on bandgap energy, *J. Electron. Mat.* 23(2), 125.
3. Nasi, L., Lazzarini, L., Salviati, G., Fermi, F., and Lenzi G. (1995) TEM, PL and PLE characterization of ordered domains in $Ga_{0.51}In_{0.49}P$ alloys, Proceedings of Microscopy of Semiconducting Materials IX, *in press on I.O.P.Conference Series.*
4. Baxter, C.S., and Stobbs, W.M. (1991) The morphology of ordered structures in III-V alloys: inferences from a TEM study, *J. Crystal Growth* 112, 373.
5. Nasi, L., Mazzer, M., Zanotti-Fregonara, C., and Salviati, G. (unpublished).
6. Horner, G.S., Mascharenas, A., Fryen, S., Alonso R.G., Bertness, K., and Olson, J.M. (1993) Photoluminescence-excitation-spectroscopy studies in spontaneously ordered $GaInP_2$, *Phys. Rev. B* 47, 4041.
7. Shinohara, M., and Inoue, N. (1995) Behavior and mechanism of step bunching during metalorganic vapor phase epitaxy of GaAs, *Appl. Phys. Lett.* 66(15), 1936.

ORDERING IN InGaP PREPARED BY MOCVD

E. DOBROČKA[†], I. VÁVRA[‡], R. KÚDELA[‡] AND M. HARVANKA[†]

[†]*Department of Solid State Physics, Faculty of Mathematics and Physics, Comenius University,*
Mlynská dolina F2, 842 15 Bratislava

AND

[‡]*Institute of Electrical Engineering, Slovak Academy of Sciences,*
Dúbravská cesta 9, 842 39 Bratislava

Since the first report of ordered structures in AlGaAs alloy in 1985 [1] atomic-scale ordering has been observed for a wide range of III-V semiconductor alloy systems [2]. It is well established that the ordering affects the optical [3] and electrical [4] properties of the material, and is therefore of significant technological interest. The most commonly observed ordered structure in III-V alloys is the CuPt or $L1_1$ structure with ordering in the $\{111\}$ planes. However, perfectly ordered material is rarely observed, structures with high density of planar defects are usually found [5]. In this paper we analyse various types of imperfections that may develop during the epitaxial growth.

The investigated samples were $In_{0.5}Ga_{0.5}P$ epitaxial layers grown in a low-pressure MOCVD equipment AIX 200 on (001) oriented semiinsulating $GaAs$ substrates at the temperature $640^{\circ}C$. Total pressure inside the reactor was $20mbar$, the flow velocity was $2.1m/s$. Specimens for TEM in orientations (001) (plane view) as well as $(1\bar{1}0)$ (cross section) were prepared by standard techniques consisting of chemical polishing and ion beam etching. TEM observations were performed in a JEOL 1200 EX microscope operating at $120kV$.

Extra diffraction spots were found in several projections around the pole [001] including the projection [001] itself. The most interesting ones are shown in Fig. 1. The spots in Figs. 1a), 1b) and 1c) are generated by the **k** vectors $\langle \frac{1}{2}, \frac{1}{2}, 0 \rangle$; $\langle \frac{1}{2}, \frac{1}{2}, \frac{1}{6} \rangle$ and $\langle \frac{1}{2}, \frac{1}{2}, \frac{1}{2} \rangle$; and $\langle \frac{1}{2}, \frac{1}{2}, \pm\frac{1}{4} \rangle$, respectively. The ordered structures associated with these vectors were found using the method of static concentration waves [6]. The arrangement of atoms in

J. Novák and A. Schlachetzki (eds.), Heterostructure Epitaxy and Devices, 217–220.

a)   b)

c)   d)

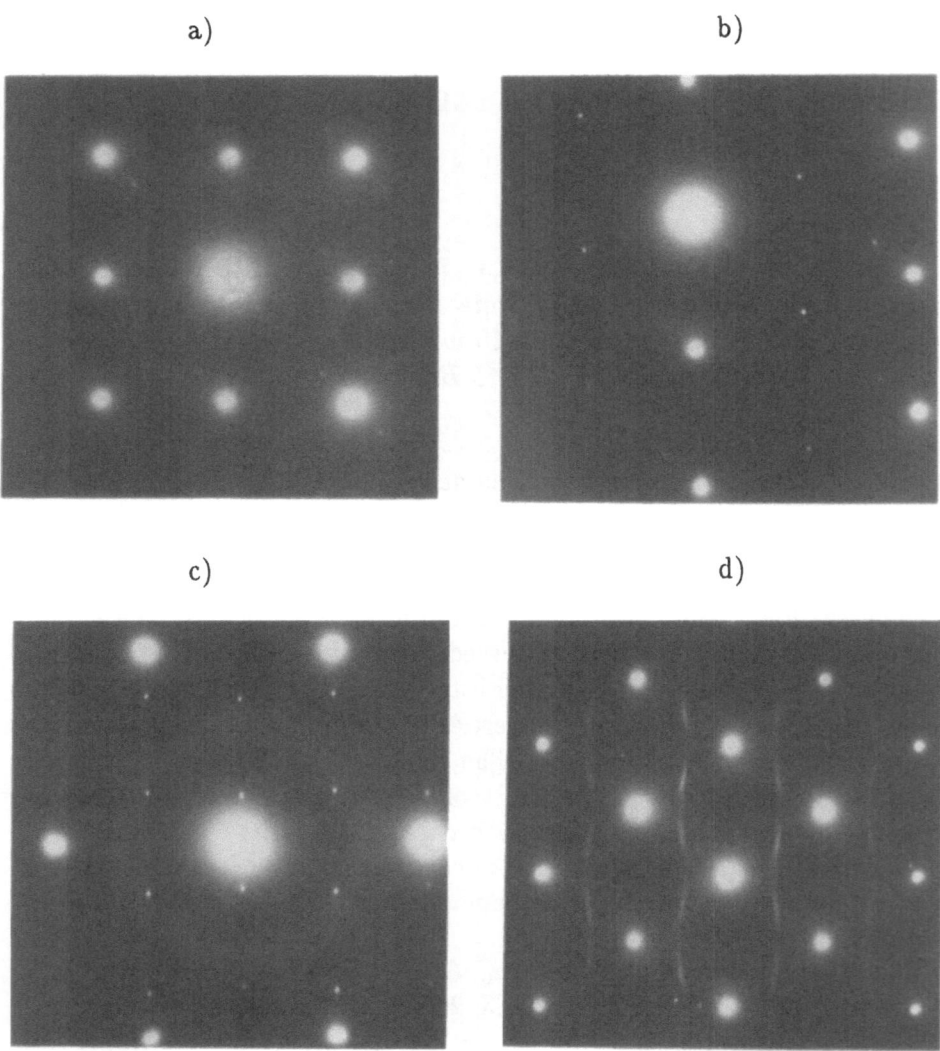

Figure 1. TEM diffraction patterns of the $In_{0.5}Ga_{0.5}P$ epitaxial layer. Specimens were prepared in orientations [001] and [1$\bar{1}$0]. a) [001] pole. b) [$\bar{1}$03] pole. c) [114] pole. d) [1$\bar{1}$0] pole.

the In-Ga sublattice in these structures is shown in Fig. 2. It can be seen that all three structures are closely related to the standard CuPt ordered structure with ordering along the [111] direction. They can be interpreted as CuPt structures with regularly spaced antiphase boundaries (APBs) oriented perpendicularly to the [001] direction. The spacing of the APBs in Figs. 2a), 2b) and 2c) is $1a$, $\frac{3}{2}a$ and $2a$, respectively, where a is the lattice parameter of the disordered structure.

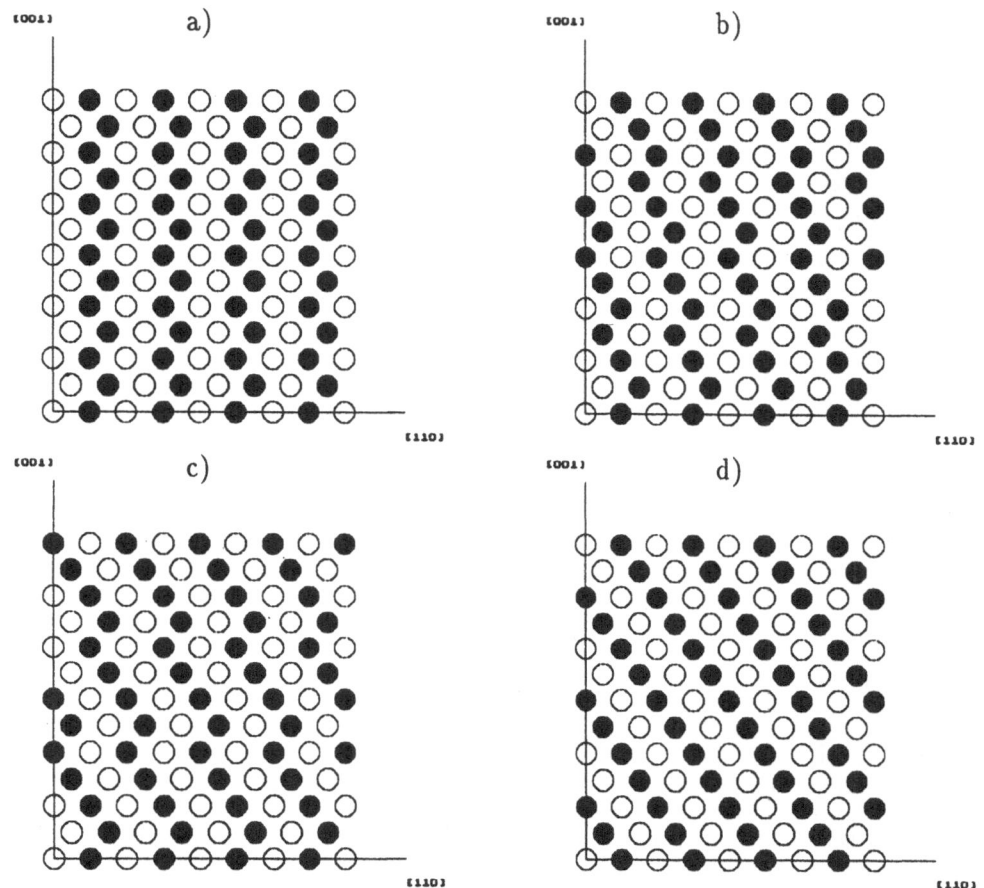

Figure 2. Projection of the ordered Ga-In sublattice in the [1$\bar{1}$0] direction. Open circles - Ga atoms, filled circles - In atoms. For the sake of simplicity P atoms are not shown. a) structure corresponding to $\mathbf{k} = \langle\frac{1}{2},\frac{1}{2},0\rangle$. The spacing of APBs - 1a. b) structure with $\mathbf{k} = \langle\frac{1}{2},\frac{1}{2},\frac{1}{6}\rangle$ and $\mathbf{k} = \langle\frac{1}{2},\frac{1}{2},\frac{\bar{1}}{2}\rangle$. The spacing of APBs - 3/2a. c) structure with $\mathbf{k} = \langle\frac{1}{2},\frac{1}{2},\pm\frac{1}{4}\rangle$. The spacing of APBs - 2a. The structures are derived from the [11$\bar{1}$] CuPt 'parent' structure - d).

Detailed inspection of various projections symmetrical with respect to the pole [001] revealed that imperfect ordering related to the [$\bar{1}\bar{1}$1] CuPt structure is also developed. On the other hand, ordering along the remaining two directions, i.e. [1$\bar{1}$1] and [$\bar{1}$11], was not found. This is in accordance with the results of other authors [4],[7].

It is easy to construct structures with even larger (but regular) spacing of APBs and to find their corresponding k vectors. Following this sequence one would arrive at the perfect CuPt ordering. As regards the size of the regions with one type of ordering, one can hardly expect that the epilayer consists of large domains with constant APB spacing inside the domains. It is more probable that the spacing of APBs varies continuously during the growth process and the resulting structure is an irregular stacking of the presented perfectly ordered structures derived from both variants. This seems to be confirmed in Fig. 1d), where the $[1\bar{1}0]$ diffraction pattern is shown. It is seen that instead of a dense array of discrete spots a continuous streaks with varying intensity along the $[001]$ direction are visible. The highest intensity is around the points $\langle \pm\frac{1}{2}, \pm\frac{1}{2}, \frac{1}{2} \rangle$ corresponding to both variants of perfect CuPt structure. It is also apparent that the streaks are not exactly aligned in the $[001]$ direction. According to [5] this misorientation can be explained by the presence of an other 'family' of APBs that have lower density and are not exactly perpendicular to $[001]$ direction.

To conclude, $In_{0.5}Ga_{0.5}P$ epitaxial layers grown in a low pressure MOCVD equipment were analysed by conventional TEM. Imperfect CuPt ordering was found in the samples grown at the temperature $640°C$. Three model structures were proposed to explain the observed extra spots in the diffraction patterns. The role of APBs in these ordered structures is discussed.

References

1. Kuan, T.S., Kuech, T.F., Wang, W.I., and Wilkie, E.L. (1985) Long-range order in $Al_xGa_{1-x}As$, Phys.Rev.Lett. **54**, 201-204.
2. Stringfellow, G.B. and Chen, G.S. (1991) Atomic ordering in III/V semiconductor alloys, J.Vac.Sci.Technol. **B 9**, 2182-2188.
3. Arent, D.J., Bode, M., Bertness, K.A., Kurtz, S.R., and Olson, J.M. (1993) Band-gap narrowing in ordered $Ga_{0.47}In_{0.53}As$, Appl.Phys.Lett. **62**, 1806-1808.
4. Friedman, D.J., Kibbler, A.E., and Olson, J.M. (1991) Cation site ordering and conduction electron scattering in $GaInP_2$, Appl.Phys.Lett. **59**, 2998-3000.
5. Baxter, C.S., Stobbs, W.M., Wilkie, J.H. (1991) The morphology of ordered structures in III-V alloys: inferences from a TEM study, J.Cryst.Growth, **112**, 373-385.
6. Khachaturyan, A.G. (1983) *Theory of structural transformations in solids*, John Willey & Sons, New York.
7. Bellon, P., Chevalier, J.P., Augarde, E., André, J.P., and Martin, G.P. (1989) Substrate-driven ordering microstructure in $Ga_xIn_{1-x}P$ alloys, J. Appl. Phys. **66**, 2388-2394.

Ti/Pt/Au OHMIC CONTACTS TO P-TYPE InGaP

K. VOGEL, V. MALINA[x], AND P. RESSEL
Ferdinand-Braun-Institut für Höchstfrequenztechnik,
Rudower Chaussee 5, D-12489 Berlin, Germany
[x]Institute of Radio Engineering and Electronics,
Czech Academy of Sciences, Chaberská 57, CZ-182 51
Prague 8, Czech Republic

1. INTRODUCTION

The great interest in InGaP is due to its superior properties for application in a variety of optoelectronic, power
and high-speed electronic devices. Because of its wide direct band-gap (1.9 eV), large valence band offset, and low
amount of deep-level traps, the InGaP can be successfully
used in light-emitting diodes and laser diodes, and has
the potential to replace AlGaAs in HEMT and/or HBT structures.

On the other hand, as a result of the wide energy
gap, careful attention must be paid to the choice of contact metallisations to ensure that reliable, low-resistan-
ce ohmic contacts can be realised. Various metals have been
proposed for ohmic [1] and Schottky [2] contacts to n-type
InGaP. In this contribution, we believe to report for the
first time on the electrical properties and thermal stability of non-alloyed Ti/Pt/Au ohmic contacts to p-type InGaP
epitaxial layers.

2. EXPERIMENTS

Epitaxial layers of p-$In_{0.5}Ga_{0.5}P$ (1/um thick, $3 \times 10^{18} cm^{-3}$)
were grown by MOCVD on SI-(100)GaAs:Cr substrates. Mesa
InGaP structures were formed by a wet-chemical etching

221

J. Novák and A. Schlachetzki (eds.), Heterostructure Epitaxy and Devices, 221–224.

to ensure one dimensional current flow. The deposition of 30nm Ti/40nm Pt/275nm Au contact system was carried out in a Leybold-Heraeus L 560 UV vacuum evaporator by means of an electron beam gun. Right before the deposition, the mesa-etched InGaP structures were treated in diluted HF. In addition, part of the structures was sputtered in-situ with low-energy (50 eV) Ar ions before the metal deposition. TLM contact patterns were then formed by a lift-off technique and annealed at different temperatures from 200 to 700°C for 30 sec in a RTA system under N_2 gas.

3. RESULTS AND DISCUSSION

Figure 1 shows that the specific contact resistance of Ti/Pt/Au contact to the p-$In_{0.5}Ga_{0.5}P$ epitaxial layer lattice-matched to GaAs substrate dropped with increasing temperature of annealing and reached a minimum value of $3-6 \times 10^{-5}$ohm.cm^2 in the temperature range of 500 to 600°C. In this range, the contacts became ohmic while at other

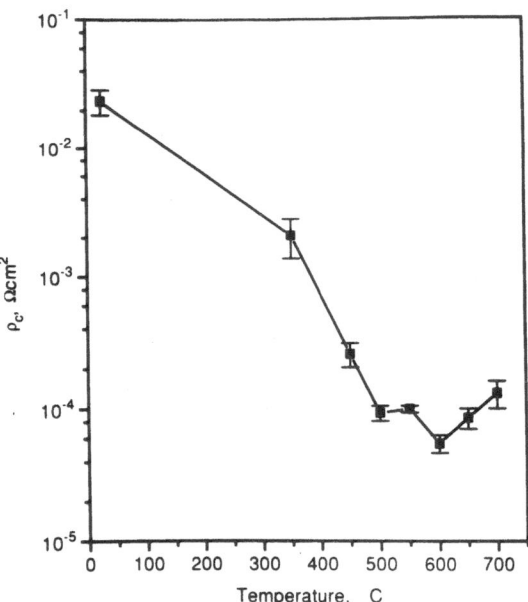

Figure 1. The specific contact resistance as a function of the annealing temperature.

temperatures they showed non-linear I-V characteristics. No marked difference was observed between the in-situ cleaned and the only wet-chemically etched samples.

To examine the long-term stability of contacts at higher temperatures, the contacts originally annealed at an optimum temperature of 500°C were subjected to an aging test at 350 and 400°C for 40 hours in N_2. As can be seen from Figure 2, such contacts exhibited good thermal stability and no increase in specific contact resistance.

The surface morphology of contacts was smooth and gold-like with sharp edge definition. It remained relatively unchanged until a temperature of about 500°C, when

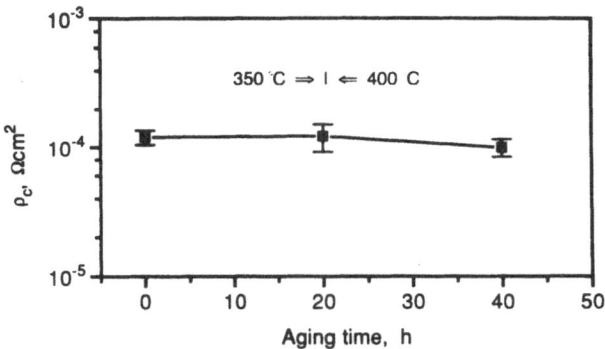

Figure 2. The specific contact resistance of contacts, annealed at 500°C, after the aging at 350 and 400°C.

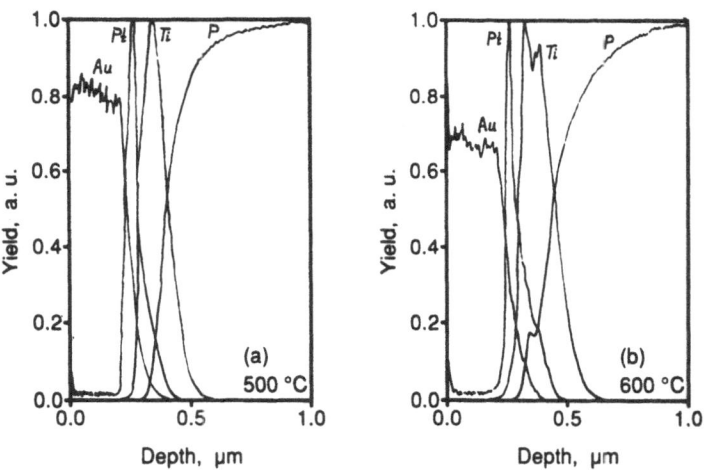

Figure 3. SIMS in-depth profiles of contacts after the annealing (a) at 500°C and (b) at 600°C.

small crystallites, homogeneously distributed over the contact surface, appeared. Practically no changes in the contact morphology were observed after the long-term aging of such contacts at 350 and 400°C. With the increasing temperature of annealing, the microstructure of contacts became more pronounced and the alloying of contacts (the formation of small molten drops) started only at a temperature of about 650°C. At the same time, the character of non-metallised surface of the p-InGaP epitaxial layer did not change and remained smooth and homogeneous in colour.

This remarkable metallurgical stability of Ti/Pt/Au/ p-InGaP system at high temperatures was confirmed by the SIMS in-depth profile measurements of samples annealed at 500 and 600°C. The results of these measurements are shown in Figure 3.

ACKNOWLEDGMENTS

This work was supported by the Bundesministerium für Bildung, Wissenschaft, Forschung und Technologie and in part by the Grant Agency of the Czech Republic under the contract No. 102/94/1059. The authors are grateful to Dr. A. Knauer of the FBH Berlin for the preparation of InGaP epitaxial layers and to A. Umbach of the HHI Berlin for performing part of the Ti/Pt/Au evaporations.

REFERENCES

1. Ren, F., Kuo, J.M., Pearton, S.J., Fullowan, T.R., and Lothian, J.R. (1992) Ohmic contacts to n-type In$_{0.5}$Ga$_{0.5}$P, J. Electron. Mater. 21, 243-247.
2. Chang, E.Y., Lai, Y.L., Lin, K.Ch., Chang, Ch.Y., and Juang, F.Y. (1993) Study of the thermal stability of the Schottky contacts on GaInP grown by LP-MOCVD, in Mat. Res. Soc. Symp. Proc. 282, Materials Research Society, Pittsburgh, pp. 253-258.

COMPARISON OF PHYSICAL PROPERTIES OF BULK CRYSTALS AND EPITAXIAL LAYERS OF GaN

H.TEISSEYRE, M.LESZCZYNSKI, T.SUSKI, P.PERLIN, J.JUN, I.GRZEGORY, M.BOĆKOWSKI, S.POROWSKI
UNIPRESS, High Pressure Research Center, Polish Academy of Science Sokolowska 29/37 01-142 Warsaw Poland
T.D. MOUSTAKAS
MBE Laboratory, Department of Electronic Computer and System Engineering, Boston University, Boston, USA
J.MAJOR
Spectra Diode Labs, 80 Rose Orchard Way, San Jose, Ca., USA
W. GĘBICKI
Institute of Physics Warsaw University of Technology 00-661 Warsaw, Poland

Abstract
The paper presents structural and optical properties of gallium nitride. We examined three kinds of samples: (i) GaN bulk crystals grown at high pressure of 10-15 kbar, (ii) GaN layers grown on sapphire, (iii) GaN layer grown on SiC. We employed various experimental techniques: high resolution X-ray diffraction, optical absorption, photoluminescence and Raman scattering. We found that GaN layer on silicon carbide is fully relaxed and contains rather small density of threading dislocations. GaN layers grown on sapphire are of a much worse crystallographic quality and have a small thermal strain. The optical properties are not strongly influenced by mosaicity of the layers, but rather by the different free-electron concentration.

Introduction

GaN and its continuous alloy systems with AlN and InN are intensively studied because of the applications in two areas: (i) optical devices, green, blue and UV light emitting diodes and lasers, (ii) high temperature and high power devices. This paper compares the crystallographic and optical properties of bulk crystals with those of epitaxial layers grown on sapphire and silicon carbide. The bulk crystals were grown at pressures of 10-15 kbar and temperatures of about 1700 K [1]. The layers were grown by molecular beam epitaxy (MBE) or metalorganic chemical vapour deposition (MOCVD).

J. Novák and A. Schlachetzki (eds.), Heterostructure Epitaxy and Devices, 225–228.
© *1996 Kluwer Academic Publishers.*

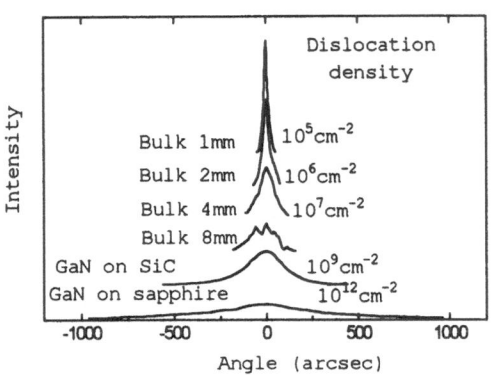

Figure 1. X-ray diffraction rocking curves for different GaN samples

Crystallographic Properties

Rocking Curves

Rocking curves of the (00.4) $CuK_{\alpha1}$ reflection were obtained by using the high-resolution X-ray diffractometer. Figure 1 shows the results for typical samples. The layers grown on sapphire substrate (16% mismatch) are characterized by broad peaks. Their full widths at half-maximum (FWHM) are of 8-15 arcmin. In the case of GaN layer grown on silicon carbide (3.4% lattice mismatch) we obtained the FWHM of 2.5-4.5 arcmin. The Bragg peaks are much narrower for bulk crystals grown at high pressure. For the crystals of size up to 1 mm the rocking curves are of 20-30 arcsec. For bigger samples the rocking curves become broader and for 6-8 mm size crystals we observe small angle (1-3 arcmin.) boundaries separating high quality crystallites of 0.5-2 mm.

Thermal expansion

The measurements of the lattice constants of bulk crystals, layer and substrates were performed by using the Bond method for the set of symmetrical and asymmetrical reflections. The samples were examined in the temperature range 295-750 K. It can be seen in Figure 2 that the lattice constant **c** (perpendicular to the interface for epitaxial layers) is the same for the bulk crystals and GaN layers. The lattice constant **a** is smaller for the layer on sapphire with respect to the bulk crystals and the layer on SiC. This means, that sapphire , which expands stronger with temperature, induces the therml strain. The relaxed lattice constants vary slightly from sample to sample, but this is not relevant for our considerations.

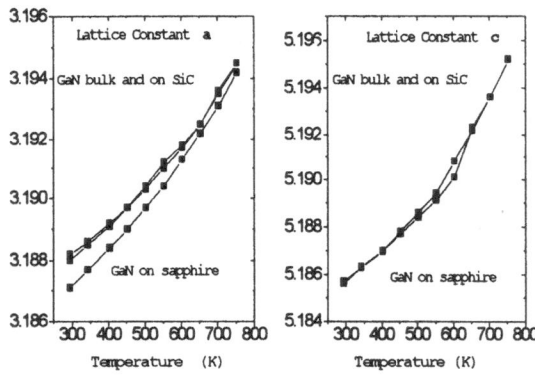

Figure 2. Temperature dependence of GaN lattice constants, a parallel to the interface and c perpendicular to the interface.

Figure 3 shows the lattice mismatch (parallel to the interface) between GaN layer and the sapphire substrate compared with the lattice constant difference between the bulk GaN crystals and sapphire. It can be seen that up to temperature of about 500 K this mismatch does not change. This means, that the layer is rigidly bound to the substrate. Above this temperature the mismatch decreases to the value for the completely relaxed layer (or bulk crystal) at growth temperature (900-1100 K).

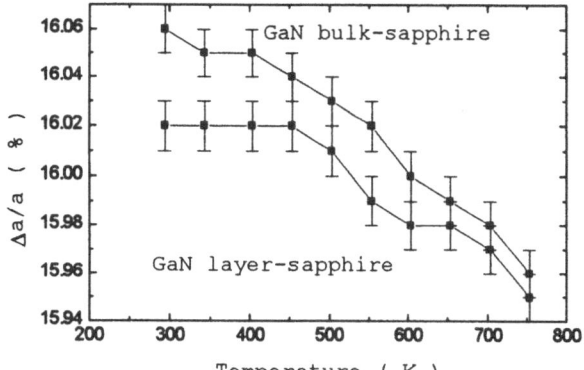

Figure 3. Temperature dependence of the parallel lattice mismatch between the GaN layer and sapphire substrate

Optical Properties

The next step of our analysis was to compare the optical properties of GaN bulk crystals and epitaxial layers. To do that we performed the absorption measurements by using 150 Watt halogen lamp, single grating Spex 500 M spectrometer and GaAs photomultiplier. All obtained spectra were normalized taking into account the spectrum of the lamp. To determine the value of the energy gap (E_g) we assumed that the absorption coefficient α is proportional to the square root of energy (parabolic conduction band approximation)[3]:

$$\alpha = \alpha_0 \sqrt{E - E_g} \qquad (1)$$

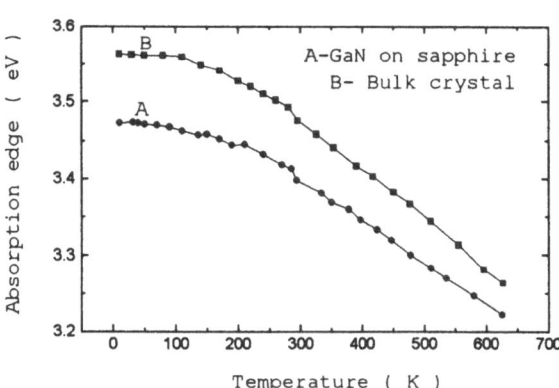

Figure 4. Temperature variation of optical energy gap in different GaN samples

where E is the photon energy and α_0 is an energy-independent parameter. Figure 4 shows the temperature dependence of the energy gap for the GaN bulk crystals and for the MBE layer grown on sapphire. We could not make the absorption experiment on the layer grown on SiC because the substrate material is not transparent in this spectral region. For bulk crystals the absorption edge is shifted to higher energies. For both kinds of samples the energy value corresponding to the absorption edge decreases with temperature, and the temperature coefficient is larger for the bulk samples. The differences in the lattice parameters and thermal expansion cannot explain such a big effect. Instead, we found it necessary to take into account that the optical transitions are from the top of the valence band to the first unoccupied state in the conduction band, i.e. to the state located around the Fermi level (Burstein-Moss effect). The free-electron concentration of the examined layer is of about 6 10^{17} cm^{-3} which causes a negligible Burstein-Moss effect in contrast to the bulk crystals. For these crystals the free-electron concentration is of about 8 10^{19} cm^{-3} (measured by Hall effect). Such a concentration should cause a blue shift of the absorption edge by about

90 meV. The larger negative temperature coefficient for the bulk crystals with respect to the epitaxial layer agrees with the calculated variation of Fermi energy with temperature.

Raman scattering

For the GaN layers grown on sapphire and on SiC the Raman scattering experiments were carried out in $Z(X,-)\bar{Z}$ configuration. In such a case at the center of Brillouin zone two modes are Raman active: $A_1(LO)$ at energy 736 cm^{-1} and E_2(high) at 568 cm^{-1}. In the case of the bulk crystal we performed the experiment also in $X(Y,-)\bar{X}$ geometry here we can see both $A_1(LO)$ and E_2(high) phonons, and in $X(Z,-)\bar{X}$ geometry try with $A_1(LO)$ and $E_1(TO)$ Raman active peaks. The selection rules and the energies of these peaks agree well with similar results obtained on different epitaxial layers[4]. However for our bulk crystals (for $Z(X,-)\bar{Z}$ geometry) $A_1(LO)$ phonon was absent and two LO phonon - plasmon coupled modes appear. The difference between Raman spectra of bulk crystals and epitaxial layers are shown in Figure 5; two branches of the phonon - plasmon mode are clearly visible for the bulk samples nd also for one of our epitaxia sample. For different samples with various free-electron concentrations we found that the high energy mode follows the plasma frequency (established from infrared reflectivity). On the other hand, the lower-frequency mode is shifted-down with respect to the standard $A_1(TO)$ phonon by about 11 cm^{-1} for the bulk samples or shifted-up to 236 cm^{-1} for one of our layers grown on sapphire The changes of the position for two phonon - plasmon branches with carrier concentration agree well with a prediction of the linear response theory for undamped phonon - plasmon modes. Taking into account these results, we expect that Raman scattering can be a contactless tool to investigate the free-carrier concentration up to about 10^{20} cm^{-3}.

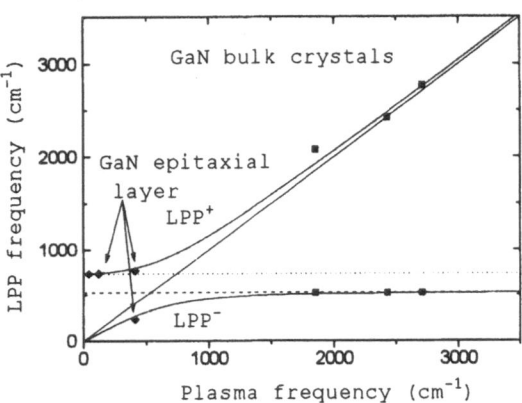

Figure 5 Frequencies of two plasmon-phonon modes as a function of plasma frequence. Experimental points are for different kinds of GaN samples.

References

1. Porowski,S., Grzegory,I., and Jun, J. (1989) *High Pressure Chemical Synthesis* ed Jurczak,J. and Baranowski,B. Amsterdam: Elsevier p 21.
2. Leszczynski,M., Grzegory,I., Bockowski,M., Jun,J., Porowski,S., Jasinski,J., Baranowski,J., Acta.Phys.Polon.1995 (in press)
3. Pankove,J.I. (1971) *Optical Proccesses In Semiconductor,* Prentice Hall,Englewood Cliffs
4. Kozawa,T., Kachi,T., Kano,H., Taga,Y., Hashimoto,M., Koide,N., Manabe,K., (1994) *J.Appl.Phys.* 75 p 1098

MOCVD GROWTH OF GA₁₋ₓINₓASᵧP₁₋ᵧ-GAAS QUANTUM STRUCTURES

M. RAZEGHI

J. HOFF, M. ERDTMANN, S. KIM, D. WU, E. KAAS, C. JELEN,
S. SLIVKEN, I. ELIASHEVICH, J. DIAZ, E. BIGAN. G.J. BROWN*,
S. JAVADPOUR

*Center for Quantum Devices,
Department of Electrical Engineering and Computer Science,
Northwestern University, Evanston, IL 60208 USA*

**Wright Laboratory, Materials Directorate, WL/MLPO, Wright
Patterson AFB, Ohio 45433-7707*

1. Introduction

Razeghi [1] has already demonstrated the feasability of Low Pressure Metalorganic Chemcial Vapor Deposition (LP-MOCVD) technology for the growth of $Ga_xIn_{1-x}As_yP_{1-y}$ - InP heterostructures, quantum wells and superlattices in electronic and optoelectronic applications and, furthermore, has shown that this system is excellent for monolithic integration on silicon substrates. More recently, Razeghi has developed $Ga_xIn_{1-x}As_yP_{1-y}$-GaAs devices by LP-MOCVD as an alternative to those of the $Al_xGa_{1-x}As$ material sysytem [2]. In this review article, we would like to present an overview of the state of the art in $Ga_xIn_{1-x}As_yP_{1-y}$-GaAs devices and to show the MOCVD as a powerful tool for their growth.

2. A brief comparison of $Ga_xIn_{1-x}As_yP_{1-y}$ and $Al_xGa_{1-x}As$

The $Al_xGa_{1-x}As$ system has been a material system of choice for optoelectronic design for many years. An alloy of GaAs (Eg=1.42 eV;a=5.653 Å) and AlAs (Eg≈3.0 eV; a=5.66 Å) [3], $Al_xGa_{1-x}As$ remains lattice matched to GaAs substrates over a wide range of bandgaps eliminating the concern over lattice mismatch. However, there are some aspects of $Al_xGa_{1-x}As$ that are less than desirable. First, $Al_xGa_{1-x}As$ has a high reactivity with oxygen [4][5] mainly due to the presence of Aluminum in the alloy. This oxidation results in processing and reliability problems and often requires extra processing steps necessary to passivate devices. Aluminum oxidation also restricts fabrication methods, such as epitaxial regrowth. Elevated deposition temperatures must be used to get device-quality $Al_xGa_{1-x}As$ material, resulting in undesirable dopant diffusion. Furthermore, the electrical properties of $Al_xGa_{1-x}As$ are affected by oxygen-related defects [6]. Therefore, while $Al_xGa_{1-x}As$ may be capable of producing optoelectronic quality devices, it is not the perfect material system.

229

J. Novák and A. Schlachetzki (eds.), Heterostructure Epitaxy and Devices, 229–245.
© *1996 Kluwer Academic Publishers.*

- It covers the same range of bandgaps as direct bandgap $Al_xGa_{1-x}As$ - from 1.42 eV for the binary GaAs to approximately 1.92 eV for the ternary $Ga_{0.49}In_{0.51}P$ [2].

- Device quality material can be deposited at a lower temperature, reducing undesirable dopant diffusion [2].

- Extremely high electron mobilities have been demonstrated in two-dimensional $GaAs$-$Ga_{0.49}In_{0.51}P$ heterostructures [7].

- Dislocation and impurity motion is much lower than in $Al_xGa_{1-x}As$ because of the large difference in atomic size between Indium and Gallium. This facilitates growth on silicon substrates and, therefore, future monolithic optoelectronic integration [2].

- Surface and interface recombination velocities are an order of magnitude lower than for $Al_xGa_{1-x}As$, translating into improved device performance[8].

- The lack of aluminum in the system greatly reduces the reactivity with oxygen which, in turn, improves electrical parameters and facilitates device processing especially for epitaxial regrowth [2].

- The high-field transport properties of $Ga_xIn_{1-x}P$ are improved over those of $Al_xGa_{1-x}As$ for $x \geq 0.3$ [5].

- Processing and fabrication techniques are aided by the ready availability of selective etchants especially for certain devices such as Master Oscillator Power Amplifier (MOPA) [1].

3. MOCVD growth technique

The growths in this paper were all carried out at a pressure of 76 Torr. Trimethylindium (TMIn)and Triethylgallium (TEGa) were the group III indium and gallium sources, respectively. Gaseous Arsine (AsH_3) and Phosphine (PH_3) were the group V arsenic and phosphorous sources. Diethylzinc and Silane provided the Zn (p-type) and Si (n-type) dopants. Hydrogen (H_2) was used as the carrier gas. The growth conditions necessary for the entire $Ga_xIn_{1-x}As_yP_{1-y}$ system lattice matched to GaAs are summarized in Table 1.

Table 1: MOCVD growth conditions for the growth of $Ga_xIn_{1-x}As_yP_{1-y}$/GaAs at the bandgap 1.42-1.92 eV

Growth Pressure	P	76 Torr
Growth Temperature	T	510°C
Mole Flow Rate (Mol/min)	TEGa	$6.6-9.0 \times 10^{-6}$
	TMIn	$2.2-5.9 \times 10^{-6}$
	AsH_3	$0-4.5 \times 10^{-4}$
	PH_3	$0-3.0 \times 10^{-3}$

Both semi-insulating (Cr doped) and n+ (Si doped) (001) oriented GaAs substrates were used during growth. All substrates had a 2° misorientation toward the (110) direction.

We used X-ray diffraction to optimize the growth of each of the different types of superlattices - the binary/ternary, the binary/quaternary, the quaternary/ternary, and the quaternary/quaternary. It was [9] determined that a symmetric (002) diffraction pattern and weak (004) superlattice satellite patterns indicate that there is no strain at either interface. Further, the greater the degree of symmetry in the (002) scan, the less integrated strain is present at the interfaces. Finally, strain in the GaInP layer was found to be distinguishable from strain at the interfaces by an asymmetric (002) scan with no significant superlattice satellite pattern in the (004) profile but a single significant peak in the (004) pattern away from the substrate peak indicating lattice mismatch between the substrate and the GaInP layers.

4. QWIPs

Quantum Well Infrared Photodetectors or QWIPs are engineered long wavelength infrared (LWIR) photodetectors. Rather than using the traditional interband transitions in narrow bandgap semiconductors like HgCdTe, QWIPs achieve narrow bandgap-like performance through intersubband transition in quantum wells formed from large bandgap III-V materials [10]. Figure 1 shows such a structure. A III-V semiconductor like GaAs (1.42 eV @ 300K) is sandwiched between two larger

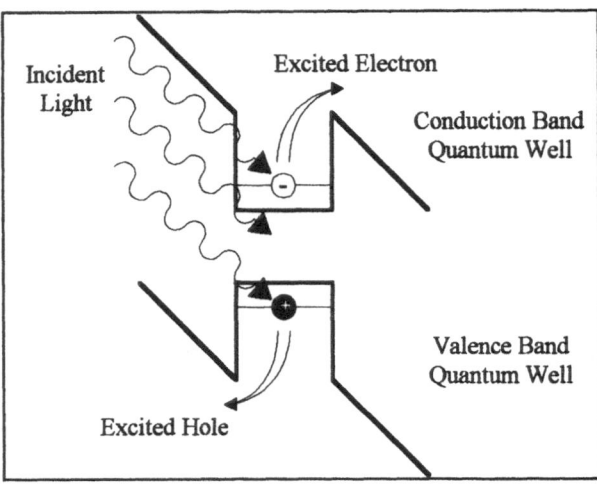

Figure 1: An illustration of conduction and valence
band quantum wells under external bias and
illumination.

bandgap materials like GaInP (1.9 eV @300K). Quantum wells are formed in the GaAs material for both the conduction and valence bands giving the designer the choice of doping the wells n-type (to use the conduction band well) or p-type (to use the valence band well). Photoresponse is achieved when incident light excites a bound charge carrier out of the quantum well and into a conducting continuum state. Once in the continuum, an applied electric field moves the charge carrier causing a photocurrent.

The cutoff wavelength of a quantum well photodetector is determined by its well width and barrier height [10] through a straightforward application of the Schrodinger Equation. The barrier height itself is determined by the band offset of the two materials used to form the well.

The choice of dopant is particularly important to quantum well infrared photodetector design. First, conduction band or n-type quantum wells cannot respond to normal incidence light due to polarization dependent quantum mechanical selection rules [10]. Grating structures or polished facets must be used with these devices to scatter normal incidence light into polarizations perpendicular to the quantum well planes. Valence band or p-type quantum wells, on the other hand allow normal incidence photoexcitation because band mixing overcomes the polarization dependent selection rules. Second but no less important, the effective mass of the charge carrier and the barrier height it experiences are not the same for both wells. Most of the work thus far in QWIPs has been with n-type devices since the lower effective mass is expected to give greater responsivity. We, on the other hand, have concentrated on p-type QWIPs since the inherent normal incidence detection and reduced dark current due to the higher effective mass of the carrier may ultimately produce better devices [11].

QWIPs are generally grown with a 25 to 50 period superlattice sandwiched between thick GaAs layers which act as top and bottom contacts. In all of our devices, the top and bottom contacts were grown to 0.5 μm and 1.0 μm respectively. Finally, in all of our devices, all of the wells and the top and bottom contacts were doped to the same level. The barriers were left undoped.

A large number of superlattices were grown for the purpose of fine tuning the cutoff wavelength of the quantum well detectors. In general, there are four types of QWIPs which can be grown from the $Ga_xIn_{1-x}As_yP_{1-y}$ system. The simplest is the binary-ternary superlattice GaAs/GaInP. Slightly more complicated are the binary-quaternary and quaternary-ternary types. The binary-quaternary type uses GaAs as the well material and then the barrier material is chosen to give the desired barrier height. The quaternary-ternary uses GaInP as its barrier material and then the well material is chosen to give the desired barrier height. Finally, the most complicated $Ga_xIn_{1-x}As_yP_{1-y}$ superlattice is the quaternary-quaternary type in which the barrier height is chosen, and then two quaternary materials are selected to meet that requirement. All four types of QWIPs have been grown and have demonstrated photoresponse.

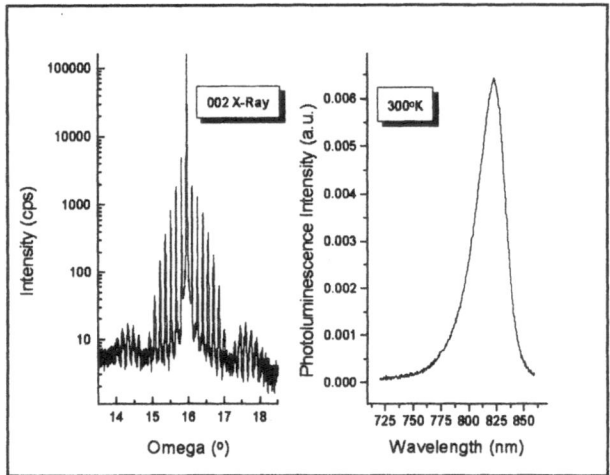

Figure 2: X-ray and photoluminescence of a GaAs/GaInP superlattice

4.1 BINARY/TERNARY QWIPS

Figure 2 shows the (002) X-ray and the room temperature photoluminescence of a typical 50 period GaAs/GaInP QWIPs grown with (nominally) 35 Å wide wells and 300 Å wide barriers . As indicated in Section 3, the high degree of symmetry in the (002) X-ray scan is indicative of a lack of strain at the interfaces. The (004) scan (not shown) showed a weak superlattice satellite pattern and no lattice mismatch between the barrier layers and the substrate which confirms the lattice quality. The room temperature photoluminescence is also where it is expected to be by a simple Kronig-

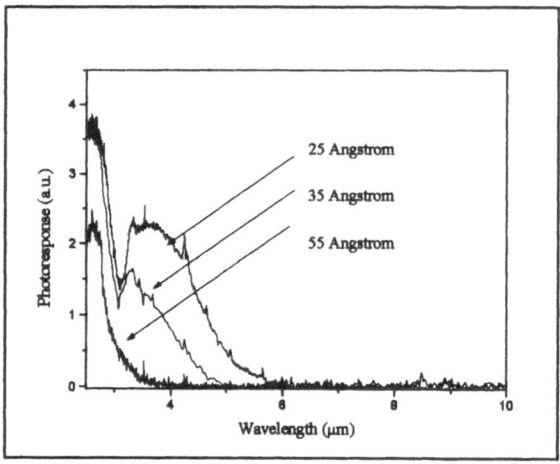

Figure 3: Photoresponses for the three different well widths in binary/ternary superlattices

Penney analysis.

A series of such binary/ternary devices were grown by LP-MOCVD all with 50 periods, 300 Å wide barriers, and wells doped 3×10^{18} cm^{-3}. The only differences among the three devices were the well widths which were 25 Å, 35 Å, and 55 Å. The normal incidence photoresponses of these devices are shown in Figure 3. Each of these devices was background limited up to sample temperatures of 120 K.

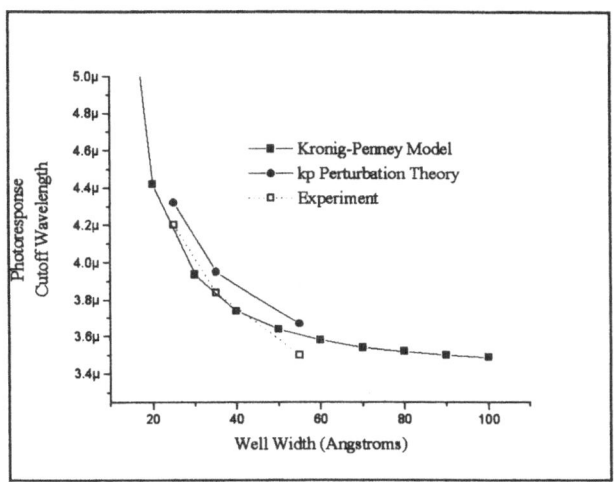

Figure 4: Experimental and theoretically predicted cutoff wavelengths for GaAs/GaInP QWIPs

The cutoff wavelengths of these photodetectors are short (3-5 μm) due to the large depth of the GaAs/GaInP valence band quantum well. These cutoff wavelength have been modeled first by a simple Kronig-Penney model and again by an 8x8 Kane Hamiltonian [12][13]. Figure 4 shows the excellent fit between theory and experiment. However, it also shows that these binary/ternary superlattices are incapable of LWIR photodetection without making the well widths extremely small. Therefore, it is essential that the barrier height be reduced to increase the cutoff wavelength. This is the motivation for growing the binary/quaternary and quaternary/ternary superlattices.

4.2 BINARY/QUATERNARY QWIPS

In this device, the quaternary material was chosen to reduce the bandgap of the barrier by approximately 100 meV under that of GaInP. A 50 period superlattice was grown with 30 Å wide wells and 300 Å wide barriers. The doping level was set to 3×10^{18} cm^{-3}. Figure 5 shows the (002) X-ray and the room temperature photoluminescence of this GaAs/GaInAsP QWIP. Both the X-ray scans and the photoluminescence again indicated non-strained superlattices of high quality.

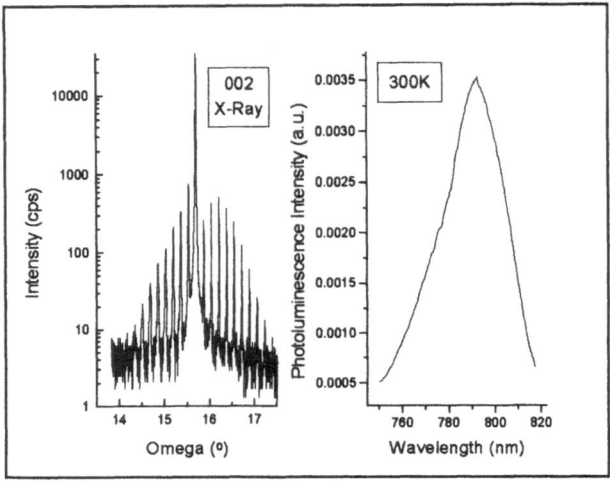

Figure 5: The x-ray and photoluminescence of a binary/quaternary QWIP

Figure 6 shows the spectral scan of the binary/quaternary detector. The photoresponse of this QWIP cutoff at a wavelength of between 6 and 7 μm. Again, the device was background limited up to 120 K sample temperature.

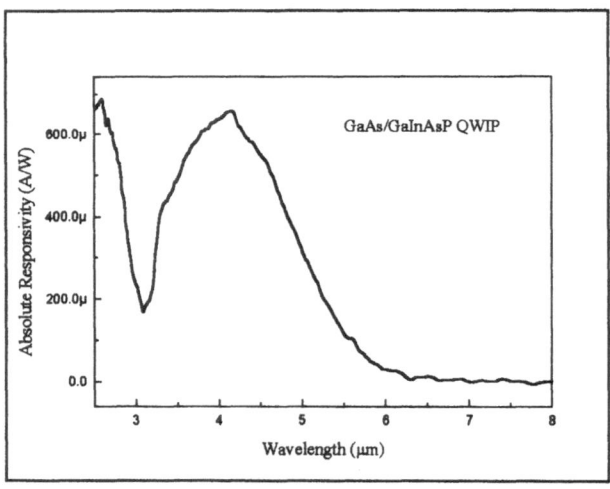

Figure 6: The photoresponse spectra of the binary/quaternary QWIP

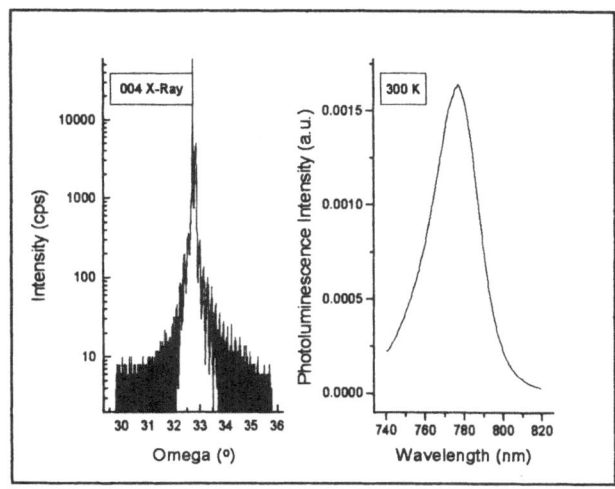

Figure 7: The (004) x-ray and photoluminescence of the quaternary/ternary QWIP.

4.3 QUATERNARY/TERNARY QWIPS

In this device, the quaternary material was chosen to increase the bandgap of the well by approximately 100 meV over that of GaAs. Again, a 50 period superlattice was grown with 30 Å wide wells and 300 Å wide barriers, and the doping level was again set at 3×10^{18} cm^{-3}. Figure 7 shows the (004) X-ray and the room temperature photoluminescence of this GaAs/GaInAsP QWIP. Both the X-ray scans and the photoluminescence again indicated high quality superlattices.

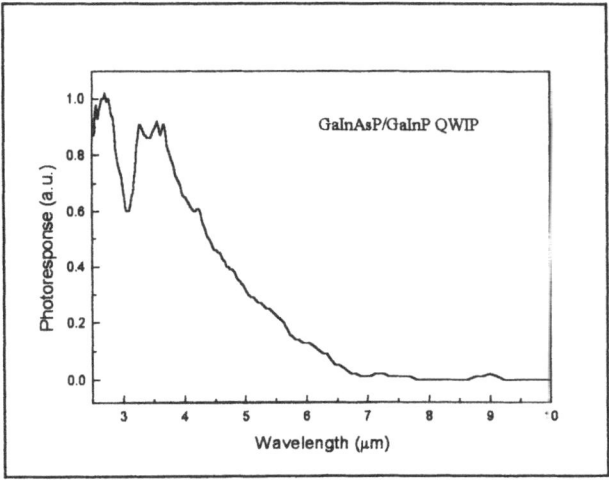

Figure 8: The photoresponse spectra of the quaternary/ternary QWIP

In this device, like the binary quaternary QWIP, the photoresponse cutoff wavelength was extended to between 6 and 7 μm and the device was background limited up to 120 K sample temperature.

The similarity between the cutoff wavelength of these two materials is important. Decreasing the bandgap of the barrier material by 100 meV increases the cutoff wavelength of a QWIP by the same amount as increasing the bandgap of the well material by 100 meV. This implicitly states that the band offset ratio is a constant in the entire quaternary material system lattice matched to GaAs. This result makes it possible to predict the materials necessary to yield 8-12 μm LWIR detectors.

5. High-power laser diodes based on InGaAsP/GaAs at 808nm and 980 nm

There is perhaps no better example of the significance of the $Ga_xIn_{1-x}As_yP_{1-y}$ material system as a replacement for $Al_xGa_{1-x}As$ than in high power lasers. Until recently, the majority of commercially produced diode lasers were based on $Al_xGa_{1-x}As$. However, aluminum oxidized easily during both fabrication and high-power operation [5][11]. Defects known as "dark lines" and "dark spots" spread through the lasers due to structural defects. [14]. The Al-free $In_{1-x}Ga_xAs_yP_{1-y}$ /GaAs system overcome these problems. It has been shown that the low non-radiative recombination rate and high device performance because it has the advantage such as selective etching and regrowth, which is essential for laser diode fabrication. The laser structure based on various $Ga_xIn_{1-x}As_yP_{1-y}$ compounds with $0.51 < x < 1$ and $0 < y < 1$, lattice matched to GaAs , which cover the same range of direct band-gap energy as AlGaAs (1.42eV-1.92eV) ,were grown and fabricated emitting at 808nm and 980nm.

5.1 GROWTH AND CHARACTERIZATION OF 808 NM LASER STRUCTURE.

The advanced separate-confinement heterostructure quantum well laser structure (fig 9), which is undoped quaternary active layer and waveguide sandwiched with Si and Zn doped InGaP cladding layer were grown and fabricated. Due to the very thin active region and quantum size effects, the energy distribution of the carriers can be narrower in this structure. The growth conditions for this structure are the same as mentioned above.

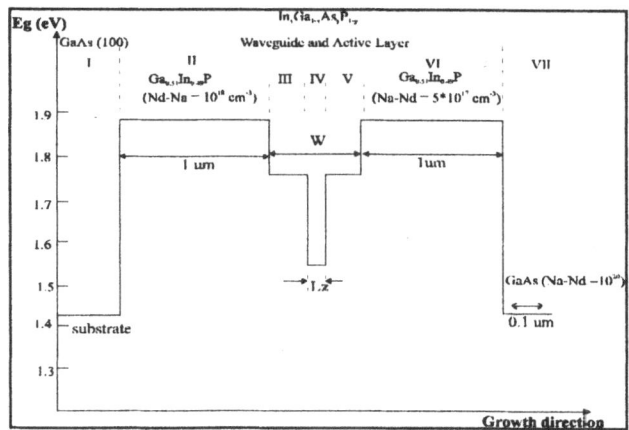

Figure 9: Schematic energy band diagram for the $Ga_xIn_{1-x}As_yP_{1-y}$ based laser structure

The typical X-ray diffraction spectrum of the epitaxial layers shows the good lattice match with $\Delta a/a = 0.0006$ range. The FWHM of 24.5 arcsec for InGaP and 25.4 arcsec for GaInAsP shows high material quality with sharp interfaces.

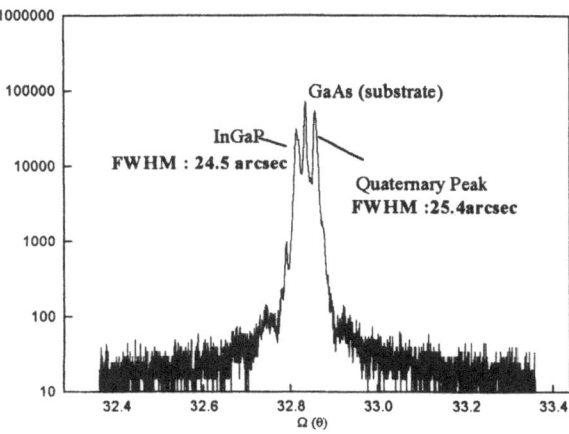

Figure 10: (004) High resolution X-ray diffraction spectrum for 808nm laser structrue grown by MOCVD

The optical properties of the grown material were characterized with photoluminescence response. The spontaneous emission from the active region is dominant as shown below. The room temperature FWHM is achieved as narrow as 26 meV which indicates the high quality of the structure.

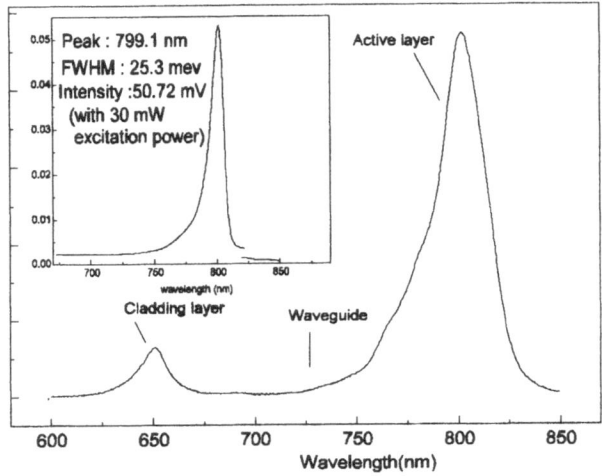

Figure 11: Photoluminescence spectrum for 808nm laser structure grown by MOCVD. The FWHM is 25.3 meV from the $Ga_xIn_{1-x}As_yP_{1-y}$ active layer.(see insert).

5.2 GROWTH AND CHARACTERIZATION OF 980 NM LASER STRUCTURE.

The strained InGaP/ GaAs/ InGaAs multiquantum well laser structures on GaAs substrate were grown by low pressure MOCVD. The growth condition is already described. .In lattice mismatched $In_xGa_{1-x}As$ on GaAs, the internal compressive strain makes the determination of optimum growth conditions critical. The x-ray diffraction spectrum shows very sharp and well resolved peaks reveals high quality strained quantum wells. The photoluminescence spectrum shows that the active region energy gap is 1.27eV which is shifted due to the biaxial elastic strain and to the quantum size effects. The FWHM of 13 meV at room temperature is the best ever reported.

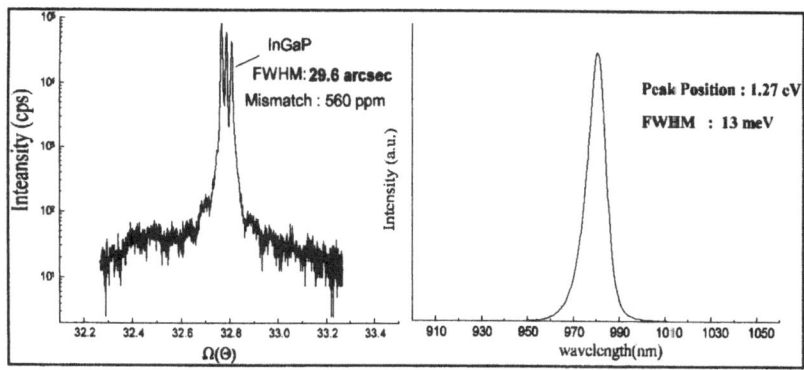

Figure 12: The X-ray diffraction and photoluminescence spectrum from 980 nm laser structure grown by MOCVD. The FWHM luminescence of 13 meV is the narrowest ever reported.

5.3 HIGH POWER LASER DIODE.

The 808nm and 980nm laser diodes were fabricated using standard photolithography and chemical etching. The threshold current density for 808nm laser was obtained as low as 220A cm^{-2} with differential efficiencies (η_d) as high as 1.3 WA^{-1} The beam divergence which is important for the optical pumping sources is as narrow as 26° which is significantly less than for AlGaAs/GaAs lasers.

The lifetesting of 808 nm laser diode with uncoated facets at 40°C over 6000 hours did not show any degradation which can be comparable with AlGaAs/GaAs lasers.

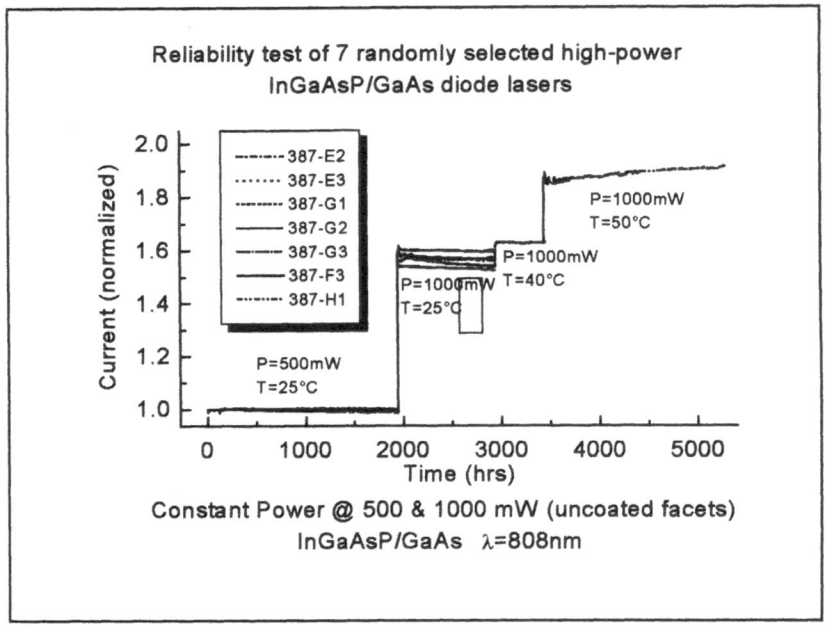

Figure 13: High power laser reliability tests

Reliable high-power lasers emitting at 980nm were fabricated. Threshold current densities as low as 80 A cm^{-2} and differential efficiencies as high as 75% were observed for 1-mm-long coated lasers. *To* ,the characteristic temperature reached 350 K which is the highest ever reported. These impressive results are shown in Table 2.

Table 2: Comparison of AlGaAs and aluminum-free GaInAsP high power laser diodes (aperture 100 μm) (references in superscript)

	$Al_xGa_{1-x}As$	$Ga_xIn_{1-x}As_yP_{1-y}$
λ (nm)	800	800
J_{th} (A/cm^2)	230 [15]-400 [16]	240-500
η_d (W/A)	1.3 [15]-0.8 [16]	1.3
T_0 (°C)	130-160 [16]	170-175
Series Resistance (Ω)	0.25 [16]	0.1
Thermal Resistance (K/W)	10 [14]	~1

Transverse beam divergence	32-48°	26°
Catastrophic Optical Damage limit for uncoated facets (MW/cm^2)	0.5-1 [15]	6
Lifetime under 1 MW/cm^2 (hours)	200 (uncoated facets)	165000 (~200 years) (uncoated facets
Degradation under 1 W (uncoated facets)	immediate	No degradation after 2000 hrs at 500 mW + 1000 hrs at 1W + 500 hrs at 1W, 40°C; test continues at 50°C.

6. Conclusions

The $Ga_xIn_{1-x}As_yP_{1-y}$ material system lattice matched to GaAs has been shown to be a viable alternative to $Al_xGa_{1-x}As$. The material system has better device and fabrication characteristics. Moreover, the growth of these materials by LP-MOCVD has matured to the point that highly successful and reproducible results have been obtained for both lasers and quantum well infrared photodetectors.

Future work in this area should be twofold. First, researchers should continue to demonstrate $Ga_xIn_{1-x}As_yP_{1-y}$ devices as replacements for $Al_xGa_{1-x}As$ devices. Second and more importantly, the transfer of this knowledge to industry should be increased greatly to allow commercial users to benefit from the performance and reliability improvements of this material system.

7. Acknowledgment

The authors would like to acknowledge Dr. Gerald Witt from the US Air Force Office of Scientific Research (AFOSR) as well as Dr. L. N. Durvasula and Dr. H. O. Everitt for their engouragement.

244

The laser work was supported by ARPA/ARO Contract No. DAAH04-93-G-0044. The QWIP work was supported by Air Force Contract No. F33615-93-C5382 throught Kopin Corporation.

8. References

[1] Razeghi, M. (1995) *The MOCVD Challenge, Vol 1: A Survey of GaInAsP-InP for Photonic and Electronic Device Applications*, Adam Hilger, Bristol

[2] Razeghi, M. (1995) *The MOCVD Challenge, Vol 2: A Survey of GaInAsP-GaAs for Photonic and Electronic Device Applications*, Institute of Physics Publishing, Bristol

[3] Landolt-Börnstein (1982), *Numerical Data and Functional Relationships in Science and Technology, Group III, Vol. 17, "Semiconductors", Subvolume a, "Physics of Group IV Elements and III-V Compounds"*, Springer-Verlag, Berlin

[4] Razeghi M., (1994) High-power laser diodes based on InGaAsP, *Nature* **369**, 631-633

[5] Besikci, C. and Razeghi, M. Electron Transport Properties of $Ga_{0.51}In_{0.49}P$ for Device Applications, *IEEE Trans. Elect. Dev.*, **41**, 1066-1069 (1994)

[6] K.L. Tsai, C. P. Lee, K. H. Chang, D. C. Liu, H. R. Chen, and J. S. Tsang, (1994), Asymmetric dark current in quantum well infrared photodetectors, *Appl. Phys. Lett.* **64**, 2436-2438

[7] M. Razeghi, M. Defour, F. Omnes, M. Dobers, J.P. Vieren, and Y. Guldner, (1989), Extremely high electron mobility in a GaAs/ $Ga_xIn_{1-x}P$ heterostructure grown by Metalorganic Chemical Vapor Deposition, *Appl. Phys. Lett.*, **55**, 278-280

[8] Fukuda, M. (1986), *J. Appl. Phys.* **59**, 4172-4175

[9] He, X. Erdtmann, M. Williams, R. Kim, S. And Razeghi, M. (1994), Correlation between x-ray diffraction patterns and strain distribution inside GaInP/GaAs superlattices, *Appl. Phys. Lett.*, **65**, 2812-2814

[10] B.F.Levine, (1993).Quantum Well Infrared Photodetectors, *J. Appl.Phys.* **74**, R1-R81

[11] Hoff, J., Kim, S., Erdtmann, M. Williams, R. Piotrowski, J. Bigan, E. Brown, G.J. and Razeghi, M. (1995), Background limited performance in p-doped GaAs/ $Ga_{0.71}In_{0.29}As_{0.39}P_{0.61}$ quantum well infrared photodetectors, *Appl. Phys. Lett*, **67**, 22-24

[12] E.O. Kane (1982), Energy Band Theory,*Handbook on Semiconductors*, Vol. 1, W. Paul, Ed. Amsterdam, North Holland, 1982, p193

[13] F. Szmulowicz and G. Brown (1995) Calculation of resonant absorption and photoresponse measurement in p-type GaAs/AlGaAs quantum wells, *Appl. Phys. Lett*, **66**, 1659-61

[14] Razeghi, M. (1995) InGaAsP-based High Power Laser Diodes, *Optics and Photonics News*, **8**, 16-22

[15] Fukuda, M. (1991) *Reliability and Degradation of Semiconductor Lasers and LEDs*, Artech House, Boston

[16] Larsson, A. (1986) High Efficiency broad area single quantum well lasers with narrow single-lobed far-field pattern prepared by MBE, *Electon. Lett.* **22**, 79

[16] Spectra Diode Labs, 1994 Product Catalog

INP–BASED HBT : PRINCIPLE, DESIGN AND TECHNOLOGY

J.L. PELOUARD
*Laboratoire de Microstructures et de Microélectronique
(L2M – CNRS)
196 Avenue H. Ravéra – 92220 Bagneux – France.*

1. Introduction to the bipolar story.

In bipolar transistors the main current is carried by electrons injected into the base by the emitter–base junction. There is a transistor effect if the collector current is nearly equal to the emitter current. Unfortunately, electron injection (J_n) into the base accompanies hole injection (J_p) into the emitter creating a parasitic current. Because the current densities through the homojunction of a Bipolar Junction Transistor (BJT) are given by the Schockley formula for short base diodes, the injection ratio (J_p/J_n) is proportional to the ratio of base (N_A) to emitter (N_D) doping levels. In order to obtain an injection ratio small enough for significant current gain, the BJT must have a much higher doping level in the emitter than in the base. This has various well known negative consequences on the behavior of the transistor as Early effect, small f_{max}, etc. As a result the BJT design is a compromise between the current gain and the speed of the transistor.

In the late 1940's W. Schockley[1] proposed to use of a wide band gap material in the emitter in order to reduce the injected hole current. Later H. Kroemer[2] published the first paper to show the promising advantages of this new concept : the heterojunction bipolar transistor (HBT). In this device the energy discontinuities in emitter–base heterojunction introduces an additional term to the injection ratio equation. In first approximation it writes :

$$\left(\frac{J_p}{J_n}\right)_{HBT} = \left(\frac{J_p}{J_n}\right)_{BJT} \exp\left(\frac{-\Delta\epsilon_v}{k_B T}\right) \tag{1}$$

By contrast of the AlGaAs/GaAs heterojunction, InP–based heterojunctions exhibit large values of the exponential term of eqn. 1. In these conditions high base doping levels can be used keeping practicable values for the current gain. In addition the abrupt heterojunction strongly modifies the electron transport in the device. In this paper the electron transport in the InP–based HBTs and its consequences on both the static and dynamic behaviors will be review. To overcome limitations in currently used HBTs, a new structure will be described and characterized.

J. Novák and A. Schlachetzki (eds.), Heterostructure Epitaxy and Devices, 247–256.
© 1996 *Kluwer Academic Publishers.*

2. Electron transport in InP–based HBTs.

2. 1. ELECTRON INJECTION BY AN ABRUPT HETEROJUNCTION.

When an electron is crossing an abrupt heterojunction, its total energy and its parallel component are conserved. Assuming the heterojunction is abrupt and the bulk properties are valid up to the interface, the perpendicular component of the wave vector ($k_{2\perp}$) is

$$k_{2\perp}^2 = k_{1\perp}^2 + k_0^2 + k_1^2 \left\{ \frac{m_2^*[1 + \alpha_2(\epsilon_1 + \Delta\epsilon_C)]}{m_2^*(1 + \alpha_1\epsilon_1)} \right\} \qquad (2)$$

where the index 1 and 2 are the initial and final materials, respectively, ϵ_i is the kinetic energy, $k_{i\parallel}$ is the parallel component of the wave vector, α_i is the non–parabolicity factor and $k_0^2 = (2m_2^* \Delta\epsilon_C)/\hbar^2 [1 + \alpha_2(\epsilon_1 + \Delta\epsilon_C)]$.

In order to first show the influence of these boundary conditions on the electron population crossing the heterojunction, flat band conditions (no space charge layer) on both sides are assumed (inset Fig. 1). From an electron population at equilibrium in the large band gap material (curve 1 Fig. 1) the velocity distributions of the injected electrons in the small band gap materials (curves 2 Fig. 1) are calculated using eqn. 2. Because of the high value of the k_0 term in eqn. 2, the perpendicular component of the velocity vector is shifted to the high positive values. The width of the distribution is reduced because of both the quadratic dispersion relationship and the reduction of the effective mass. All these velocity distributions are far from equilibrium, even for a conduction band discontinuity as small as $\Delta\epsilon_C$=0.12eV. As a result, the injected electrons in the base develop a strongly forward–oriented, quasi–monkinetic population.

Because the main interaction in the range of the injected energy is the emission of a polar optic phonon which is highly non–isotropic and inelastic, the probability for the injected electrons to get back to the emitter is negligible. These conditions correspond to those described by Bethe[5] for thermionic current in Schottky diodes. Monte–Carlo simulations[6] have shown that this theory is applicable to the case of semiconductor heterojunctions.

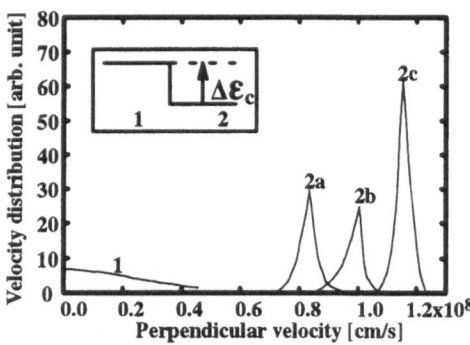

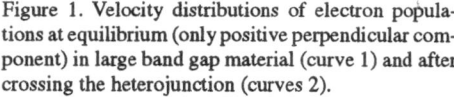

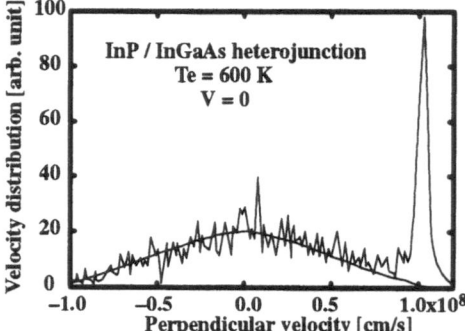

Figure 1. Velocity distributions of electron populations at equilibrium (only positive perpendicular component) in large band gap material (curve 1) and after crossing the heterojunction (curves 2).

Figure 2. Distributions of the perpendicular component of the electron velocity in the base at 5nm from emitter–base heterojunction.

2. 2. ELECTRON TRANSPORT IN THE BASE.

In BJTs and HBTs with a compositionally–graded interface, the diffusion current through the emitter–base junction leads to an excess of minority electron density in the base. Because the reverse biased base–collector junction maintains the electron density at the equilibrium level near the edge of the space charge layer, there is an electron density gradient in the base which creates a diffusion current. In HBTs with an abrupt emitter–base heterojunction, the injected electron population is far from equilibrium with high mean velocity (Fig. 1). The effect of any concentration gradient on such population is negligible. Thus, a correct description of electron transport in the base must take into account the initial conditions and the relaxation of both the energy and momentum of the hot electron population injected into the base.

The velocity distribution of the electron population has been computed by Monte–Carlo simulation[7] into a InP/InGaAs diode. In the base, at 5nm from the emitter–base interface the distribution (Fig. 2) can be split in two parts : 1) a narrow peak centered around a high mean velocity and 2) a broad peak centered around a very low mean velocity. This behavior suggests that some electrons (narrow peak) have experienced no interactions or else interactions with a small deviation angle leading to a ballistic or quasi–ballistic transport, and some other electrons (broad peak) have experienced one or more isotropic interactions leading to a velocity–relaxed population for which the transport is dominated by the diffusive component. In order to obtain an analytical description of the evolution of these electron populations in the base a simple model has been developed.

Lets consider only two types of interactions. The first one put together the isotropic interactions which randomize the orientation of the velocity vector (probability λ_{iso}). In the second one, there are the non–isotropic interactions which induce a small deviation angle keeping the same velocity distribution (probability λ_{n-iso}). The flux evolution, between x and x+dx, of the electron population which has undergone "i" non–isotropic interactions is :

$$\phi_i(x + dx) = \phi_i(x) - \phi_i(x)\frac{\lambda_{iso} + \lambda_{n-iso}}{v_i(x)}dx + \phi_{i-1}(x)\frac{\lambda_{iso}}{v_i(x)}dx \qquad (3)$$

where v_i is the velocity of the quasi–ballistic population i. In order to simplify the following calculations a constant velocity is assumed for all i populations. Because the velocity varies quadratically with the energy, this assumption leads to an error of less than a few percents. The integration of Eqn. 3 gives the evolution of the current density carried by each of the i quasi–ballistic populations (Fig. 3) :

$$J_i(x) = \frac{J_0(0)}{i!}\left(\frac{x\lambda_{n-iso}}{V_{QB}}\right)^i \exp\left[\frac{-x(\lambda_{iso} + \lambda_{n-iso})}{V_{QB}}\right] \qquad (4)$$

After integration, the total current density due to the quasi–ballistic electrons is :

$$J_{QB}(x) = \sum_{i=0}^{\infty} J_i(x) = J_0(0) \exp\left(\frac{-x}{\Lambda}\right) \qquad (5)$$

where $\Lambda = V_{QB}/\lambda_{iso}$. The quasi–ballistic current decreases exponentially with the distance from the launching heterojunction. The characteristic length (Λ) of this decay depends only on the non–isotropic interaction probability and is always longer than the mean free path. Because the mesured and calculated Λ values[8] are in a accessible range for device processing, the ballistic and quasi–ballistic transport in the base reduces the base transit time in InP–based HBTs.

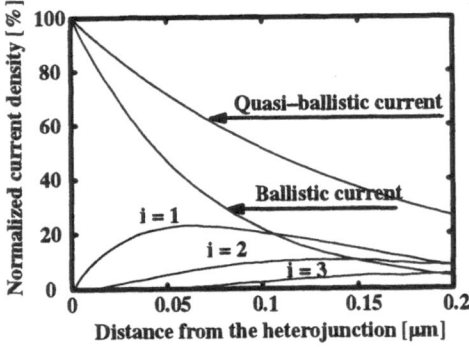 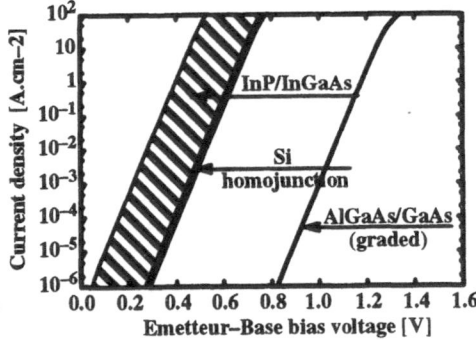

Figure 3. Contribution of the total electron current in the base due to ballistic (i=0) and quasi–ballistic (i=1,2,3) electrons.

Figure 4. Emitter current versus emitter–base bias voltage for InP/InGaAs, AlGaAs/GaAs and Si junctions.

2. 3. ELECTRON COLLECTION BY THE BASE–COLLECTOR JUNCTION.

2. 3. 1. Collection conditions.

In the reverse biased base–collector junction, the electric field is high (>100kV/cm), then the electrons entering in the space charge layer (SCL) are strongly accelerated and have a ballistic transport where they obtain a very high mean velocity ($\sim 10^8$ cm/s). During this ballistic flight the electrons increase their kinetic energy up to a range of energy where the transfer to the lateral valleys is the predominant interaction. Behind that point, the mean electron velocity reaches very rapidly the saturation velocity ($\sim 10^7$ cm/s). In addition, the electrons in the lateral valleys can significantly transfer back to the Γ–valley only in the low electric field region[9]. Thus the electrons stay at the saturation velocity until they reach the n^+ collector region. At that point they transfer back to Γ–valley by means of isotropic interactions. Thus only one half of the population is directly collected, the other part goes back toward the base. These last electrons reduce their energy and velocity against the electric field. Finally, they are accelerated toward the collector and collected. This builds a trapped domain at the collector edge and corresponds to an additional delay in the transit time of the base–collector SCL. Introducing new potential profiles in the space charge layer, many attempts[10] have been done to reduce this transit time without real success. A new structure where the base–collector transit time is reduced, will be presented in the last section of this paper.

2. 3. 2. Collection by an abrupt heterojunction.

In order to reduce the leakage currents and the avalanche processes many authors have used a base–collector heterojunction. This symmetric structure is, in addition, well suited for ECL or I2L integration because it can be used in both collector–up and emitter–up modes. However the heterojunction spike in the conduction band provides a selective collection of electrons with high enough perpendicular velocities to cross over the barrier. This effect has been used in different types of transistors[11] to demonstrate the existence of ballistic and quasi–ballistic transport through the active region. Because this selective collection introduces both a dependence of the current gain on the base–collector voltage and an additional power dissipation, it has to be avoided in the HBT design. Because compositionally–graded materials are difficult to grow in the InP–based system, various structures[12] have been proposed including the introduction of spacers, delta–doped plans or composite material in the base–collector junction. Even with reasonable collection conditions such structures exhibit parasitic as excess of leakage current or charge trapping in the heterojunction notch.

3. Static behavior of InP–based HBTs.

In this section the static behavior of the InP–based HBT will be revisited taking into account the non–stationary electron transport describe in the previous section. The main static feature in the bipolar transistor is the current gain. In a common base configuration this gain is given by : $a_0 = \gamma BM$ where γ is the injection efficiency, B the transport factor and M the multiplication factor in the base collector SCL.

3. 1. THERMIONIC CURRENT.

As it has been shown in section 2. 1. the current through an abrupt heterojunction is a thermionic current described by :

$$J(V_{BE}) = A^* T^2 \exp\left(\frac{-q\phi_B}{k_B T}\right)\left[\exp\left(\frac{-qV_{BE}}{nk_B T}\right) - 1\right] \qquad (6)$$

where A^* is the modified Richardson's constant and ϕ_B is the equilibrium barrier height:

$$\phi_B = \Delta\epsilon_C + E_{g2} + k_B T \ln\left(\frac{N_A N_D}{N_{C_1} N_{V_2}}\right) \qquad (7)$$

where E_{g2} is the band gap of the base material and N_D and N_A (N_{C_1} and N_{V_2}) are the doping levels (density of states) of the emitter and the base respectively. Also n is a coefficient which takes into account the electrostatic potential distribution between the base and the emitter. In the complete depletion approximation, n is given by : $n = 1 + (\epsilon_1 N_D)/(\epsilon_2 N_A)$. Thus the conduction band discontinuity introduces two disadvantages in the injected current behavior : 1) reduction of the saturation current and 2) introduction of a non–ideality factor larger than 1. In order to estimate the influence of these effects, the injected current is plotted (Fig. 4) versus the emitter–base bias voltage for various heterojunctions. Because the heterojunction is not always perfectly abrupt and because tunnel effect and image force reduces the effective barrier height, current through InP/InGaAs junctions have been calculated for both abrupt and graded heterojunctions. It can be seen that the abrupt InP/InGaAs heterojunction has a turn–on voltage smaller by a factor of 2 than that of the fully graded AlGaAs/GaAs and is comparable to that of the Si homojunction.

3. 2. TRANSPORT FACTOR.

Because the recombination probability for hot electrons is lower than that of thermal ones and because the transit time of ballistic and quasi–ballistic electrons (~ 0.1ps) is much shorter than their lifetime (>10ps), the recombination of the quasi–ballistic electrons can be neglected, and the transport factor is given by :

$$B = 1 - \frac{W_B^2}{2L_D^2}f \qquad (8)$$

where $\qquad f = 1 + \frac{2\Lambda}{W_B}\left\{\exp\left(\frac{-W_B}{\Lambda}\right) + \frac{\Lambda}{W_B}\left[\exp\left(\frac{-W_B}{\Lambda}\right) - 1\right]\right\} \qquad (9)$

It can be noticed that for a base width of one third of the quasi–ballistic length the current gain limitation due to the transport factor is improved by a factor of 5. On the other hand, because the velocity distribution of the injected electrons is strongly forward–oriented, the ratio of injected electrons into the extrinsic base is smaller than for a graded hetero-junction where the injection is isotropic. This improvement is significant when the lateral dimensions are reduced in order to improve the dynamic behavior.

4. Dynamic behavior of InP–based HBTs.

The HBT dynamic behavior is mainly characterized by two typical frequencies. First, the cutoff frequency (f_t) at which the dynamic current gain is equal to one :

$$f_t = \frac{1}{2\pi\tau_{EC}} \tag{10}$$

where $\tau_{EC} = \tau_E + \tau_C + t_B + t_{BC}$ is the total emitter–to–collector transit time. τ_E and τ_C are the charging times for the emitter–base and base–collector junctions, respectively, and t_B and t_{BC} are the transit times in the base and base–collector SCL, respectively. The other characteristic frequency is the maximum oscillation frequency (f_{max}) at which the dynamic power gain is equal to one :

$$f_{max} = \sqrt{\frac{f_t}{8\pi R_B C_C}} \tag{11}$$

where R_B is the base resistance and C_C the collector capacitance.

In order to increase the values of f_t and f_{max} the reduction of any parasitic effect has been intensively studied through the reduction of contact resistances[13], the reduction of base resistance[14], the reduction of collector capacitance[15], by device scaling[16] and by the development of self–aligned technology[17]. Parasitic effects have been reduced so effectively that delays due to electron transport ($t_B + t_{BC}$) are now the dominant limita-tions. Let's consider, as an example the InP/InGaAs HBT done at Fujitsu[18] which is one of the most advanced HBT ever reported, exhibiting both f_t and f_{max} larger than 160 GHz. The delay due to electron transport represents about 60% of the total emitter–to–collector transit time τ_{EC}. On the other hand, ballistic and quasi–ballistic transport in the base make the base transit time (t_B) as short as 0.1ps[8] for a base width of 500nm. In this conditions the transit time in the collector SCL (t_{BC}) is about one half of the total delay in these tran-sistors. It is clear that a further improvement of the dynamic behavior means the reduction of the transit time (t_{BC}) in the base–collector SCL. In the following we present a transistor structure based on a Schottky collector where the dynamic behavior can be significantly improved by reducing t_{BC}.

4. 1. COLLECTION BY A SCHOTTKY JUNCTION..

Metal–semiconductor junction (Schottky junction) has already proposed to be used, as opposed to a pn junction, to form the collector in HBTs[19]. Main arguments have been about the charging time and minority carrier charge storage in the collector under satura-tion conditions. Here we propose to use a Schottky barrier as a collector for the following reasons. (1) In a Schottky junction the maximum value of the electric field is located at the metal–semiconductor interface, that is at the end of the electron path in the SCL. By contrast this maximum is located close to the base in pn junctions because the doping level is much higher in the base than in the collector SCL. In these conditions there is no low electric field region in SCL and the base–collector transit time is reduced. (2) The electrons are collected by the metal directly from the lateral valleys of the semiconductor. Because they experience, in the metal, a lot of inelastic interactions with the thermal electrons, they have no chance to get back–scattered toward the semiconductor. In these conditions there is no trapped domain at the collector edge and the base–collector transit time is again reduced. (3) Because the metal resistivity is smaller than that of semiconductor, the collector access resistance (R_C) is smaller with a Schottky barrier than with a pn junction as collector. (4) Because it is difficult to grow semiconductor on metal, the Schottky barrier used as a collector is on top of the structure (collector–up HBT). Thus the base–collector capacitance is smaller than that of emitter–up HBT due to the smaller junction area. In these conditions, the scaling trade–off about the collector SCL thickness –base–collector transit time (t_{BC}) versus the base–collector capacitance (τ_C)– is shifted toward the thinner SCLs, reducing the base–collector transit time. Because the collector access resistance has also been reduced the shift is even larger. (5) Because leakage currents are smaller in Schottky diodes than in InGaAs homojunctions[20] and because there is no spike in the conduction band blocking the electron transport, Schottky junction provides good collection conditions as an pn homojunction and high breakdown voltage as for a pn heterojunction. (6) Finally the metal can be used as a sink for the excess of energy due to electron relaxation in the collector layer. Because HBTs are fast when they drive large current densities and because the thermal conductivity of the metal is much higher than that of semiconductor, it possible to build circuits with a much higher integration density using HBT with Schottky collector than with semiconductor collector. For the same reason HBT with Schottky collector is suited for power applications.

4. 2. COLLECTOR–UP HBT WITH A SCHOTTKY JUNCTION (MHBT).

Introducing a semiconductor–metal junction in the HBT structure, a new design has been developed in order to take advantage, as described in the previous section, of the Schottky junction without introducing new parasitic. This design is based on six main points. (1) To reduce lateral parasitic features (resistances and capacitances), a fully self–aligned process has been developed. The whole transistor (except pad connections) is done with only one masking level. (2) To reduce the base transit time, thin base layers (down to 30 nm) must be used. (3) To reduce the access resistance, base ohmic contacts are done on thick (>0.15 μm) extrinsic base layer which has been highly doped by Mg implantation. (4) To reduce the capacitance due to the extrinsic part of the emitter–base junction, emitter layer is under–etched using selective wet etching. This under cut gives the same area for both emitter–base and base–collector junctions. (5) To reduce extrinsic capacitance, collector and base are connected to pads by air–bridges.

4. 2. 1. Process.
A schematic cross section view of the MHBT is shown on Fig. 5. The top layer is done by a deposition of Tungsten which is used as a collector. Tungsten has been chosen because of its electrical and thermal qualities and because it may be used as a mask for ion implantation. There are two components in the base layer : the intrinsic part which is defined at the epitaxy and the extrinsic part which is thicker than the intrinsic one and is highly doped by Mg implantation. The former part is thick enough (>0.15 mm) to have both good mechanical strength after the under cut of the InP layer and low contribution to the base access resistance. The under cut of the InGaAs layer allows to line up the inner

254

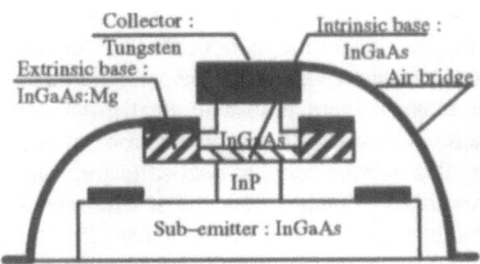

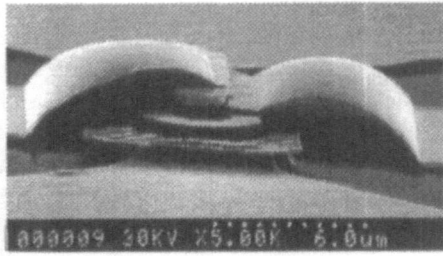

Figure 5. Schematic cross–section view of the MHBT.

Figure 6. SEM view of the MHBT.

edge of the base contact with the edge of the Tungsten layer. Electric isolation of these contacts is provided by the thickness of the base–collector transition region.

The extrinsic region of the emitter–base junction is removed by selective under cut of the InP layer. A good control of the etching process allows to equalize emitter–base and base–collector junction area in order to minimize extrinsic capacitances. Additionally this under–cut removes also damages in the emitter–base junction due to base contact annealing and ion implantation annealing. The discontinuity between the base layer and the emitter layer ensures both alignment of the emitter edge contact with the outer base edge and electrical isolation of these two layers. The whole process described up to now is done using only one masking level. An isolation mesa is etched around each transistor. Finally base contacts and collector contacts are connected to pads via air–bridges in order to reduce extrinsic capacitances. After process, SEM observations (Fig. 6) have shown the efficiency of this self–aligned process which has been entirely done by conventional technics such as 400 nm lithography, wet etching, ion milling, etc.

4. 2. 2. Demonstrator

In order to demonstrate, the feasibility of the MHBT fabrication a simplified structure has been processed. Using a base thick enough (0.15 µm) the Mg implantation in the extrinsic base has been avoided. Also the base–collector transition layer has been chosen thick enough (0.45 µm) in order to reduce short cuts between base and collector contacts. Because the expected values of the base transit time and base–collector transit time are about 1 ps and 1.2ps respectively, the vertical scaling of this structure is not optimized. However fast behavior has been measured, demonstrating the big potential of the MHBT.

Epitaxial layers were grown by gaz source molecular beam epitaxy (GSMBE) at the Tampere University. The structure consisted of a 500–nm–thick InGaAs sub–emitter ($n=3x10^{19}$ cm^{-3}), a 300–nm–thick InP emitter ($n=1x10^{19}$ cm^{-3}), a 100–nm–thick InP emitter ($n=1x10^{17}$ cm^{-3}), a 150–nm–thick InGaAs base ($p=1x10^{19}$ cm^{-3}), a 450–nm–thick undoped InGaAs ($n=5x10^{15}$ cm^{-3}). A 5–nm–thick undoped InGaAs spacer has been introduced between the base and the emitter. The dopants for p– and n–type were Be and Si, respectively.

4. 2. 3. Static characterization of MHBT.

Typical Gummel plot is shown in Fig. 7. From the statistical study based on a large number of transistors (~ 200), it has been shown a good agreement of both saturation current val-

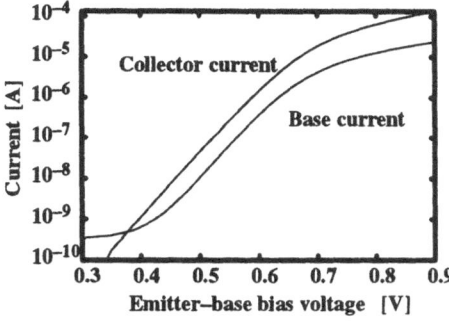

Figure 7. Typical Gummel plot of MHBT.

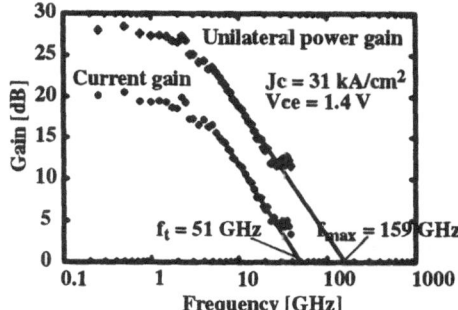

Figure 8. Frequency dependence of the current gain and unilateral power gain of MHBT

ues and ideality factor of the collector current with the thermionic theory. No significant values can be found for a component current proportional to $\exp(V_{BE}/2k_BT)$. Thus recombination current in the n–InP/p–InGaAs heterojunction is negligible when the InGaAs layer is grown on the InP layer as it has already been reported[21] for InP on InGaAs. As a result the current gain is almost constant over more than 8 decades of the collector current. The W/p–InGaAs junction used as a collector junction exhibits a very small leakage current (1nA at $V_{BC}=-10V$) in agreement with previous results[20]. The leakage reduction observed in Schottky junctions is attributed to a smaller generation current than those observed in pn homojunctions. As a result, Ic–Vce characteristics have small output conductance up to more than 3V.

4. 2. 4. *Dynamic characterization of MHBT.*
The current gain, h_{21}, and unilateral power gain U of a typical 0.5x50 μm^2 are plotted in Fig. 8 as a function of frequency. The extrapolated values of the cutoff frequency f_t from h_{21} and the maximum oscillation frequency f_{max} from U are 51 GHz and 159 GHz, respectively. The f_t value is mainly limited by transport delays in both the base ($t_B=1ps$) and the base–collector SCL ($t_{BC}=1.2ps$). However the f_{max} value is, for this non–optimized structure, in the range of the best values reported so far[18,22]. In addition the effective delay time $(R_CC_{BC})_{eff} = f_t/(8\pi f_{max}^2)$ is only 78 fs, which is smallest value ever reported for any bipolar transistor. These results demonstrate the feasibility of the MHBT technology. They also indicate a promising potential for ultra–fast behavior, reducing the base–collector transit time with a metallic collector.

5. Conclusions
The non–stationary electron transport generated by the conduction band discontinuity in the abrupt emitter–base heterojunction strongly modifies the device behavior of the InP–based HBTs. The electron current injected in the base is a thermionic current. Ballistic and quasi–ballistic currents in the base reduce the base transit time, leading to higher values of both the current gain and the cutoff frequency. In order to reduce the transit time in the base–collector SCL which represents more than one half of the total emitter–to–collector delay time in the most advanced InP–HBTs, a new HBT structure, based on metallic collector (MHBT), has been proposed. A non–optimized demonstrator has been realized and characterized, exhibiting a very high speed behavior. The results indicate a promising potential for an improvement by a factor of 2 or 3 of the best performances to date.

256

6. References

1. Schokley W. U.S. Patent 2569347 filed June 26, 1948 and issue Spetember 26, 1951.
2. Kroemer H. (1957) Theory of a wide–gap emitter for transistors, *Proc. IRE* **45**(11), 1535.
3. Hayes J.R., Capasso F., Malik R.J., Gossard A.C., and Wiegmann W. (1983) Optimum emitter grading for heterojunction bipolar transistors, *Appl. Phys. Lett.* **43**, 949.
4. Marty A., Rey G., and Bailbe J.P. (1979) Electrical behavior of an npn GaAlAs/GaAs heterojunction transistor, *Solid–St. Electron.*, **22**(6), 549.
5. Bethe B.A. (1942) M.I.T. Radiation Lab. Repport 43/12.
6. Pelouard J.L., Castagné R., and Hesto P. (1987) Monte–Carlo study of ballistic transport in heterojunction bipolar transistors and in high electron mobility transistors, SPIE proceedings **795**, 41.
7. Hesto P., Pône J.F., Mouis M., Pelouard J.L., and Castagné R. (1985) Monte–Carlo modelling of semiconductor device, Proceedings of NASECODE IV Conference (Dublin, Ireland), 315.
8. Levi A.F.J. (1988) *Electron. Lett.* **24**, 1273. Bardyszewski W. and Yevick D. (1989) *Appl. Phys. Lett.* **54**, 837. Baquedano J.A., Levi A.F.J. and Jalali B. (1994) Forward delay in scaled AlInAs/InGaAs heterojunction bipolar transistors, *Appl. Phys. Lett.*, **63**(16), 2231.
9. Baranger H.U., Pelouard J.L., Pône J.F., and Castagné R. (1987) Ballistic peaks in the distribution function from intervalley transfer in a submicron structure, *Appl. Phys. Lett.*, **51**(21), 1708.
10. Tomizawa K. and Pavlidis D. (1990) Transport equation approach for heterojunction bipolar transistor, *IEEE Trans Electron Devices*, **37**(3), 519. Morizuka K. et al (1988) Electron space–charge effects on high–frequency performance of AlGaAs/GaAs HBT under high current density, *IEEE Electron Device Lett.*, **9**(11), 570.
11. Heiblum M., Nathan M.I., Thomas D.C., and Knoedler C.M. (1985) Direct observation of ballistic transport in GaAs, *Phys. Rev. Lett.*, **55**(20), 2200. Berthold K., Levi A.F.J., Walker J., and Malik R.J. Extreme nonequilibrium transport in heterojunction bipolar transistors, *Appl. Phys. Lett.*, **52**(26), 2247.
12. Su L.M., Grote N.G., Kaumanns R., and Schroeter H (1985) N–n–p–N double heterojunction bipolar transistor on InGaAsP/InP, *Appl. Phys. Lett.*, **47**, 28. Pelouard J.L., Hesto P., Praseuth J.P., and Goldstein L. (1986) Double heterojunction GaAlInAs/GaInAs bipolar transistor grown by molecular beam epitaxy, *IEEE Electron Device Lett.*, **EDL–7**(9), 516. Kurishima K. et al. (1993) *Appl. Phys. Lett.*, **62**(19), 2372. Feygenson A. et al. (1993) Proceeding of the 5th conference on InP and related materials, 572.
13. Rao M.A. et al. (1987) An (AlGa)As/GaAs heterojunction bipolar transistor with non–alloyed graded–gap ohmic contacts to the base and the emitter, *IEEE Electron Device Lett.*, **EDL–8**(1), 30. Nagata K. (1987) Improved AlGaAs/GaAs HBT performance by InGaAs emitter cap layer, *Electron. Lett.* **23**(11), 566.
14. Schuitemaker P. et al. (1986) InP/InGaAs double heterostructure bipolar transistors grown by MBE, *Electron. Lett.* **22**(15), 781. Fisher R. and Morkoc H. (1986) Reduction of extrinsic base resistance in GaAs/AlGaAs heterojunction bipolar transistor, *IEEE Electron Device Lett.*, **EDL–7**(6), 359.
15. Lee W. and Fonstad C.G. (1987) Application of O+ implantation in inverted InGaAs/InAlAs heterojunction bipolar transistors, *IEEE Electron Device Lett.*, **EDL–8**(5), 217. Fukano H. et al. (1988) High–speed InAlAs/InGaAs heterojunction bipolar transistors, *IEEE Electron Device Lett.*, **EDL–9**(6), 312.
16. Nottenburg R.N. at al. (1988) High–current–gain submicrometer InGaAs/InP heterostructure bipolar transistor *IEEE Electron Device Lett.*, **EDL–9**(10), 524. Hayama N. (1987) Submicrometer fully self–aligned AlGaAs/GaAs heterojunction bipolar transistor, *IEEE Electron Device Lett.*, **EDL–8**(5), 246.
17. Chang M.F. et al. (1986) GaAs/(GaAl)As heterojunction bipolar transistors using a self–aligned substitutional emitter process, *IEEE Electron Device Lett.*, **EDL–7**(1), 8. Mishra U. et al. (1989), *IEEE Electron Device Lett.*, **EDL–10**(10), 467.
18. Shigematsu H. et al. (1995) Ultrahigh ft and fmax new self–alignement InP/InGaAs HBT's with a highly Be–doped base layer grwon by ALE/MOCVD, *IEEE Electron Device Lett.*, **EDL–16**(2), 55.
19. Emeis M. and Beneking H., (1985) Fabrication of widegap–emitter Schottky–collector transistor using GaInAs/InP, *Electron. Lett.* **21**(1), 85. Vleck J.C. and Fonstad C.G. (1986) InGaAs/InAlAs heterojunction Schottky transistors grown by MBE, *Electron. Lett.* **22**(20), 1088.
20. Pelouard J.L., Matine N., Pardo F., Sachelarie D., and Benchimol J.L. (1993) Fully self–aligned InP/InGaAs heterojunction bipolar transistors grown by chemical beam epitaxy with a Schottky collector, Proceeding of the 5th conference on InP and related materials,
21. Nottenburg R.N. et al. (1987) High–speed InGaAs/InP heterojunction bipolar transistors, *IEEE Electron Device Lett.*, **EDL–8**(6), 282.
22. Chau H.F. et al. (1993) High–speed InP/InGaAs heterojunction bipolar transistors, *IEEE Electron Device Lett.*, **EDL–14**(8), 388.

TECHNOLOGY AND PROPERTIES OF ALUMINIUM-FREE PSEUDOMORPHIC HEMTs BASED ON InP/InGaAs STRUCTURES

P. KORDOŠ

Institut für Schicht- und Ionentechnik,
Forschungszentrum Jülich (KFA),
D-52425 Jülich, Germany

Abstract. Pseudomorphic high-electron-mobility transistors based on III-V heterostructures show superior high-frequency and low-noice properties. InAlAs/InGaAs-based pHEMTs grown on InP substrates exhibit highest current gain cutoff frequency measured on three-terminal device. However, their performance needs to be improved in order to establish wide application of pHEMTs. One possibility towards this goal could be the preparation of pHEMTs without using an Al-containing multilayered heterostructure. From this reason Al-free InP/InGaAs pHEMTs have been studied. (p)InGaAs or (p)InP barrier enhancement layers can be used instead of InAlAs layer and φ_B up to 0.68 eV can be obtained. Optimized pseudomorphic InP/InGaAs 2-DEG structures grown by LP MOCVD exhibit channel conductivities comparable to conventional InAlAs/InGaAs structures. On InP/InGaAs pHEMTs with 0.25 µm gate length (no T-gates have been used) the cutoff frequencies f_T = 85 GHz and f_{max} = 180 GHz have been evaluated from the high-frequency S-parameter measurements. These are only slightly lower values than reported on InAlAs/InGaAs pHEMTs and an improvement of their properties can be expected on T-gate devices. Pseudomorphic InP/InGaAs structures are suitable for monolithic integration with double-Schottky barrier capacitors and high speed MSM photodetectors.

1. Introduction

The mobility enhancement in modulation-doped heterostructures has given birth to a new class of high-frequency and low-noice devices called the High Electron Mobility Transistor (HEMT) and Modulation-Doped Field-Effect Transistor (MODFET) [1]. Lattice-matched AlGaAs/GaAs and InAlAs/InGaAs/InP heterostructures with two-dimensional electron gas (2DEG) are used for HEMT preparation. GaAs-based HEMTs offer better combination of performance and

257

J. Novák and A. Schlachetzki (eds.), Heterostructure Epitaxy and Devices, 257–266.
© *1996 Kluwer Academic Publishers.*

cost and have now replaced the GaAs FETs for a wide range of microwave and millimeterwave device and circuit applications. InP-based HEMTs, on the other hand, offer higher cutoff frequencies and lower noice figures.

Recently, extensive research and development of more advanced HEMT technology, based on pseudomorphic layers, began. Lattice-mismatched heterostructures, AlGaAs/InGaAs/GaAs or InAlAs/InGaAs/InP, with "pseudomorphically" grown strained InGaAs channel, are used. This approach results in further mobility enhancement (due to the lower band-gap of strained channel) and better high-frequency performance of pseudomorphic HEMTs in comparison with lattice matched HEMTs. On 50 nm self-aligned-gate InAlAs/InGaAs/InP pHEMT the highest current gain cutoff frequency reported up to now on three-terminal device, $f_T = 340$ GHz, has been demonstrated (however with power gain cutoff frequency $f_{max} = 250$ GHz $< f_T$) [2].

The research activity on HEMTs is today directed toward the solution of various problems which hinder their wide application. Among them are high gate leakage current, high DC output conductance, low breakdown voltage and kink in the I_d–V_d characteristics, as well as poorer gate-recess etching control and shorter long-term stability [3]. Practically all these limitations of AlGaAs/GaAs and InAlAs/InGaAs HEMTs can be contributed to aluminium-containing barrier and buffer layers. It is known that InAlAs layers exhibit high concentration of deep traps and are highly oxidizing and that does not exist a selective etchant for InAlAs and InGaAs. One possibility to solve these problems might be the preparation of HEMTs without using an Al-containing layer [4]. However, the Schottky barrier height of InGaAs can be enhanced by InAlAs interlayers very efficient and the conduction band discontinuity at the InAlAs/InGaAs heterointerface is high, $\Delta E_c = 0.55$ eV.

In this study results concerning the technology and properties of InP/InGaAs pseudomorphic HEMTs, prepared on Al–free multilayer structures on InP substrates, are presented. At first, the Schottky barrier height enhancement by thin (p)InGaAs, (n)InP and (p)InP layers is described. Further, results obtained at the optimization of the channel conductivity in InP/InGaAs 2-DEG structures are shown. Finally, DC and high frequency properties of prepared HEMTs are demonstrated and compared with conventional InAlAs/InGaAs pHEMTs.

2. Schottky Barrier Enhancement on (n)InGaAs

It is well known that the Schottky barrier height on (n)InGaAs is only 0.2 eV. Barrier enhancement is therefore needed for preparation of good gate contacts [5]. InAlAs "barrier" layers, which enable to reach values of $\varphi_B = 0.68$ eV, are commonly used at the HEMT preparation.

We have studied the Schottky barrier enhancement on (n)InGaAs using various Al-free layers like lattice matched (p)InGaAs and InP, and lattice mismatched GaAs and InGaP [6,7]. The barrier enhancement by (p)InGaAs layers increases with the layer thickness and dopant density, respectively, and barrier heights up to 0.68 eV have been measured. The barrier enhancement can be described by the two-carrier model. InP surface layers can be used as an alternative to enhance the barrier on (n)InGaAs. The barrier height increases with the (n)InP layer thickness up to 0.55 eV, i.e. up to values reported on (n)InP Schottky diodes. Additional barrier enhancement can be achieved by acceptor doping of the InP surface layer and barrier heights of 0.66 eV have been obtained by 30 nm thick (p)InP enhancement layer on (n)InGaAs. Barrier heights measured on (n)InGaAs with (p)InGaAs, (n)InP and (p)InP enahncement layers of different thickness are shown in Fig. 1.

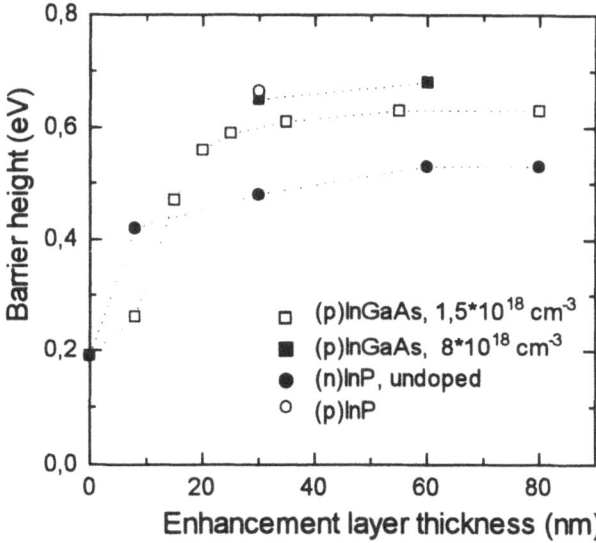

Figure 1. Schottky barrier height enhancement on (n)InGaAs by (p$^+$)InGaAs, (n)InP and (p)InP layers of different thicknesses.

On structures with barrier enhanced (n)GaAs layers, a remarkable decrease of the reverse current density is observed if the layer thickness is reduced to the critical layer thickness, but the barrier height is very low due to the small (n)GaAs thickness. For structures with slightly lattice-mismatched (n)InGaP ($x_{GaP} = 0.11$) measured barrier heights are similar to those for (n)InP enhancement layers of the same thickness.

3. Channel Conductivity Optimization in InP/InGaAs 2-DEG Structures

One important feature of high frequency operated HEMTs is the 2-DEG channel conductivity $G_{ch} = e^*n_s^*\mu_n$ which should be high as possible. Both, the carrier mobility (due to the lower ΔE_g of the channel) and the carrier density (due to the higher ΔE_c at the channel/spacer interface) can be enhanced by pseudomorphic 2-DEG structures. On the other hand, the critical layer thickness of strained channel which depends on the layer composition, and the quality of spacer/channel heterointerface are the limiting factors in improvement of the channel conductivity.

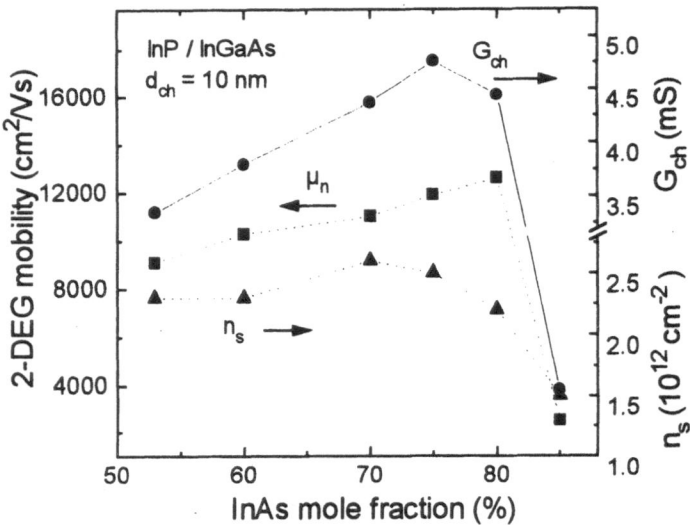

Figure 2. Electron mobility, sheet carrier density and channel conductivity in 2-DEG InP/InGaAs structures as a function of the channel composition.

Chough et al. [8] have studied the influence of the channel composition (d_{ch} = 10 nm) on the performance of MBE grown InAlAs/InGaAs HEMTs. The carrier mobility of 9500 cm^2/Vs and sheet carrier density of 2.5*10^{12} cm^{-2}, i.e. G_{ch} = 3.8 mS, have been obtained on lattice matched structures. The mobility increases continuously up to 12800 cm^2/Vs (G_{ch} = 5 mS) with increased x_{InAs} up to 0.85 and maximal DC transconductance is observed on this structure. However, highest channel conductivity (G_{ch} = 5.9 mS) and highest current-gain cutoff frequency f_T are obtained on structures with x_{InAs} = 0.77. This illustrates the role of the channel conductivity optimization on the HF performance of HEMTs.

In contradiction to InAlAs/InGaAs, the conduction band discontinuity at the spacer/channel InP/InGaAs interface is much smaller, ΔE_c ~ 0.3 eV, which indicates that lower n_s values can be obtained. Channel conductivity optimization is therefore more important in this case. Figure 2 presents our results obtained on LP MOCVD grown InP/InGaAs 2-DEG structures (d_{ch} = 10 nm) with different channel composition [9]. The carrier mobility increases similarly as in InAlAs/InGaAs structures, from 9100 to 12600 cm^2/Vs if x_{InAs} increases from 0.53 to 0.8. The sheet carrier density exhibit a maximum of 2.6*10^{12} cm^{-2} at x_{InAs}

= 0.7, but the channel conductivity is maximal at x_{InAs} = 0.75. Further optimization of the layer structure by decreasing the channel thickness to 8 nm and the spacer thickness to x nm an increase of the channel conductivity has been obtained. Resulting value of G_{ch} = 5.5 mS is comparable to this obtained on InAlAs/InGaAs 2-DEG structures. However, Nguyen's structures with record f_T value exhibit channel conductivity of 7.3 mS [2].

4. Preparation of InP/InGaAs HEMTs

The layer structures for HEMTs are prepared by low pressure MO CVD growth on semi-insulating InP substrates. The channel consists of 8 nm thick strained InGaAs layer with x_{InAs} = 0.7-0.8. 10 nm thick InP carrier supplying layers and 7 nm thick InP spacers are situated below and above the channel. On the top of this 2-DEG multilayer a barrier enhancement double-layer, which consists of 15 nm undoped InP and 15 nm (p)InP, and a 30 nm thick (n^+)InGaAs contact layer are grown.

contact	(n)InGaAs	30 nm	$n = 6*10^{18}$ cm^{-3}
barrier	(p)InP	15 nm	$p = 1.5*10^{18}$ cm^{-3}
barrier	(n^o)InP	15 nm	
carrier	(n)InP	10 nm	$n = 2.7*10^{18}$ cm^{-3}
spacer	(n^o)InP	7 nm	
channel	(n)InGaAs	8 nm	
spacer	(n^o)InP	7 nm	
carrier	(n)InP	10 nm	$n = 1.5*10^{18}$ cm^{-3}
buffer	(n^o)InP	300 nm	
substrate	SI InP		

Figure 3. Layer sequence of InP/InGaAs HEMT.

Device fabrication consists of standard optical lithography, metal evaporation and lift-off processes. As a first step mesa etching is performed to place the contact pads on the semi-insulating substrate. The ohmic source and drain contacts are realised using an Ni/AuGe/Ni metal system with subsequent annealing at 380 °C for 90 s. The (n^+)InGaAs contact layer is selectively removed by wet etching. Ti/Au is used for the gate metallisation and polyimid for mesa isolation. Cr/Au metallisation performs the microstrip interconnections and the contact pads. Devices with different gate length between 1.2 and 0.25 µm are prepared. Electron beam lithography is used to define the gate with a length below 0.75 µm.

5. Properties of InP/InGaAs HEMTs

DC measurements on InP/InGaAs HEMTs show that the drain-to-source break-down voltage is higher than 5 V and no kink effects can be observed. An example of measured I–V characteristics is shown in Fig. 4. Extrinsic transconductances g_m = 520 mS/mm and 870 mS/mm are evaluated on devices with 0.7 μm and 0.25 μm gate length, respectively. The channel width is 100 μm.

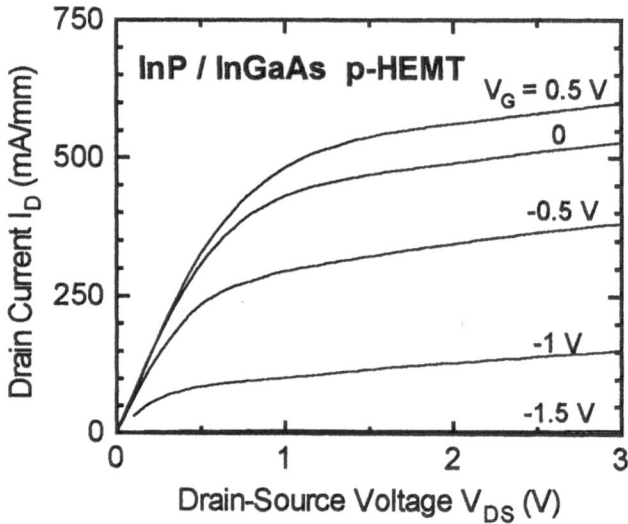

Figure 4. Typical I–V characteristics of InP/InGaAs pHEMTs.

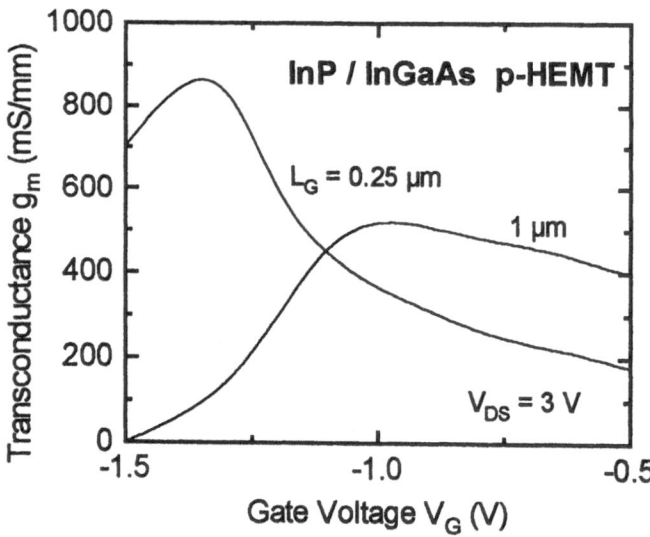

Figure 5. Transconductance as a function of gate voltage for InP/InGaAs pHEMTs.

The extrinsic transconductance as a function of the gate voltage $g_m = f(V_G)$ for HEMTs with 0.7 µm and 0.25 µm gate length are shown in Fig. 5. The transconductance maximum is shifted to higher gate voltages if the gate length is reduced. Similar effect can be observed on InAlAs/InGaAs pHEMTs.

High frequency S-parameter measurements up to 26.5 GHz are performed using on-wafer cascade microwave prober. The frequency dependence of the current gain $|h_{21}|^2$ and unilateral gain GU for InP/InGaAs pHEMT with $L_G = 0.25$ µm are shown in Fig. 6. Extrapolation of the measured data with a slope of -20 dB/decade gives the current gain cutoff frequency $f_T = 85$ GHz (on some devices f_T up to 104 GHz has been found) and the power gain cutoff frequency $f_{max} = 180$ GHz. Devices with the gate length of 0.5 µm show the cutoff frequencies $f_T = 55$ GHz and $f_{max} = 117$ GHz. The f_T*L_G–product is smaller for devices with shorter gates, 27.5 GHz*µm for devices with 0.5 µm gates and only about 22–25 GHz*µm for 0.25 µm gates. Values of about 29 GHz*µm have been found recently on similar InP/InGaAs pHEMTs with $L_G = 1$ µm [10]. This effect is caused by lower gate aspect-ratio in devices with shorter gate length. Similar situation occurs at the conventional InAlAs/InGaAs HEMTs [2]. However, our devices are prepared without using T-gates (this process is now under development) and therefore an improvement in f_T and f_{max} can be expected.

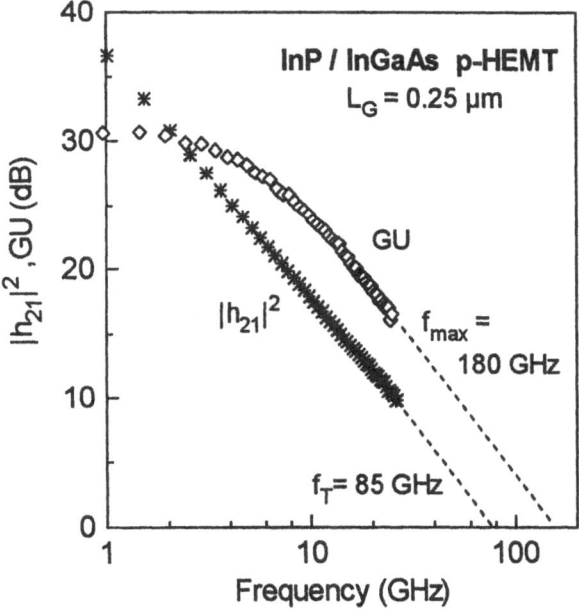

Figure 6. Current and unilateral power gain against frequency for InP/InGaAs pHEMTs.

Fig. 7 shows the current gain cutoff frequency f_T as a function of the gate length evaluated on our Al-free InP/InGaAs pHEMTs with different gate length ranging from 1.2 μm down to 0.25 μm. Record values of f_T published on InAlAs/InGaAs pseudomorphic devices with 50 and 100 nm gate length (open dots) [2] are shown for comparison too. Al-free InP/InGaAs HEMTs offer comparable microwave properties in comparison to conventional InAlAs/InHaAs HEMTs. On the other hand, on Al-free HEMTs higher breakdown voltages and no kinks in I-V characteristics are observed. This can be an important result in order to improve the long-term stability of HEMT performance.

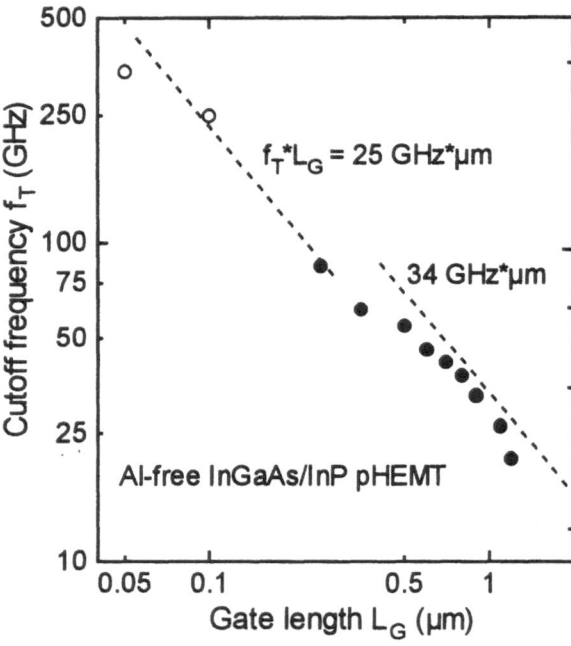

Figure 7. Current gain cutoff frequency as a function of gate length for Al-free InP/InGaAs and conventional InAlAs/InGaAs pHEMTs.

6. Integration Possibilities of InP/InGaAs HEMT

InP/InGaAs 2-DEG structures offer a good possibility to integrate monolithically HEMT with other discrete devices. We have studied the properties of the Schottky diodes with the transport along the 2-DEG [11]. Double-barrier MSM diodes show excellent varactor properties with the capacitance ratio C_{max}/C_{min} up to 86 and the varactor sensitivity $S = (dC/C)(V/dV)$ up to 11 at 1 V bias [12]. RC-limited cutoff frequency of 135 GHz have been measured on these diodes. An example of the capacitance variation with bias is shown in Fig. 8. Further, we found that InGaAs MSM 2-DEG diodes exhibit good photoelectric properties with FMHM at 1.3 μm less than 60 ps (Fig. 9) [13]. This gives the possibility to realize a monolithic integrated photoreceiver based on pHEMT structures [14].

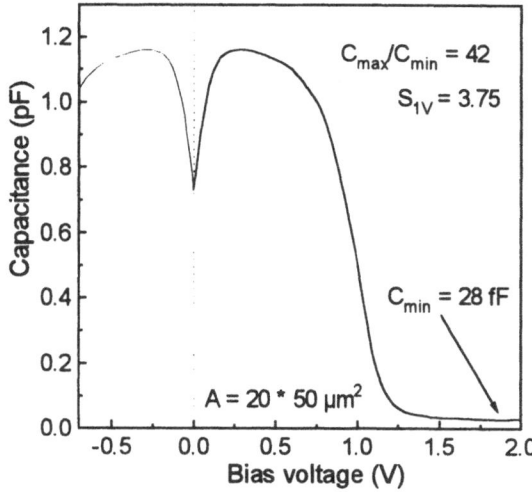

Figure 8. Capacitance variation with bias in InP/InGaAs pHEMT-like MSM diode.

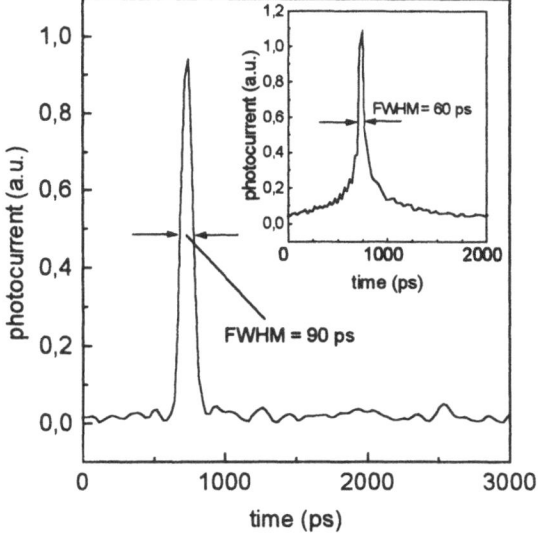

Figure 9. Pulse response of InGaAs MSM-2DEG photodetector.

7. Conclusions

We have shown that Al-free InP/InGaAs can be a good alternative to conventional InAlAs/InGaAs structures for HEMT preparation with improved reliability. Devices with 0.25 μm gate length exhibit $f_T = 85$ GHz and $f_{max} = 180$ GHz which can be increased by using T-gates. InP/InGaAs 2-DEG structures offer a possibility to integrate monolithically pHEMTs with microwave varactors ($f_{RC} = 135$ GHz) and high-speed photodetectors (FWHM ≤ 60 ps).

266

8. References

1. Nguyen, L.D., Larson, L.E., and Mishra, U.K. (1992) Ultra-high-speed modulation-doped field-effect transistors: a tutorial review, *Proc. IEEE* **80**, 494-518.
2. Nguyen, L.D., Brown, A.S., Thompson M.A. and Jelloian L.M. (1992) 50-nm self-aligned-gate pseudomorphic AlInAs/GaInAs high electron mobility transistors, *IEEE Trans. Electron Devices* **39**, 2007-2014.
3. Inoue, K. (1993) Recent advances in InP-based HEMT/HBT device technology, (1992) *Proc. 4th Intern. Conf. InP and Related Materials*, 10-13.
4. Mesquida Küsters, A., Puls, C., Wüller, R., Behres, A., Kohl, A., Somer, V. and Heime, K. (1995) High-performance Al-free InGaP/InP/InGaAs/InP backside-doped split-channel HFETs with 0.25μm T-gates, *Electron. Lett.* **31**, 409-411.
5. Kordoš, P. and Marso, M. (1993) Schottky barriers and ohmic contacts on InGaAs, in P. Bhattacharya (ed.) *Properties of lattice matched and strained Indium Gallium Arsenide*, INSPEC London, 145-151
6. Kordoš, P., Marso, M., Meyer R., and Lüth, H. (1991) Barrier height enhancement of n-InGaAs Schottky diodes grown by MOCVD technique, *Electron. Lett.* **27**, 1759-1761.
7. Kordoš, P., Marso, M., Meyer R., and Lüth, H. (1992) Schottky barrier height enhancement on n-InGaAs, *J. Appl. Phys.* **72**, 2347-2355.
8. Chough K.B., Chang T.Y., Feuer M.D., and Lalevic B. (1992) Comparison of device performance of highly strained GaInAs/AlInAs MODFETs, *Electron. Lett.* **28**, 329-330.
9. Meyer, R., Hardtdegen, H., Leuther, A., Marso, M., Kordoš, P., and Lüth, H. (1993) Optimization of Strained GaInAs/InP heterostructures towards high channel conductivity for HEMT application, *Proc. 5th Intern. Conf. InP and Related Materials*, 485-488.
10. Mesquida Küsters, A., Kohl, A., Brittner, S., Sommer, V., and Heime, K. (1994) Effect of indium mole fraction on charge control, DC and RF performance of single quantum-well InP/InGaAs HEMTs, *Proc. 6th Intern. Conf. InP and Related Materials*, 323-326.
11. Kordoš, P., Marso M., Fox, A., Hollfelder, M., and Lüth, H. (1992) n-InGaAs Schottky diode with current transport along 2-DEG, *Electron. Lett.* **28**, 1689-1690.
12. Marso M., Kordoš, P., Fox, A., Meyer, R., Hardtdegen, H., and Lüth, H. (1993) A novel InGaAs Schottky-2DEG diode, *Proc. 5th Intern. Conf. InP and Related Materials*, 397-400.
13. Horstmann, M., Marso, M., Fox, A., Rüders, F., Hollfelder, M., Hardtdegen, H., Kordoš, P. (1995) InP/InGaAs photodetector based on a high electron mobility transistor layer structure: its response at 1.3 μm wavelength, *Appl. Phys. Lett.* **67**, 983-985.
14. Horstmann, M., Marso, M., Schimpf, K., Rüders, F., van der Hart, A., Hollfelder, M., Hardtdegen, H., Kordoš, P., Lüth, H. (1995) Novel InP/InGaAs MSM photodetector for integration in HEMT circuits, *Proc. 25th ESSDERC'95*, Frontieres, 443-446.

TUNNELING HETEROSTRUCTURE DEVICES

HANS LÜTH
Institut für Schicht- und Ionentechnik (ISI)
Forschungszentrum Jülich GmbH
D-52425 Jülich, Germany

1. Introduction

Resonant tunneling devices, in their simplest form resonant tunneling diodes (RTDs), are the only quantum devices so far which function at room temperature. RTDs are fabricated as multiple heterostructure layer systems, where a material with larger band gap (e.g. AlGaAs) forms two thin barriers (thickness approximately 2 nm), about 5 nm apart from each other, within a semiconductor with smaller band gap (e.g. GaAs). The device is fabricated as a lithographically structured mesa structure with two ohmic contacts; alternatively the device might be isolated from the surrounding by ion implantation. Biasing the diode between top and bottom contacts causes current flow. For conduction electrons, which propagate wave like, the two barriers can be penetrated by tunneling; they act as semitransparent mirrors for the electron waves similarly as mirrors do for light waves in a Fabry-Perrot interferometer. When multiples of half the electron wavelength $(n\lambda/2)$ match the distance between the two inner sides of the barriers (well region), the transparency of the double barrier structure reaches high values (in the ideal case one) due to so-called resonance tunneling. A convenient bias between the two contacts rises the energy of electrons on the emitter side such that electrons near the Fermi energy E_F fulfill the matching condition $d = \lambda/2$ and a strong current through the RTD results. Further increase of the applied voltage brings the electrons on the emitter side out of resonance, the current breaks down; but it rises again when the effect of the second barrier becomes negligible. The current voltage (I-V) characteristic of a RTD is N shaped and exhibits a negative differential resistance (NDR). This highly non-linear I-V characteristic with NDR is the basis for the application interest in resonance tunneling devices. Apart from oscillators and rectifyers the bistability of the characteristics

J. Novák and A. Schlachetzki (eds.), Heterostructure Epitaxy and Devices, 267–276.
© *1996 Kluwer Academic Publishers.*

can be used in logic circuits, flip-flop memories etc. In this respect RTD devices have a considerable potential for a reduction in circuit complexity. With much less single devices and lower circuit complexity considerably higher logic complexity can be achieved. Faster and more complex circuits can be designed with significantly relaxed lithographic requirements for lateral structuring.

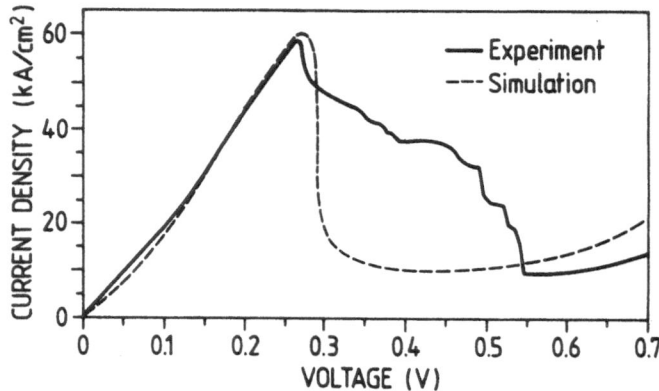

Figure 1. Measured (full line) and calculated (broken line) current-voltage (I-V) characteristic of an AlAs (6 monolayers thick) double barrier RTD embedded in $In_{0.17}Ga_{0.83}As$; well width 5nm, diode mesa area 4 μm × 4μm [2]

2. Coherent and Sequential Tunneling

The quantum mechanical transmission of an electron wave through the double barrier structure is described mathematically by a coherent solution of the Schrödinger equation outside the well region (left and right) and in between the two barriers, in the well [1, 2]. Due to the thin barrier regions (couple of monolayers) the quantized states in the well "leak out" through the barriers and couple coherently to moving plane wave states, left and right. As a result the states in the well can be considered as metastable and the lifetime of the electron within the well is finite. The corresponding quasibound states between the barriers are therefore described apart from their quantum energy E_i, by a lifetime τ during which a tunneling electron occupies this state. Through the uncertainty relation $\tau\Gamma_0 = \hbar$ the lifetime is related to the energetic halfwidth Γ_0 of the quasibound state. For typical AlAs/GaAs RTDs with barrier widths of 6 monolayers (6 ML = 1.7 nm) AlAs and a GaAs well width of 5 nm typical halfwidths Γ_0 are 2 meV and 25 meV for the first and second quasibound state, respectively.

If one calculates the tunneling current solely based on the described coherent tunneling [2], the simulated I-V curves exhibit a much too sharp resonance peak and the peak-to- valley ratio (PVR) exceeds measured data by more than a factor of 100. The reason is found in inelastic scattering events within the double barrier region which lead to sequential transmission of electrons through the barriers. A coherent wave function does no longer exist over the whole RTD. Without accounting for details of the scattering mechanism one can formally describe the combination of resonant tunneling and sequential or incoherent tunneling (inelastic scattering) by a generalisation of the Lorentzian form of resonant transmission based on the Breit-Wigner formalism. For a symmetric RTD and an electron with energy E Stone and Lee [3] derive a total transmission probability in the presence of inelastic scattering as

$$T_{tot} = \frac{\frac{1}{4}\Gamma_0\Gamma}{(E - E_r)^2 + \frac{1}{4}\Gamma^2} \tag{1}$$

where E_r is the energy of the quasi-bound state, and Γ_0 the inherent half width of the quasi-bound eigenstate (energy E_r) in purely coherent transmission. $\Gamma = \Gamma_0 + \Gamma_i$ is the total half width of the resonance which is obtained by folding the resonance peak by a Lorentzian with halfwidth Γ_i, which describes the inelastic scattering probability. Within this formalism inelastic scattering, i.e. incoherent or sequential tunneling diminishes the PVR according to

$$PVR = (PVR)_0 \frac{\Gamma_0}{\Gamma_0 + \Gamma_i} \tag{2}$$

where $(PVR)_0$ is the peak-to valley ratio as calculated without scattering.

According to Fig. 1 these simulations allow a statisfying formal description of measured I-V curves, eventhough there is no direct physical explanation of the scattering mechanisms.

A detailed theoretical study of various scattering effects within RTDs, including acoustic and optical phonon, impurity, alloy disorder and interface roughness scattering has been given by Chevoir and Vinter [4]. For RTDs prepared from a combination of a direct and an indirect semiconductor as AlAs/GaAs Γ to X valley scattering plays an important role. In AlAs/GaAs diodes, e.g. AlAs acts as a potential barrier for Γ electrons, but as a potential well for X electrons. Γ electrons from the supply (emitter) region of the RTD might therefore tunnel through X states in the barriers if they gain k-vector (from Γ to X) due to a scattering event. The relevance of Γ to X valley scattering for sequential tunneling has been shown from the pressure dependence of the I-V curves of RTDs by Mendez et al. [5] and Brugger et al. [6].

The effect of different types of interface roughness on interface scattering has been demonstrated by Förster et al. [7]: Using growth temperatures

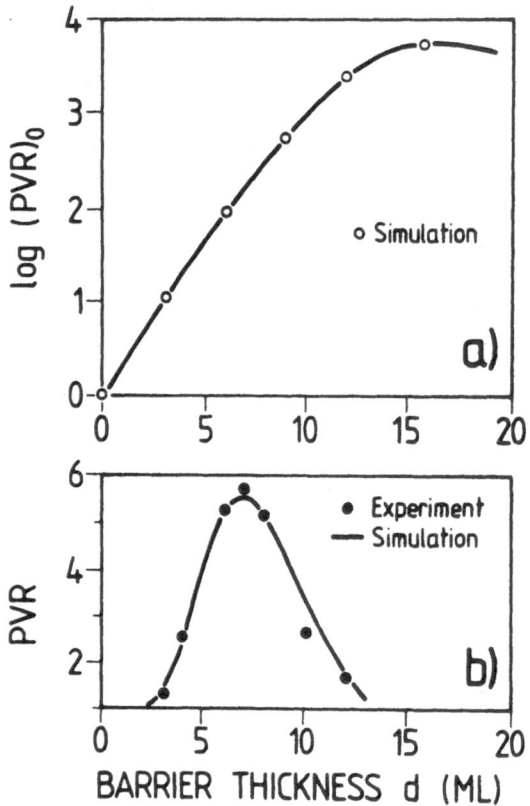

Figure 2. a) Logarithmic plot of the calculated peak–to–valley ratio (PVR) for a RTD consisting of AlAs barriers with varying thickness (in monolayers ML) embedded in $In_{0.17}Ga_{0.83}As$; well width 5nm. Only resonant tunneling is taken into accout.
b) PVR measured on RTD's (dark circles) as described in a); diode area 7×7 μm^2. The simulation (solid line) includes beside resonant tunneling also sequential tunneling processes [2]

between 480° C and 680° C barrier interfaces on AlAs/GaAs RTDs could be prepared, which varied in their morphology between short range lateral roughness (step distance $\sim$ 8 to 18 nm) at 480° C, long range roughness (step distance $\sim$ 100 nm) at 580° C and interdiffusion broadening at 680° C. Accordingly a significant increase of the PVR from diodes grown at 480° C to those grown at 580° C is observed, while the PVR decreases again for growth temperatures between 600 and 680° C [7]. Sequential tunneling contributions also explain the characterisitic dependence of the PVR on barrier thickness as found experimentally on AlAs/GaAs RTDs (Fig. 2). In contrast the $(PVR)_0$ calculated on the basis of purely resonant tunneling (Fig. 2a) the measured PVR is smaller by more than a factor of 100 and strongly peaked around a barrier thickness of 6 ML AlAs (1.7 nm) [2]. This characteristic behaviour can be simulated using eq. (1) and (2). The strong increase of the PVR between 3 and 6 ML is mainly caused by scattering on long scale roughness whereas the decrease between 6 and 13 ML involves mainly Γ to X valley scattering by phonons.

3. High Frequency Behaviour

The important figure of merit of a RTD is the peak-to-valley ratio I_{peak}/I_{valley}, which defines the extension of the negative differential resistance (NDR). For the use in RF circuits the persistence of the NDR up to highest frequencies is necessary: One defines a maximum cut-off frequency f_{max}, up to which NDR, i.e. a non-negligible PVR is observed.

The simplest equivalent circuit for a RTD which one would expect, must at least consist of a resistance R_d ($-|R_d|$ because of NDR) in parallel with a diode capacitance C_d and again in series with a series resistance Rs, mainly due to contacts. This simple equivalent circuit describes indeed the RF behaviour up to frequencies of 26 GHz. As is obvious from Fig. 3 the complex impedance Z of an AlAs/GaAs RTD as determined by a four- port scattering (S) parameter measurement is well described by a simulation using the simple RTD equivalent circuit consisting of the three parameters R_d, C_d, Rs. Based on this circuit the maximum cut-off frequency is obtained as

$$f_{max} = \frac{1}{2\pi |R_d C_d|} \sqrt{\frac{-R_d}{R_s} - 1} \qquad (3)$$

From the simulation (Fig. 3) one derives f_{max} values between 65 and 110 GHz, depending on the particular operating point, for a diode with a mesa area of 4×4 μm^2 and a PVR of 4.3 at 300 K. Scaling down the mesa area by a factor of 10, therefore, should easily lead to f_{max} values of a couple of hundred GHz, of course at lower current and power values.

4. Tunnel Transistors

RTDs are two-port devices. In an integrated circuit additionally amplifying three-port devices, transistors, are necessary. More compact circuits could be built by devices in which the transistor function is integrated with the RTD function within one and the same device. Several types of resonant tunneling transistors (RTT) have been suggested and fabricated. One approach consits in incorporating a double barrier diode into the emitter-base barrier of a unipolar hot electron transistor (HET). This resonant hot electron transistor (RHET) exhibits a non monotonic form of the I-V characteristics thus enabling a multiple state transistor. Realised in the

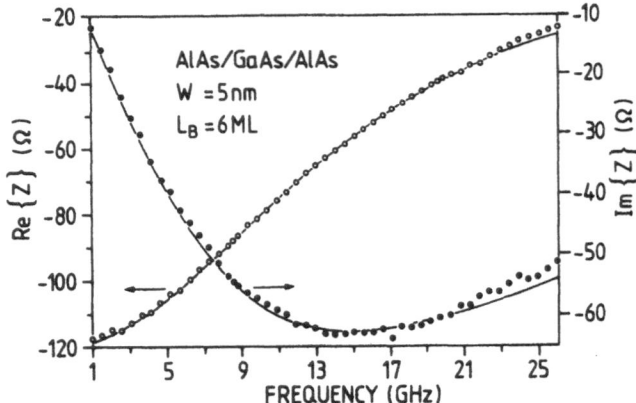

Figure 3. Real and imaginary part of the complex impedance Z of an AlAs/GaAs/AlAs double barrier RTD (well width 5nm, barrier thickness 6ML) measured by an S-parameter set-up. The full lines through the data points are a fit based on a simple equivalent circuit (see text) [18]

GaAs/AlGaAs material system [8] the device suffered from a poor current gain of 3 to 4 and a poor PVR in the transfer characteristics. In the In-GaAs/In(GaAl)As system, where higher and thinner barriers are possible and intervalley scattering can be reduced, the peak current densities are increased by a factor of about four, the current gain rises to approx. 17 and a PVR of about 10 can be achieved [9]. Recent redesigned RHETs were able to achieve cut-off frequencies of about 120 GHz [10].

A conceptually new approach is the two-dimensional to two-dimensional Tunnel HEMT (2D-2D T-HEMT) investigated by Leuther et al. [11]. According to Fig. 4 two modulation doped $In_{0.75}Ga_{0.25}As$ channels of different thickness (different subband energies) are separated by a 10 nm thick InP barrier, through which tunneling is possible between the two two-dimensional electron gases (2 DEGs) in the channels. The electron concentration within both 2 DEGs is controlled by a metallic gate, isolated by a SiO_2 layer The upper channel is contacted by non-alloyed metallic source and drain contacts which are filled in into lithographically opened holes. This type of ohmic contact is possible only to InGaAs channels, where Fermi-level pinning occurs in the conduction band for sufficiently high In concentrations. The device structure in Fig. 4 can be considered as a HEMT, where a second 2 DEG channel is coupled by tunneling to the normal upper channel. Compared with other RTT structures there are several advantages: The excellent noise and high frequency features of the HEMT are combined with tunnel characteritics. Tunneling can easily be controlled by a surface gate since there is essentially no screening between

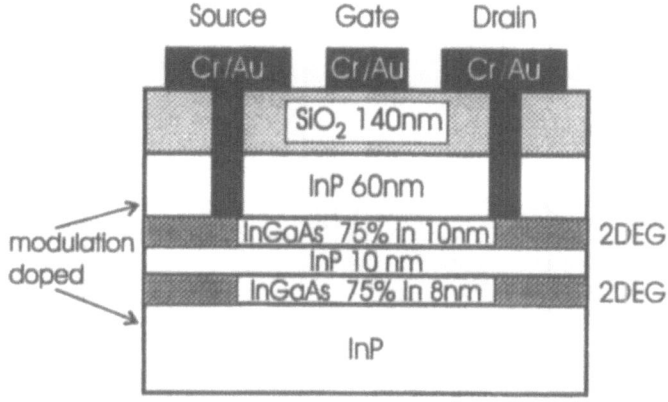

Figure 4. Scheme of a Tunnel-HEMT [11]

the channels and the gate. In contrast to standard RTDs, where tunneling occurs between 3D-2D-3D regions, tunneling in the T-HEMT is a 2D-2D process. The condition for a tunneling process between two 2D states is the conservation of the total energy and of the in-plane momentum. Given the parabolic energy dispersion in the 2 DEGs tunneling occurs only for one particular fixed voltage accross the barriers.

Depending on the height of the voltage drop resonance occurs between both ground states in the channels or if the first subband-level in one channel energetically equals the ground state of the other one. Depending on the height of the gate voltage V_G two types of I-V characteristics (source-drain current I_{DS} versus source-drain voltage V_{DS}) are found for the T-HEMT (Fig. 5). For lower V_G (-8 to -10 V) there is a sudden increase in the drain current. This increase shifts to higher V_{DS} with lower gate voltages (Fig. 5, top). Upon reversing the source-drain voltage a strong hysteresis appears, which is not completely understood so far. For higher gate voltages V_G (-13.2 to 13.8 V) there is a distinct NDR with a PVR of up to 3.7 at 4 K The onset of the NDR shifts to higher drain voltages V_{DS} for increasing gate voltage V_G.

The performance of the T-HEMT is understood as follows: A finite gate voltage separates the source and drain region by depletion below the gate spatially from each other, such that two separate tunnel resistances below the drain and below the source region have to be considered. They connect the two 2 DEG channels with each other. In the case of low gate voltages the 2 DEG channel resistances are of the same order and considerably lower than the tunnel resistance across the InP barrier. Therefore the current I_{DS} only flows through the upper channel. At a particular hig-

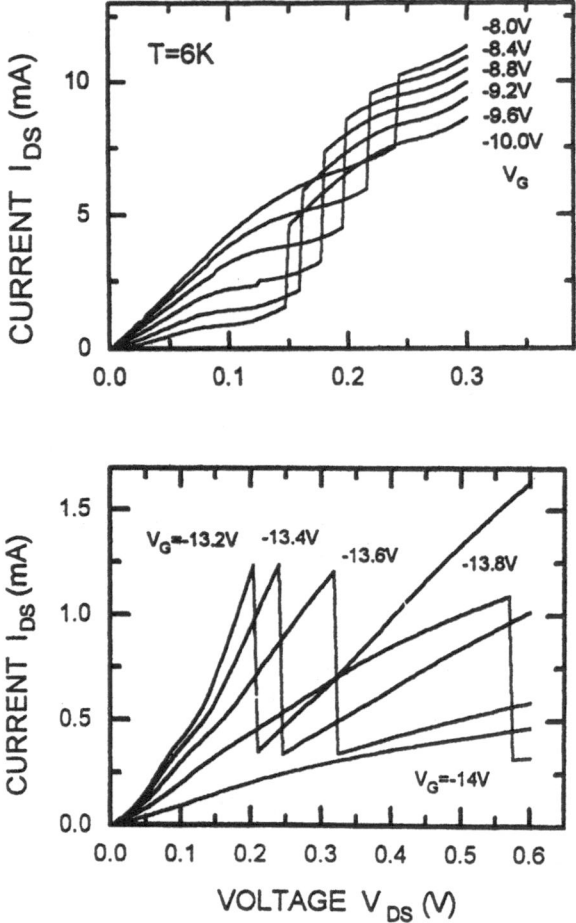

Figure 5. Drain-source current-voltage (I_{DS}-V_{DS}) characteristics of a Tunnel-HEMT in the InGaAs/InP material system, for different gate voltages V_G [11].
Upper part: gate voltage regime below -10 V
Lower part: gate voltage regime above -13 V

her source-drain voltage V_{DS} the tunnel diodes come into resonance and the current I_{DS}, starts to flow through the lower channel, too; a steep increase in I_{DS} results (Fig. 5, top). The shift of this step with increasing V_G can be understood by the corresponding variation of the voltage drops across the different resistances. At higher gate voltages (Fig. 5, bottom) the upper channel is completely depleted and the current I_{DS} flows through the second 2 DEG channel only and the two tunnel resistances in series. In this case the I-V characteristics is similar to that of a normal RTD. With higher gate voltages the resistance of the lower channel increases, which leads to a lower voltage drop at the tunnel diodes. A higher V_{DS} is required to bring the T-HEMT into resonance. This explains the shift of the NDR region with varying V_G. This qualitative interpretation of the I-V characteristics can be made quantitative at least for dc performance using a selfconsistently calculated band scheme for the tunneling process [11]. A deeper understanding of the T-HEMT performance requires a detailed stability analysis of the whole complex system which has not been made so far. The presented T-HEMT data result from recent investigations of this novel device concept. An improved technology based on experience with HEMTs will certainly bring further progress. In particular, more sophisticated layer structures enabling higher tunnel barriers and better Schottky gates will allow room temperature performance of the T-HEMT and also gate control by considerably lower gate voltages.

5. Applications

Since resonant tunneling transistors are in a preliminary state of development, applications of tunneling devices are known so far mainly for RTDs. The application interest in RTDs is based on the N-shaped I-V characteritics which yields depending on the operating points regimes of stability, instability and bistability. As a detailed analysis of the corresponding differential equations shows, the regimes of instability with selfgeneration of oscillations, of stability and bistability sensitively depend on the circuit parameters diode capacity, and series resistance and inductance [2]. Using the instability in the NDR regime high speed oscillators for frequencies up to several hundred GHz have been built [12]. By integrating a number of RTDs either in parallel [13] or in series [14] a structure with multiple peaks in the I-V characteritics can be made. Such devices, with multiple stable operating points, have been shown to be capable of a number of interesting circuit applications, such as frequency multiplier and parity generator. Lakhani et al. [15] have demonstrated an 11-bit parity generator using a single device with 5 RTD structures integrated vertically, replacing ten conventional exclusive-OR gates. This example shows an interesting feature of resonant tunneling devices in general, namely to reduce circuit complexity but simultaneously enhance the logic complexity and ability of the circuit.

A number of further applications have also been discussed by Capasso et al. [16] and Luryi [17].

6. Conclusions

As multifunctional devices resonant tunneling diodes and transistors certainly have a great potential for future high frequency electronics, when quantum effects severely limit further miniaturisation in "main-stream" CMOS technology.

But beside the development of novel device concepts based on resonant tunneling also new ways of circuit design must go hand in hand with the research on single devices in order to gain the full benefit of multifunctionality. We are only at the beginning of this development.

References

1. Ricco, B. and M. Ya Azbel (1984): Phys. Rev. **B 29**, 1970
2. Förster, A. (1994): "Resonant Tunneling Diodes: The Effect of Structural Properties on their Performance", in: Festkörperprobleme/Advances in Solid State Physics, Vol. **33**, ed. by R. Helbig (Vieweg, Braunschweig/Wiesbaden), pp. 37
3. Stone, A. D. and P. A. Lee (1985): Phys. Rev. Lett. **54**, 1196
4. Chevoir, F. and B. Vinter (1989): Appl. Phys. Lett. **55**, 1859
5. Mendez, E. E. , W. I. Wang, E. Calleja, and C. E. T. Goncalves da Sil (1987): Appl. Phys. Lett. **50**, 1263
6. Brugger, H. , U. Meiners, C. Wölk, R. Deufel, A. Marten, M. Rossmanith, K. v. Klitzing, and R. Sauer (1991): "Pseudomorphic Two-Dimensional Electron-Gas-Emitter Resonant Tunneling Devices", Proc. of the 21st European Solid State Device Research Conference, ESSDERC '91, Lausanne, Microelectronics Engineering **15**, 663
7. Förster, A. , J. Lange, D. Gerthsen, Ch. Dieker, and H. Lüth (1994): J. Phys. D: Appl. Phys. **27**, 175
8. Yokoyama, N. and K. Imamura (1986): Electron. Lett. **22**, 1228
9. Yokoyama, N. , S. Muto, H. Ohnishi, K. Imamura, T. Mori, and T. Inata (1990): in Physics of Quantum Electron Devices, ed. F. Capasso (Springer, Berlin), p. 253
10. Mori, T. , T. Adachihara, M. Takatsu, H. Ohnishi, K. Imamura, S. Muto, and N. Yokoyama (1991): Electron. Lett. **27**, 1523
11. Leuther, A. , M. Hollfelder, H. Hardtdegen and H. Lüth to be published
12. Brown, E. R. , T. C. L. G. Sollner, C. D. Parker, W. D. Goddhue, and C. L. Chen (1989): Appl. Phys. Lett. **55**, 1777
13. Sen, S. , F. Capasso, A. Y. Cho, and D. L. Sirco (1987): IEEE Trans. Electron. Devices ED- **34**, 2185
14. Potter, R. C. , A. A. Lakhani, D. Beyea, H. S. Hier, E. Hempfling, and A. Fahtimulla (1988): Appl. Phys. Lett. **52**, 2163
15. Lakhami, A. A. , R. C. Potter, and H. S. Hier (1988): Electron. Lett. **24**, 681
16. Capasso, F. , S. Sen, F. Beltram, and A. Y. Cho (1990): "Resonant Tunneling and Superlattice Devices: Physics and Circuits", in Physics of Quantum Electron Devices, ed. F. Capasso, Springer Series in Electronics and Photonics **28** (Springer Berlin, Heidelberg, New York)
17. Luryi, S. (1988): "Electronic Devices Using Multilayer Structures", in Physics, Fabrication and Applications of Multilayered Structures, ed. P. Dhez and C. Weisbuch, NATO ASI Series Vol. **B 182** (Plenum Press New York and London), p. 241
18. Scheuermann, Th. (1994): Diploma Thesis at RWTH Aachen

InGaAs/GaAs PSEUDOMORPHIC DOUBLE δ-DOPED HEMTS

Some Limitations of Design.

Ľ. MALACKÝ, J. KUZMÍK, Ž. MOZOLOVÁ, M. KUČERA
Institute of Electrical Engineering, Slovak Academy of Science
Dúbravská 9, 842 39 Bratislava, Slovakia
K. LÜBKE
Microelectronics Institute, University of Linz, A-4040 Linz, Austria
H.- H. WEHMANN
Institute for Semiconductor Technology, University of Braunschweig,
Hans Sommer Str. 66, 38106 Braunschweig, Germany

1. Introduction. The superior transport properties of electrons confined in a strained InGaAs channel make pseudomorphic InGaAs/GaAs high electron mobility transistors (HEMTs) a subject of great interest. Aluminium free devices show their advantages especially in applications at low temperatures. Various structures with excellent high frequency and/or low noise properties have already been reported by many authors [1]. From the physical principles of a HEMT's operation it follows that minimising the gate - quantum well (QW) distance together with minimising the series resistance are the most progressive ways of improvement of the device's parameters. Consequently, a high sheet carrier concentration n_S, introduced e.g. by the pulse doping of the GaAs barrier, is desirable to obtain a small gate-QW distance. Further, it is known that the electron/donor spatial separation reduces carrier scattering. Thus as many carriers as possible must remain confined in the channel. This is provided by the band discontinuity ΔE_C, keeping the mobility high and R_S low. However, the InGaAs/GaAs HEMTs have some design limits resulting from the relatively low ΔE_C at the heterointerface. In our work, a HEMT structure was

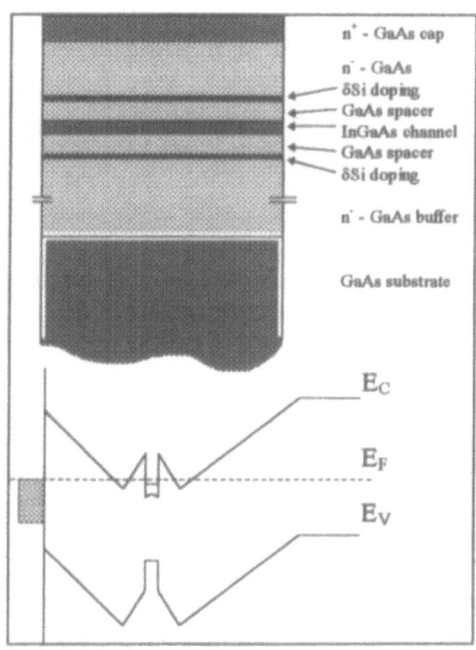

Figure 1. Cross-section of the InGaAs/GaAs HEMT structure

J. Novák and A. Schlachetzki (eds.), Heterostructure Epitaxy and Devices, 277–280.
© *1996 Kluwer Academic Publishers.*

278

studied from the point of view of an optimal confinement, maximal n_S and mobility.

Table 1.

Technology	LP MOCVD (AIXTRON)
Growth temperature	650°C
Pressure	20mbar
Gas flow [ccm/min] AsH₃	45ccm
TMG	13ccm for GaAs
	5ccm for InGaAs
TMI	120ccm for x = 0.28
2% SiH₄	1 - 12ccm for δ-doping
	10ccm for cap layer
Growth rate GaAs	0.81nm/s
InGaAs x = 0.28	0.44ccm/s

2. Device preparation. An extremely high 2DEG sheet concentration can be obtained by δ-doping the GaAs barrier and buffer layers symmetrically on both sides of the strained InGaAs channel [2]. A cross-section and a simplified band diagram of the structure analysed are depicted on Figure 1. Samples were grown on (001) SI-GaAs substrate by low pressure MOCVD. All growth data are summarised in the Table 1. The δ-doping of the GaAs layers was performed by switching the TMG flow off, introducing a second's growth interruption, than turning on the silane (2%) flow for 20seconds while the AsH₃ flow was kept constant. A growth interruption for 5seconds was also used with the lower and upper interfaces of the InGaAs SQW. Photoluminescence and Hall measurements were performed on the samples prepared in order to obtain electric and crystallographic properties of the structure. HEMTs were processed on structures with x=0.28

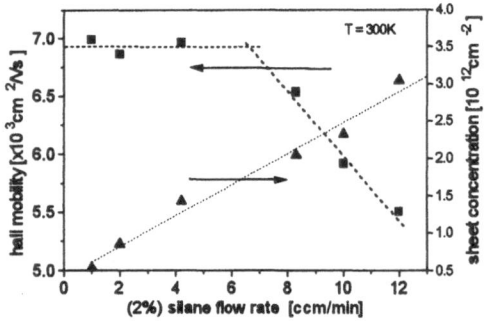

Figure 2. Hall mobility and sheet concentration vs. silane flow rate

and δ-doping yielded from a silane flow rate of 10ccm/min. The Ti/Pt/Au gate was 2μm long and 70μm wide, gate recess was 90nm. The d.c. characteristics of the HEMTs processed were investigated and correlated with the structure properties.

3. Results and Discussion. The limited height of ΔE_C ($\cong 0.2eV$) causes the saturation of sheet carrier concentration in the InGaAs channel, which independent of doping.

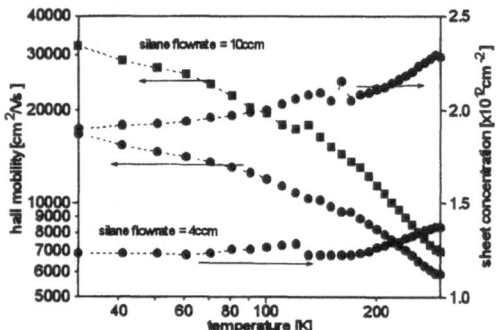

Figure 3. Mobility and sheet concentration vs. temperature

Figure 2 shows the dependence of the Hall mobility and total sheet concentration of our structure with x=0.28 on silane flow rate. While the increase of n_S with the flow rate is linear, two regions of the mobility can be distinguished. First, for silane flow below approximately 7ccm/min the mobility saturates at a value of 6900cm²/Vs. The further increase in doping leads to a reduction of the Hall mobility. The inflex point corresponds to the total sheet concentration of

1.8×10^{12} cm⁻², which may be the upper limit for n_S keeping at maximum the mobility for our 2δ-doped structure and given technological data. Although, the value is relatively low, it is easy to show that our results agree well with the analytical model [3]. Figure 3 supports our conclusions. The low temperature (20K) mobility of the highly doped sample (silane f.r. = 10ccm/min) was about 17000cm^2/Vs whereas the low doped samples (silane f.r. = 4ccm/min) exhibit mobility above 30000cm^2/Vs. This may be explained by parallel conduction through the GaAs barrier layer. Figure 4 shows PL spectra of the sample L1904 (28% In content in channel SQW, silane f.r. = 10ccm/min) measured at varied excitation power at 5Kelvin. The PL signal in the doped SQW arises

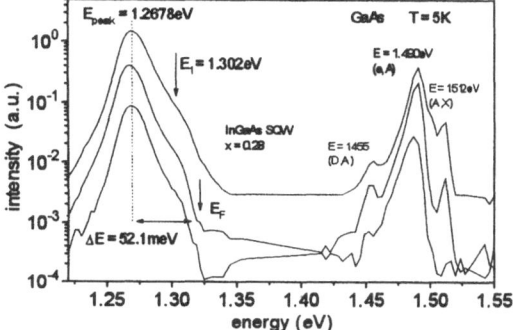

mainly from the recombination of electrons confined in the well with optically generated holes. The measurements yield an intensive PL signal peaking at 1,267eV, which clearly indicates a good quality of the strained InGaAs channel without relaxation [4]. A weaker signal detected at 1.49 - 1.512eV indicates that some carriers recombine through GaAs, which is a sign of the presence of carriers not confined in the channel. The GaAs peak disappears at

Figure 4. Photoluminescence spectra of the HEMT structure

room temperature. The PL line corresponding to transitions in InGaAs has a shape that is typical for doped SQW[5]. The abrupt cut-off clearly seen on the high energy side of the linewidth indicates the position of the electron Fermi energy. Assuming a simple model [6], the SQW sheet concentration is given as $n_S = (4\pi m^*/h^2)\Delta E$ and it is 1.72×10^{12}cm^{-2}, taking in to account $m^* = 0.071m_0$ and $\Delta E = 52.1$meV (Fig. 4). This value is in close agreement with the predicted maximal sheet concentration in the channel, given by the inflexion point in Figure 2. Finally, on Figures 5, 6 we present the results of the d.c. characterisation of the HEMTs. The output characteristics exhibit good saturation behaviour with a threshold voltage of -2V. The transfer and

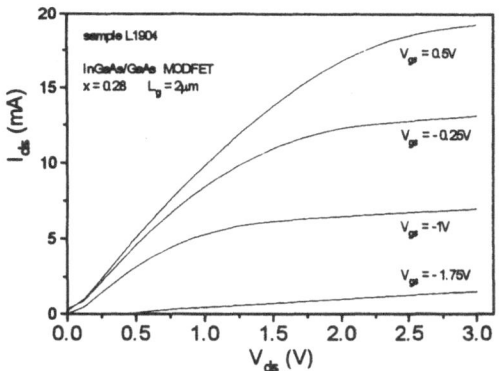

transconductance characteristics are interesting from several points of view. The maximal transconductance around 135mS/mm keeps its steady value in a broad range of gate biases. This can be attributed to the technology approach based on the use of the double delta doping which enables us to obtain the high g_m even for a low noise figure biased HEMTs. The double peak of g_m characteristics between 0 - -0.7V gate bias was reproducibly measured for all devices. Due to the small distance between the gate and the upper δ-

Figure 5 The output I-V characteristics of a double delta doped InGaAs HEMT

layer highly sensitive depletion of carriers with gate bias fairly detects both the parallel

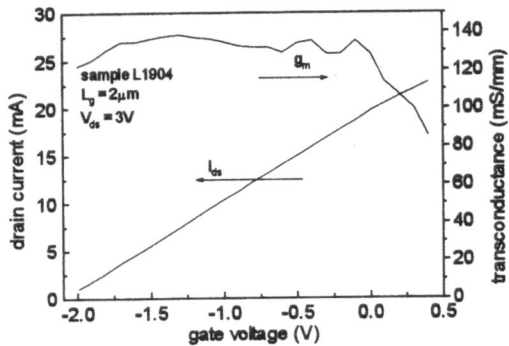

Figure 6. Transconductance as a function of gate voltage

(GaAs) and truth (InGaAs) channels. The lower δ-layer is depleted by gate bias close to threshold and helps to keep the high values of g_m. The carrier confinement in the InGaAs channel may be improved using the wider channel layer with higher In content. Apparently, the low critical layer thickness h_C of such highly strained layers seems to limit such access to this problem. However, as it is shown in Figure 7, the value of h_C of the strained InGaAs quantum wells buried in sufficiently thick barriers increases up to several tens of nm. This is true also for $x = 0.3$ or more [7]. The strain dependence of the effective electron mass in InGaAs SQW remains of marginal impact.

4. Conclusion. We present InGaAs/GaAs pseudomorphic double δ-doped HEMT structures grown by LP MOCVD. Measurements on the structure and HEMTs indicate limits in carrier confinement on heterointerface. It is shown that the maximal sheet concentration for given the technological design is $1.8 \times 10^{12} \mathrm{cm}^{-2}$. For higher doping fair reduction of the mobility can be expected. Problems caused by the low ΔE_C can be solved using highly strained InGaAs channel layers.

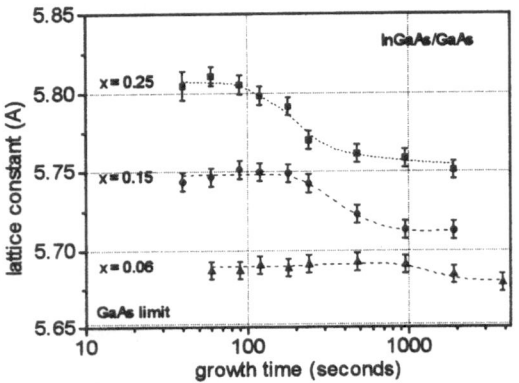

Figure 7. Lattice constant of strained InGaAs layers vs. growth time.

5. Acknowledgement. This work was supported by the Slovak Grant Agency under project 119094, by the BMBF/Germany under project X. 263.1 and by the Action Slovakia/Austria.

References.
1. Daembkes H. editor; *Modulation-Doped Field-Effect Transistors, Principles/Design/Technology*, IEEE Press Inc. New York 1991.
2. Hsu W.-Ch., Shieh H.-M., Wu Ch.-L., Wu T.-Sh.; A High Performance Symmetric Double δ-doped GaAs/ InGaAs/GaAs Pseudomorphic HFET's Grown by MOCVD, *IEEE Trans. Electr. Dev.*, 1994, 41, pp. 456 - 457.
3. Shur M.; *GaAs Devices and Circuits*, Plenum Press, New York 1987.
4. Elman B., Koteles E.S., Melman P., Ostereich K., Sung C., Low substrate temperature molecular beam epitaxial growth and the critical layer thickness of InGaAs grown on GaAs. *J. Appl. Phys.*, 1991, 70, pp. 2634 - 264.
5. Saker M.K., Skolnick M.S., Claxton P.A., Roberts J.S., Kane M.J.; The effects of free carriers on the photoluminescence and photoluminescence excitation spectra of InGaAs-InP quantum wells; *Semicond. Sci. Technol.*, 1988, 3, pp. 691 - 700.
6. Dodabalapur A., Sadra K., Streetman B.G.; Relationship between photoluminescence spectra and low-field electrical properties of modulation-doped AlGaAs/GaAs quantum wells; *J.Appl. Phys.*, 1990, 68, pp. 4119 - 4126.
7. Wang C.A., Groves S.H., Reinold J.H., Calawa D.R.; Critical Layer Thickness of Strained-Layer InGaAs/GaAs MQWs Determined by Double-Crystal X-Ray Diffraction, *J. Electron. Mat.*, 1993, 22, pp. 1365 - 1368.

ALPHA PARTICLE RADIATION EFFECTS IN HIGH ELECTRON MOBILITY TRANSISTORS

G.J.PAPAIOANNOU,M.J.PAPASTAMATIOU,N.ARPATZANIS,
P.DIMITRAKIS, C. MICHELAKIS* and Z.HATZOPOULOS*
*Solid State Physics Section, University of Athens, Panepistimiopolis
Zografos, Greece 157 84*
* *IESL, FORTH, P.O. Box 1527, Heraklion, Greece 711 10*

Abstract
    The effects of alpha particle irradiation have been investigated in High Electron Mobility Transistors. Devices with different layer structures have been employed for the better understanding of the failure mechanism sources. Finally, a charge control model allowed the determination of buffer layer degradation.

Introduction

    The AlGaAs/GaAs High Electron Mobility Transistor (HEMT) is an important component for applications involving high speed digital and microwave integrated circuits for data and signal processing and communication systems. Since GaAs based HEMTs and integrated circuits are finding applications in systems which have to function in a hostile environment, such as space applications, it is appropriate to assess their radiation hardness and identify the degradation mechanisms. Thus the understanding of the failure mechanisms of HEMTs exposed to radiation is required. A significant effort has been made to determine the failure sources in these devices. In all previous studies the effort was concentrated, mainly, on the determination of the dependence of the threshold voltage on the radiation fluence. Other device parameters such as the transconductance, the drain saturation current, the noise and frequency response were also considered. Regarding the modeling of the device degradation, that has been introduced relatively recently in the literature [1-3]. The aim of the present work is to investigate the alpha particle radiation effects in various layer structure HEMTs.

Experiment

    The layer structure of the HEMTs used in the present work is summarized in Table I. In addition to the conventional structures (A) we have also investigated HEMTs (structure B) in which the AlGaAs spacer and donor layers as well as the GaAs contact layer were grown at lower temperature, Tg=510°C, as in the case of GaAs pseudomorphic HEMTs (PM-HEMTs).
    All samples have been irradiated at room temperature with 5MeV average energy alpha particles obtained from an Am source. The irradiation fluence covered the range of $10^{10}cm^{-2}$ to about $10^{13}cm^{-2}$. All devices were assessed by

J. Novák and A. Schlachetzki (eds.), Heterostructure Epitaxy and Devices, 281–284.
© 1996 *Kluwer Academic Publishers.*

obtaining the drain current-voltage and C-V characteristics on long gate devices. Additionally deep level transient spectroscopy (DLTS) measurements were performed in order to monitor the background trap concentration and the introduction of new traps by irradiation. The dependence of the channel drift mobility on the sheet carrier concentration was determined.

Results and Discussion

In HEMTs the heavy ion radiation is manifested through the degradation of their drain current-voltage characteristics which in turn is the result of the degradation of the mobility and the carrier concentration in the semiconductor, the decrease of the electric field in the junctions and finally the variation of other parameters such as the source and drain parasitic series resistance.

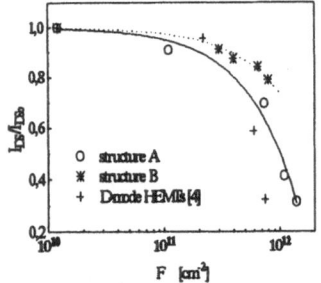

Figure 1 Relative degradation of drain saturation current vs alpha paticle fluence.

Table I. HEMT layer structures

Layer	Doping	Structure A	Structure B
GaAs	10^{18} cm^{-3}	10nm	10nm
AlGaAs	10^{18} cm^{-3}	45nm	45nm
AlGaAs	undoped	3nm	3nm
GaAs	undoped	1μm	1μm
AlGaAs	undoped	- - -	- - -
GaAs	SI	substrate	substrate

(A) AlGaAs with Al mole fraction 28%
(B) AlGaAs growth temperature was
510°C and with Al mole fraction 28%

Recent investigation [4] on heavy ion radiation has shown that the HEMT drain current degrades by 20% at fluences as high as 3×10^{11}cm^{-2} of 3Mev He^4 ions. In fact the device performance is not significantly affected for fluences lower than 10^{11}cm^{-2}. Beyond this level the device degradation becomes significant and the drain current drops by almost 20% of its pre-irradiation level at a total fluence of about 3.6×10^{11}cm^{-2} (Fig.1), which is in good agreement with the fluence reported in [3,4]. The replacement of the conventional AlGaAs donor and spacer layers with another, grown at lower temperature (510°C) furthere termed as LT-AlGaAs, lead to a significant improvement in radiation hardness. It was observed that these devices can withstand 5MeV alpha particle fluences as high as 6×10^{11}cm^{-2} for a change of 20% in the drain current (Fig.1). The improved performance of these structures is obviously caused by the presence of additional background bulk and interface defects, introduced during the lower temperature growth, which require a larger fluence in order to increase the total defect concentration and hence to affect the device characteristics. The dependence of the normalized drain saturation current (I_{DS}/I_{DSo}) on the total radiation fluence in various type HEMTs including depletion mode ones, which data were obtained from [4], are presented in Fig.1. Assuming a linear degradation of the drain saturation current on the irradiation fluence and fitting the experimental data to this model [3], we found that the corresponding degradation parameters were $(2.6\pm0.2)\times10^{13}$ cm^{-2}

for structure B devices and $(4.9\pm0.4)\times10^{13}$ cm^{-2} for all the others. This clearly shows that the devices with a LT-AlGaAs donor layer exhibit an improved radiation hardness over other HEMT structures. In addition the degradation rate of conventional structures does not seem to depend significantly on the device operation mode, enhancement or depletion, but on other more determinative parameters such as the donor layer and heterojunction interface "quality".

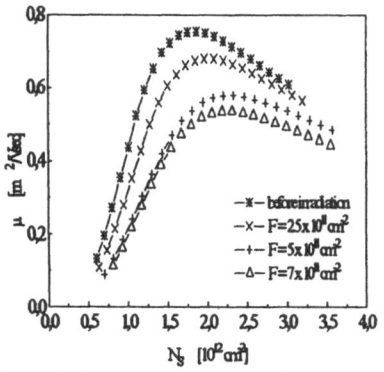

Figure 2. Dependence of the mobility on sheet carrier concentration

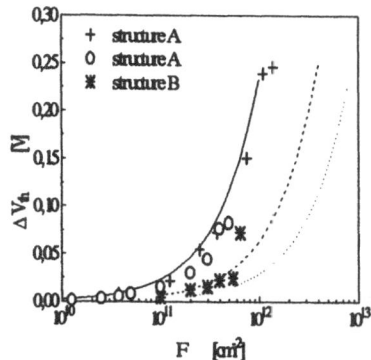

Figure 3. Variation of the threshold voltage.

The electron mobility is sensitive to ionized impurity scattering even at room temperatures. This is more pronounced in HEMTs where the modulation doping technique separates the donors and the conducting electrons thus minimizing the Coulombic interaction. Further screening is achieved by the 2DEG itself. So the issue of the screening efficiency to the radiation induced defects in HEMTs constitutes a critical parameter for the device hardening. For this reason the two dimension electron gas average mobility has been determined [3,5]. A common feature encountered in all μ-N$_S$ characteristics is that the mobility increases with increasing N$_S$, attains a maximum and beyond this the average mobility starts to decrease (Fig.2). The increase of the average mobility with the 2DEG density has been partially attributed to screening of the electron-phonon interaction while the decrease at large N$_S$ values can be attributed to parallel conduction in the AlGaAs donor layer, where the mobility is very low, about 500cm²/Vsec [6]. Results from HEMTs with a conventional buffer layer (Fig.2) and from those with low temperature AlGaAs donor layers showed that the latter exhibit an improved tolerance to alpha particles due to the larger concentration of defects at the interface and in the spacer layer, as expected from the lower temperature growth. In these structures low radiation fluences do not alter the population of the defects and much larger fluences are needed to increase the scattering rate on charged defects and hence to reduce the electron mobility. The mobility degradation parameter was found to be about $(4.25\pm0.21)\times10^{13}$ cm^{-2} and $(1.31\pm0.03)\times10^{13}$ cm^{-2} in conventional structures and structure B ones respectively. The investigation of such structures leads to the conclusion that a compromise can be done regarding the material quality, from the point of view of drain saturation current and carrier mobility, and the HEMT radiation tolerance.

284

Radiation introduces electron and hole traps in the buffer layer. The charge state of these traps depends on their position in the band gap and the distance of Fermi level from the conduction band. The charge state of these traps determines the width and net charge of the buffer layer space charge region. The reduction of the net acceptor concentration increases the depletion region in the GaAs buffer layer. This in turn decreases the electric field at the heterointerface and lowers the sub-bands into the quantum well. In addition the defects that are introduced into the donor layer compensate the donors, decrease the free carrier concentration and lower the Fermi level into the donor layer band-gap.

In order to investigate the contribution of these mechanisms to the shift of the device threshold voltage, a charge control model was introduced, which took into account the changes that are introduced by irradiation into both the buffer and donor layer respectively. A detailed description of this model is presented in [3]. The theoretically calculated shift of the threshold voltage is presented in Fig.3 assuming a typical structure such as structure A. The carrier removal rate in the AlGaAs donor layer was used as fitting parameter and theoretical curves were obtained using values of (dotted line) 6×10^4 cm^{-1} , (dashed line) 10^5 cm^{-1} and (solid line) 3×10^5 cm^{-1}. The degradation parameter of the byffer layer was assumed to be $\beta_A = 0.014$.

Conclusions

The He ion radiation introduces defects in all layers of a HEMT structure. Mechanisms related to these defects affect different device parameters, such as the drain saturation current and electron mobility. Regarding the shift of the threshold voltage, that is the result of a compensation between the subbands lowering into the triangular quantum well and the shift of the Fermi level, towards the conduction band, in the donor layer. The latter is in fact the dominant effect. An important degradation source in HEMTs is the increase of the parasitic series resistances. Finally the donor layer growth temperature plays an important role on the device radiation hardness. It has been shown that the use of a lower temperature grown AlGaAs donor and spacer layers improves the device hardness to He ion irradiation.

Acknowledgments

The present work has been partially supported by the Hellenic GSRT under grant No.87EΔ71 and the U.S. Air Force under grant AFOSR-90-0086.

References

1. R.J. Krantz, W.L. Bloss and M.J. O'Loughin (1988) High Energy Neutron Irradiation Effects in GaAs Modulation Doped Field Effect Transistors: Threshold Voltage, IEEE Trans. Nuclear Science NS-35, 1438-43
2. B.K. Janousek, R.J. Krantz, W.L. Bloss, W.E. Yamada, S. Brown, R.L. Remke and S. Witmer (1989) Characteristics of GaAs Heterojunction FETs and Source Follower Logic Inverters exposed to High Energy Neutrons, IEEE Trans. Nuclear Science NS-36, 2223-7
3. G.J. Papaioannou, M.J. Papastamatiou and A. Christou (1995) He Ion Radiation Effects in HEMTs, J. Appl. Physics, in press
4. Anderson, W.T, Knudson, A.R., Meulenberg, A., Hung, H.L., Rousos, J.A., and Kiriakidis, G. (1990) Heavy Total Fluence Effects in GaAs Devices, IEEE Trans. Nuclear Science NS-35, 1438-70
5. M.S. Shur in "GaAs Devices and Circuits", Ch.10, Plenum Press N.Y., 1986
6. D.C. Look and G. B. Norris (1986) Classical Magnetoresistance Measurements in Al$_x$Ga$_{1-x}$As /GaAs MODFET Structures: Determination of Mobilities, Solid State Electronics 29, 159-65

SINGLE VERSUS DOUBLE CURRENT BISTABILITY IN RESONANT–TUNNELLING DEVICES

T. FIGIELSKI, T. WOSIŃSKI and A. MĄKOSA
Institute of Physics, Polish Academy of Sciences
Al. Lotników 32, 02-668 Warszawa, Poland

Resonant tunnelling of electrons in double-barrier AlGaAs-based hetero-structures attracts a great deal of interest as a fascinating research subject and a powerful research tool in mesoscopic physics. Moreover, resonant-tunnelling devices (RTDs), fabricated by advanced epitaxial-growth methods, have become of a technological importance as they can exhibit negative differential resistance (NDR) region in the current-voltage, $I(V)$, characteristics. This property, together with fast tunnelling transport, brings about a possibility to design electronic devices operating at frequencies up to the Terahertz range.

Actually, however, RTDs very seldom display NDR! Instead, the dominant majority of RTDs can be classified in one of two categories according to their apparent $I(V)$ characteristics. Their resonant maxima in the tunnel current versus bias voltage end in either a *single-step* [1] or a *double-step* [2] feature (compare Figures 1 and 3). Each step represents a current bistability where the apparent current follows the load line of measuring circuit. So, the desired NDR is not exhibited in such devices. Only RTDs with a very low current density display at the resonance the $I(V)$ curve which is smooth, bell-shaped and has well pronounced NDR region. It is noteworthy that the both alternative features can appear in the same device when it exhibits several resonance peaks. In such a case the peaks appearing at lower bias display the single-step feature while those appearing at higher bias - the double-step feature [3]. There is a great deal of confusion in the literature about these features and associated effects that has been the genesis of this paper.

First, we discuss the single-step feature. It is surely a consequence of significant charge accumulation in quantum well of the device. The predicted effect of this charge accumulation is to tip the normally bell-shaped resonance in the $I(V)$ curve over the right to give a Z shaped curve [4, 5]. Such Z shape cannot be revealed by conventional measuring technique, i. e. using a circuit with positive load resistance, since it gives rise to the appearance of a bistable and

J. Novák and A. Schlachetzki (eds.), Heterostructure Epitaxy and Devices, 285–288.
© 1996 *Kluwer Academic Publishers.*

discontinuous characteristic. Instead, it can be done if to apply specially developed technique with an "active load line" equivalent to a voltage source and negative series resistance [5, 6]. Nevertheless, the lower arm of the Z can be revealed with conventional measuring technique by including a large capacitance in parallel with the device, as it is shown in Figure 1. Then, a hysteresis loop appears whose area rises with increasing capacitance. Obviously, the included capacitance cannot affect the static I(V) characteristic of the device but it influences the transient between two stable states.

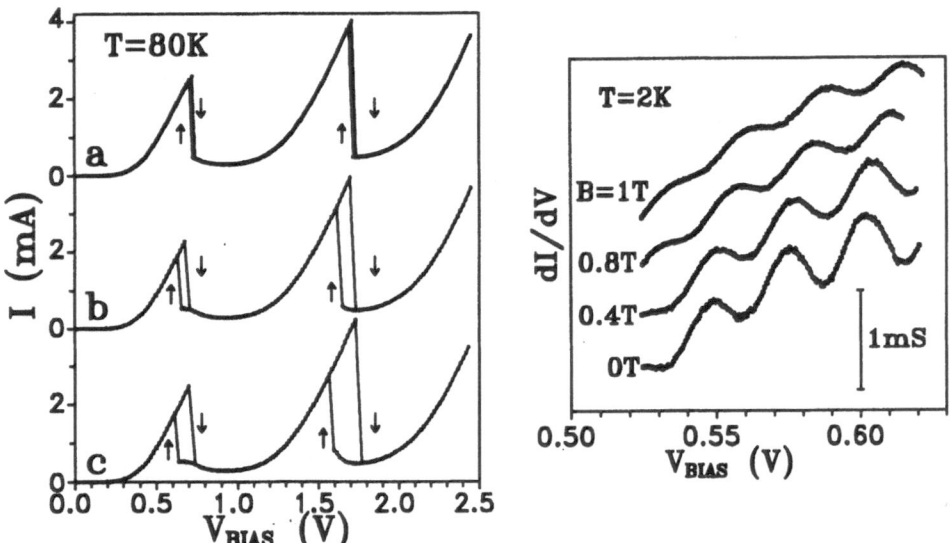

Figure 1. Current-voltage characteristics of resonant-tunnelling device, labelled DBH No 1 in Ref. [1], displaying the single-step feature beyond two resonance peaks. Apparent hysteresis loops rise with increasing capacitance, C, included in parallel with the device. (a): C = 2 nF, (b): C = 25 nF, (c): C = 400 nF.

Figure 2. Fine oscillatory structure in the first derivative of the tunnel current corresponding to the increasing slope of the resonance curve, measured in the same device as in Figure 1. Notice a phase shift of the oscillations with a magnetic field, B, applied normally to the current flow.

In RTDs displaying the single-step feature, we have discovered a fine oscillatory structure of the tunnel current on the increasing slope of the resonance curve [1, 7]. This fine structure behaves in a very specific way in a magnetic field applied normally to the current flow [8]. Positions of the oscillation maxima shift on the bias-voltage scale with increasing magnetic field and their amplitude is rapidly suppressed; Figure 2. It has been explained by a quantum interference of ballistic electrons which escaped from the quantum well and have been partially reflected at a potential step in the collector part of the device [8]. The necessary condition for the appearance of such interference effects is the coherence of wave functions of individual electrons during the resonant tunnelling. This points to *coherent* tunnelling [9] as the dominant process responsible for this kind of *I(V)* curve.

Sequential tunnelling has also been proposed [10] as an alternative process to the coherent one in order to explain NDR in RTDs. The sequential process consists of two independent tunnelling events, i. e. the electron incoming to and outgoing from the quantum well, separated by scattering events in between (both elastic and inelastic) which destroy the electron wave-function coherence in the well. Coherent and sequential interpretations of resonant tunnelling lead to the same predictions for dc current [11] and therefore they can be hardly distinguished in experiment.

Let us now turn to the double-step-feature *I(V)* characteristics; Figure 3. Usually, the shoulder between the two steps is explained as a kind of artifact being due to self-induced oscillations in the region of NDR. Such a view has been, however, criticised for several reasons[2, 12]. We have shown that each step represents a bistable behaviour of the device and the main resonance peak is tipped over the right. This indicates, on one hand, that some charge is accumulated in the quantum well and, on the other hand, that the half-width of the resonance peak is much smaller than the shoulder extension. Moreover, we have shown that LO phonons contribute to this shoulder. We have inferred from these experiments that the shoulder consists of satellite resonances occurring through the coupled electron-phonon states in the well [2]. Oscillations which develop within the shoulder range would then be the result and not the cause of this part of the *I(V)* curve. Evidently, the sequential tunnelling has to be the dominant process of resonant tunnelling in the latter case.

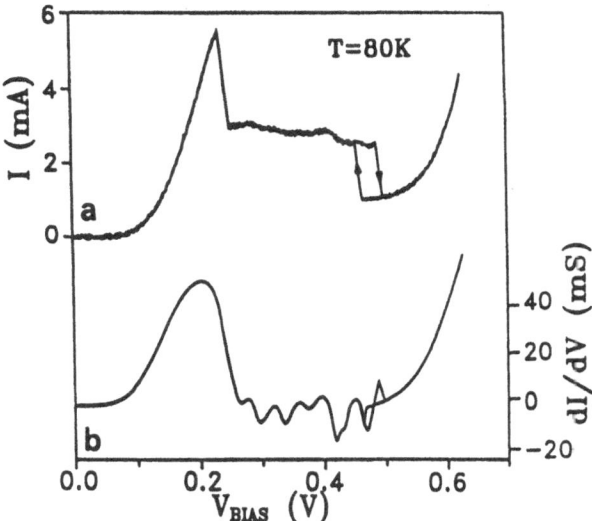

Figure 3. (a) current-voltage characteristic of a device described in Ref. [2] showing the double-step feature. (b) the first derivative of the tunnel current with respect to the bias voltage revealing periodic structure with a period of 36 mV.

The apparent correlation between the dichotomy in the $I(V)$ characteristics of RTDs and the two distinct mechanisms of resonant tunnelling can be, however, misleading. We may only state that the appearance of the fine oscillatory structure indicates the coherent resonant tunnelling while the appearance of the two-step feature indicates the sequential tunnelling.

The dichotomy in question can be tentatively explained by a competition between the charge accumulation and the electron-phonon coupling strength in the quantum well. A crucial parameter would be the escape time of an electron from the well, τ. When τ is sufficiently short neither substantial charge can be stored in the well nor has the electron enough time to emit an LO phonon while dwelling in the well. Only in such a case, the $I(V)$ curve is smooth and has well pronounced NDR region. As τ becomes longer the electron can emit several phonons before it escapes from the well. Then, the resonant tunnelling through a number of coupled electron-phonon states of the well gives rise to a characteristic broad shoulder in the $I(V)$ curve beyond the main resonance peak. As τ increases further the charge accumulated in the well screens the electron-phonon interaction and gives rise to a pronounced Z shape of the resonance peak.

A part of the results presented here has been obtained in collaboration with Prof. A. E. Belyaev and Dr. S. A. Vitusevich from the Institute of Semiconductor Physics in Kiev. This work has been partially supported by the Committee for Scientific Research of Poland under Grant 2 P03B 035 08.

References

1. Figielski, T., Vitusevitch, S.A., Mąkosa, A., Dobrowolski, W., Belyaev, A.E., Wosiński, T., Konakova, R.V., and Kravchenko, L.N. (1995) *Solid State Commun.* **94**, 93.

2. Figielski, T., Mąkosa, A., Wosiński, T., Harness, P.C., and Singer, K.E. (1994) *Solid State Commun.* **91**, 913.

3. Alves, E.S., Leadbeather, M.L., Eaves, L., Henini, M., and Hughes, O.H. (1989) *Solid State Electron.* **32**, 1627.

4. Sheard, F.W. and Tombs, G.A. (1988) *Appl. Phys. Lett.* **52**, 1228.

5. Lerch, M.L.F., Martin, A.D., Simmonds, P.E., Eaves, L., and Leadbeather, M.L. (1994) *Solid State Electron.* **37**, 961.

6. Lerch, M.L.F., Fischer, D.J., Martin, A.D., Zhang, C., and Eaves, L. (1995) EP2DS XI, Nottingham '95, Workbook, p. 177.

7. Vitusevich, S.A., Figielski, T., Mąkosa, A., Wosiński, T., Belyaev, A.E., Konakova, R.V., and Kravchenko, L.N. (1995) *Acta Phys. Polon. A* **87**, 377.

8. Figielski, T., Vitusevich, S.A., Wosiński, T., Mąkosa, A., Belyaev, A.E., and Dobrowolski, W. (1995) submitted to *Phys. Rev. B*.

9. Ricco, B. and Azbel, M.Ya. (1984) *Phys. Rev. B* **29**, 1970.

10. Luryi, S. (1985) *Appl. Phys. Lett.* **47**, 490.

11. Weil, T. and Vinter, B. (1987) *Appl. Phys. Lett.* **50**, 1281.

12. Fu, Y. and Willander, M. (1993) *J. Appl. Phys.* **74**, 3264.

HETEROSTRUCTURE LASERS BASED ON GaSb AND InAs FOR SPECTROSCOPY

HULICIUS,E., ŠIMEČEK,T., HOSPODKOVÁ,A. AND OSWALD,J.
Institute of Physics, Academy of Science,
Cukrovarnická 10, Prague, 162 00, Czech Republic

ALIBERT,C. AND BARANOV,A.N.
Equipe de Micro-optoélectr. de Montpellier, Unité de Recherche
Associé au C.N.R.S. N° 392, Université de Montpellier II,
Montpellier, France

YAKOVLEV,YU.P. AND LITVAK,A.M.
A.F.Ioffe Physical Technical Institute, 194021 St. Petersburg,
Russia

AND

WERLE,P. AND MÜCKE,R.
Frauenhofer Institute für Atmosphärische Umweltforschung,
Kreutzeckbahnstr. 19, 82467 Garmisch-Partenkirchen, F.R.G.

Tunable Diode Laser Absorption Spectroscopy (TDLAS) is an extremely sensitive, precise and reproducible method offering at the same time very good time resolution for the detection of a wide range of trace concentrations of gases.

There are some TDLAS detectable volatile organic compounds like HCHO and HCOOH. There is also urgent need for quick, frequent, cheap and precise measurement of the most important pollutants.

Table 1 from [1] is showing some examples of TDLAS detectable gases like oxides, hydrides, free radicals and other reactive species and their detection limits.

There are of course other methods :

— Cryogenic sampling with electron spin resonance detection - this method requires very long measuring time.
— Chemical chain reactions giving from RO_2 (R = radical e.g. HO_2)in the presence of NO many NO_2 molecules - this method is nonspecific.
— Laser induced fluorescence - nonspecific and less reproducible method.

J. Novák and A. Schlachetzki (eds.), Heterostructure Epitaxy and Devices, 289–292.

TABLE 1. Calculated 1σ-Detection Limits DL for selected molecules and molecular absorption lines for a minimum detectable absorbance of $OD_{min} = 1.10^{-6}$. Assumed are 30mbar pressure in 25m multi path cell @ 296K. $1pptv = 10^{-12}$ volume mixing ratio.

gas	λ [μm]	DL [pptv]	DL^{STP} [$\mu g/cm^3$]
CO_2	4.405	160	0.31
N_2O	4.472	5	0.01
CO	4.610	11	0.014
CH_4	3.260	36	0.026
HCl	3.396	12	0.02
HBr	3.775	96	0.35
H_2CO	3.596	65	0.087
HCN	2.997	23	0.028
CH_3Cl	3.289	3100	7.1
C_2H_2	3.067	33	0.038

- Fourier Transform IR Spectroscopy (FTIR).
- Differential Optical Absorption Spectroscopy (DOAS).
- Light Detection And Ranging (LIDAR).

Besides lower sensitivity all these methods have some other disadvantages.

TDLAS is a general method working with long optical path (25m) using often a low pressure (30 mbar) multiple pass optical cell. The remaining problems of this method are :

- Relatively complicated calibration.
- Laser light sources and their noise, stability and lifetime.

This paper will be devoted to the key problem of the TDLAS method and according to our opinion the most challenging - the tunable diode lasers for the extended near IR part of spectrum. These lasers have to be small, reasonably cheap, sufficiently powerful, stable, easily tunable, highly coherent, with very narrow single mode emission line and reasonable life expectancy.

At present the most frequently used lasers for TDLAS are lead salt diode lasers (e.g. PbSnTe family) making use of the conventional derivative modulation technique or high frequency modulation.

Only the development of heterostructure GaSb based semiconductor lasers with emission wavelength round $2\,\mu$m [2], [3] and recently also improved parameters of InAs based lasers for wavelength range $3 - 4\,\mu$m [4] have opened wide perspectives for the exploitation of these lasers in TD-

LAS. There is an indication, that band to band transitions at heterojunctions of the type II may be used for laser emission especially at wavelengths above 4 μm in near future.

Extended near IR injection Semiconductor Diode Lasers (SDL) have several common features, a positive one being their durability. The long wavelength photons have low energy $(0, 3 - 0, 6\,eV)$ compared with the activation energy of defect migration in the vicinity of the active area of the laser. No degradation of our GaSb based lasers was observed after few years of operation.

On the other hand an important issue of these lasers is the temperature dependence of their parameters. We have measured the temperature dependence of both laser threshold and electroluminescence intensity [5], [6] since these narrow gap materials are more sensitive to the temperature changes. The nonradiative Auger recombination starts to prevail at higher temperatures. The best value of T_o obtained in our experiments up to now is more than 100 K.

GaSb based SDL's for the wavelength range $1.8 - 2.2\,\mu m$ have been studied for a number of years [5] while InAs based SDL's that seem especially suitable for TDLAS have been incorporated in our COPERNICUS project more recently [7]. Examples of the results are shown in the following pictures.

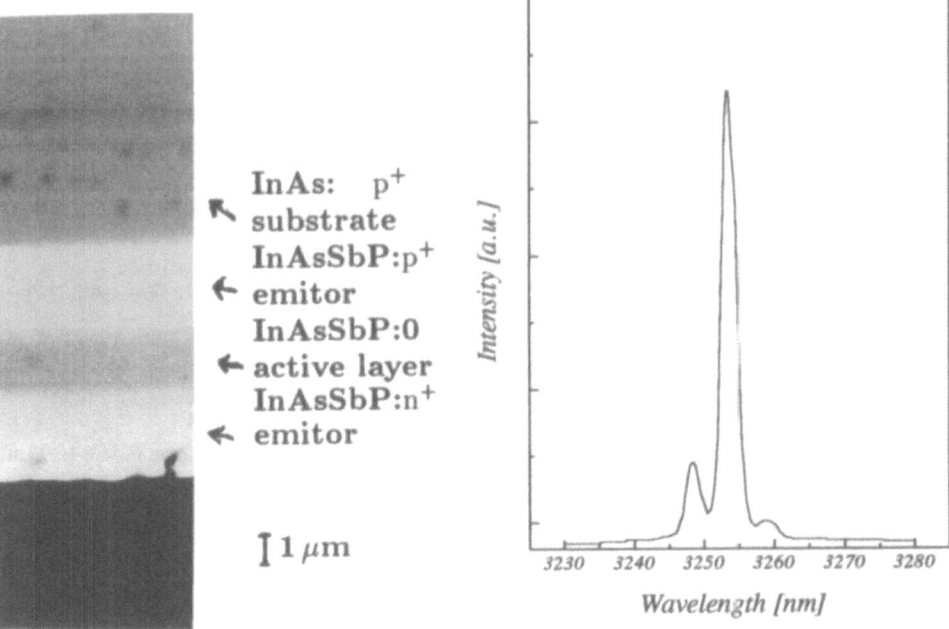

Fig. 1: SEM picture of laser diode structure no. S-277

Fig. 2: $77\,K$ laser electroluminescence spectrum of InAsSbP cw laser no. S 268-1

The up to now prevailing technology in laboratories round the world (USA, Japan, Russia, France) is still LPE because of problems with the preparation of quaternary compounds in modified double heterostructures in materials containing P or Sb. Recently some succesful experiments with growing these structures by MBE and MOCVD have been reported.

Conclusions

GaSb and InAs based heterostructure diode lasers can work continuously at temperatures even above $77\,K$. Under certain conditions narrow line monomode lasers may be prepared which are tunable by the injection current and have reasonable lifetime. Decreasing of the temperature dependence of the laser threshold current and increasing of the minimum temperature for laser operation to at least "Peltier" temperatures, the noise parameters, wavelength stability and monomode behavior have to be improved in order to make TDLAS, using these lasers, the most versatile, sensitive, reliable and cheap trace gas analytical technique.

This work has been supported by the COPERNICUS 1994 project under contract N^o ERBCIPACT940/58.

References

1. Werle,P.(1993) High frequency modulation spectroscopy: a sensitive detection technique for atmospheric pollutants, SPIE, **Vol. 2092** *Substance Detection Systems*, 4-15

2. Dolginov,L.M., Druzhinina,I.N., Eliseev,P.G. et. al. (1978) Injekcionnyj geterolazer na osnove czetyrechkomponentnovo tverdovo rastvora InGaAsSb *Kvantovaja elektronika* **5/3**, 703-704

3. Baranov,A.N., Dzuhartov,B.E., Imenkov,A.N., Shernyakov,Yu.M, Yakovlev (1986) *Sov. Tech. Physs. Lett.*, **12**, 228

4. Baranov,A.N., Imenkov,A.N., Sherstnev,V.V., Yakovlev,Yu.P. (1994) $2.7-3.9\,\mu m$ InAsSb(P)/InAsSbP low treshold diode lasers *Appl. Phys. Lett.* **64**, 2480-2482

5. Hulicius,E., et. al. (1989) Sources and detectors of radiation based on GaSb for $2\,\mu m$ range, *Crystal Properties & Preparation* **19 & 20**, 353-356

6. Joullie,A., Alibert,C., Mani,H., Pitard,F., Tournie,E., Boissier,G. (1988) Characteristic temperature T_0 of $Ga_{0.83}In_{0.17}As_{0.15}Sb_{0.85}/Al_{0.27}Ga_{0.73}As_{0.02}Sb_{0.98}$ injection lasers, *Electronic Letters* **24**, 1076

7. Alibert,C. (1995) Control of enviromental pollution by tunable diode laser absorption spectroscopy in the spectral range $2-4\,\mu m$, *progress report 1 of COPERNICUS 1994, ERB 3512PL940813 (COP-813)*, Montpellier, France

AlAs AND InGaP POTENTIAL BARRIER PHOTODETECTOR GROWN ON VICINAL SURFACES

R.REDHAMMER, J.KOVÁČ, Š.NÉMETH, V.GOTTSCHALCH*[1],
B.RHEINLÄNDER*[2], A.KOVÁČIK AND J.ŠKRINIAROVÁ

*Dept. of Microelectronics, Faculty of Electrical Engineering, Slovak
Technical University, SK-81219 Bratislava, Slovak Republic
*1 Fakultät für Chemie und Mineralogie, Uni. Leipzig, D-04103
Leipzig, Germany
*2 Fakultät für Physik und Geowissenschaften, Uni. Leipzig, D-04103
Leipzig, Germany*

1. Introduction

Recently, a concept of tunnelling photo detector reported [1]. It was based on the modulation of a potential barrier height by the charge optically generated and retained on a potential barrier. Such photodiode works as Camel diode, only p-type doping is replaced by potential barrier forming triangular charge-retaining quantum well [2]. More detailed study on structures with barriers of different materials, such as AlAs and InGaP shows higher optical gain for structures with barrier layer of material with higher hole potential barrier [3]. Here we demonstrate basic behaviours of detector with combination of two barriers of different materials AlAs and InGaP as shown in Figure 1. Also dependence on angle of substrate misorientation is shown.

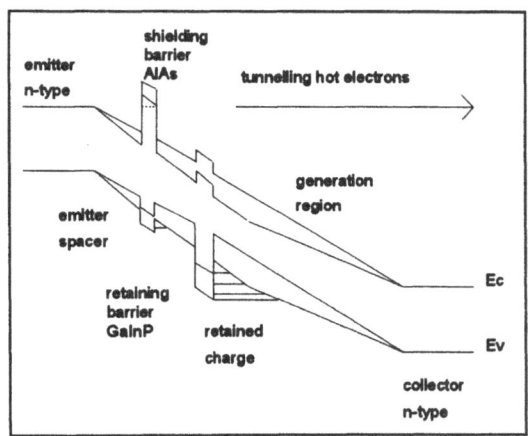

Figure 1. Schematic band gap diagram of a potential barrier photodetector.

293

J. Novák and A. Schlachetzki (eds.), Heterostructure Epitaxy and Devices, 293–296.
© 1996 *Kluwer Academic Publishers.*

2. Structure concept

The original structure was based on a shielding barrier layer placed asymmetrically in a GaAs intrinsic region between two n+ GaAs contacts. Under the applied voltage the electrons are emitted from emitter contact through an emitter spacer and they tunnel through or flow above the shielding barrier. Under illumination, light generates electron - hole pairs mainly in wider collector spacer and the electric field separates them. The generated electrons increase the total electron collector current, while the holes, moving in opposite direction, are retained by valence band edge potential barrier. The hole accumulation changes the potential profile across the structure and reduces the height of the electron shielding potential barrier against the energy of emitted electrons. In this way the accumulation of retained hole charge modulates the emitter current and amplification is obtained. The electron charge retaining effect at the conduction band edge potential barrier [3] is significantly smaller than the hole charge retaining effect due to different effective masses of holes and electrons. The structures with two barriers enable to enlarge the hole charge retain.

3. Samples

Three different types of structures, with barrier of AlAs and InGaP, were grown by MOCVD [4]. Structures NIN4 and NIN6 have one barrier of AlAs and GaInP material respectively and was grown on various vicinal surface substrates (100) $\pm$ 0°, 2° and 6° off towards <110>. Structures HET1 have two potential barriers of both AlAs and GaInP. All orientations of each structure were grown in one growth process. Table 1 shows the actual layer sequences. The thickness of the layers was checked using TEM on samples 2° off. Standard device processing [5] with MESA [6] and area of 0,0026 mm^2 was used.

TABLE 1 Sample layer structure. Numbers in brackets are form the TEM measurements

	NIN4		NIN6		HET1	
	d [nm]	N_d [cm^{-3}]	d [nm]	N_d [cm^{-3}]	d [nm]	N_d [cm^{-3}]
contact layer	300	n+	300	n+	300	n+
precontact layer	50	n = 5x10^{16}	50	n = 5x10^{16}	50	n = 5x10^{16}
spacer	30	i	30	i	30	i
barrier	8,2	i-AlAs	4	i-InGaP	5,1	i-InGaP
spacer II					100	i-GaAs
barrier					7,4	i-AlAs
generation region	70	i	70	i	30	i
contact layer	50	n = 5x10^{17}	50	n = 5x10^{17}	50	n = 5x10^{17}
substrate		n+		n+		n+

4. Results and discussion

4.1 DARK CURRENT CHARACTERISTICS

The I-V characteristics of several diode systems of each structure were measured. The structures polarized with contact layer closer to the barrier layer as emitter shows lower current and higher opening voltage than in opposite polarisation. The emitted electrons overcome the potential barrier by thermoemission or by tunnelling process. The further potential barrier from emitter the lower electric field necessary to overcome the potential barrier by thermoemission. Figure 2 show I-V dark characteristics of NIN4 structures. The InGaP barrier NIN6 structures show much higher dark current comparing to AlAs barrier NIN4 structure. Here extremely small conduction band discontinuity of InGaP material on GaAs gives potential barrier of practically zero height for electrons. HET structure is combination of both barrier layers and therefore the current is not as high as in NIN6 structures.

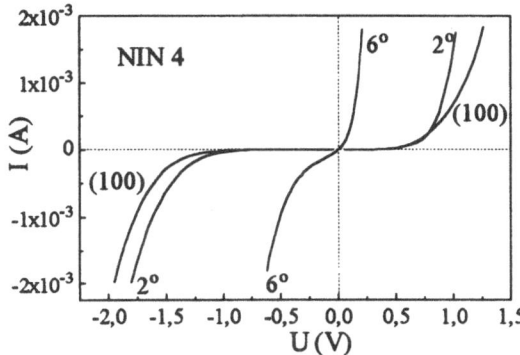

Figure 2. I-V characteristics of structures NIN4 grown on substrates with various vicinity.

4.2 INFLUENCE OF SUBSTRATE MISORIENTATION

Interesting is dependence on substrate misorientation Figure 2. The structures with one potential barrier NIN4 (similarly as other structures) shows strong dependence on misorientation angle. The higher misorientation, the higher current. The higher off orientation, the higher current for the same voltage, or lower voltage for constant current. Origin of this behaviour is not fully clear. It may results from natural anisotropy of band structure of used materials, but may also results from various interface quality, from different deviation of growth speed for different materials on vicinal surfaces, or from different charging of various density of interfaces traps.

4.3 LIGHT CHARACTERISTICS

The I-V characteristics for different incident optical power intensities for each structure were measured. We used 1 mW laser CW lasing on 850 nm wavelength. Substrate misorientation influence was not very evident comparing to dark current characteristics. The most interesting feature observed was transistor-like characteristics of HET structures. The higher incident optical power the higher current plateau in I-V characteristics, as shown in Figure 3. The plateau is more evident in polarisation when the InGaP barrier layer is closer to the emitter and higher current is emitted. HET structure can be understood as a photo transistor with induced base formed by retained hole charge at the potential barrier.

296

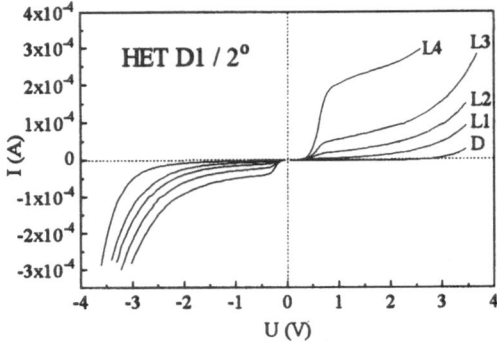

Figure 3. I-V characteristics of HET1 structure for various optical incident power

4.4 LIGHT SENSITIVITY

Sensitivity of these structures was evaluated as a ratio of photocurrent and incident optical power. We used a calibrated LED source of 850 nm wavelength and of 8,2 μW power in optical fibre. The sensitivity is dependent on applied voltage. NIN4 structure with AlAs barrier reaches maximum of 80 A/W. The HET structure shows high sensitivity of approx. 100 A/W for (100) orientation.

5. Conclusion

The charge retaining effect has a significant influence on the injected current and the choose of material used for barrier layer is important. The structure with barrier layer of GaInP shows the highest sensitivity due to large hole charge retaining. HET structures with both types of barrier materials combine the electron barrier AlAs layer for reducing the dark current and InGaP layer with large hole retain barrier. This structure shows transistor like I/V characteristics, that corresponds to the idea of phototransistor with induced base. Physics of strong misorientation current dependence remains unclear.

6. References

[1]   Redhammer, R., Kováč, J. and Németh, Š. (1993) Proc. of Int. Symp. GaAs and Related Compounds 1993, Freiburg Germany, 831-832; IOP Publishing.
[2]   Redhammer, R. and Allsopp, D. W. E. (1992) Proc. of ESSDERC '92, Leuven Belgium pp 899-902; Microelectronics Engineering 19, 899.
[3]   Redhammer, R., Kováč, J., Németh, Š., Gottschalch, V., Rheinländer, B., Kováčik, A., Jakabovič, J., Tomaška, M. and Škriniarová, J. (1994) 4th European Heterostructure Technology, 15-16 Sept. 1994 Schloss Reinsemburg, Gunsburg, Germany.
[4]   Gottschalk, V., Franzheld, R., Keller, St., Kriegel, St., Wagner, G. and Berndorf, G (1993) Fifth European Workshop on Metal-Organic Vapour Phase Epitaxy and Related Growth Techniques, 2-4 June 1993 Malmo, Sweden.
[5]   Kováč, J., Uherek, F., Šatka, A., Smánek, R., Jakabovič, J. (1992) Proceedings of International Conference of Microelectronics, Warsaw, SPIE Vol. 1783, 640-650.
[6]   Gregušová, D., Eliáš, P., Malacký, L., Kúdela, R. and Škriniarová, J. (1995) Phys. stat. sol. (a) 151, 113-118.

GaP-BASED DIODES FOR ELECTROMETRIC APPLICATIONS

I.A.BOCHAROVA, S.A.MALYSHEV, L.N.SURVILO,
Yu.V.TROFIMOV
Institute of Electronics, Academy of Sciences of Belarus
22 Logoiskii tract, Minsk-90, 220841, Republic of Belarus

1. Introduction

When measuring low and extra low currents or voltages the specific requirements are imposed on input stages of electrometric instrumentation. One of them is necessity to protect elements of the input stages from incidental voltages exceeding the levels allowed. This problem can be overcome with the use of protective semiconductor diodes having very low (approx. 10^{-15} A) currents under normal working conditions (0.01-0.1 V) both at forward and reverse bias [1].

Gallium phosphide as a wide gap semiconductor is very suitable for designing such diodes. As a rule, there are two protective diodes being switched in opposite directions in the input stage. The direct current pulse density may be very high during the protection procedure, thus causing the change in p-n junction parameters or full degradation of the diode. Therefore, another important requirement is very high reliability of the diodes.

In this report GaP p-n junction current has been studied as a function of forward and reverse bias voltages. Some of the GaP diodes parameters which would be applicable as informative ones for failure prediction are reported here.

2. Experiment

Devices examined were planar p-n junctions formed in n-type GaP wafers doped with Te (100) to $3.2 \cdot 10^{17}$ cm^{-3} concentration. P-type regions of 250 μm diameter and doping density of $N_a = 6.1 \cdot 10^{17}$ cm^{-3} were formed in GaP at a depth of about 0.8 μm through windows in silica mask by local Zn-diffusion in open gas flow system [2]. A 0.25 μm silica layer was used as a Zn-diffusion mask. It was formed by chemical vapor deposition of SiO_2 from tetraethoxysilane at 330°C. Ohmic contact to p-type GaP was formed on the Zn-enriched semiconductor surface after diffusion process. Palladium contact was used in case of p-type GaP, and indium-nickel alloy in case of n-type. Front and back contacts were formed using standard

297

J. Novák and A. Schlachetzki (eds.), Heterostructure Epitaxy and Devices, 297–300.
© 1996 *Kluwer Academic Publishers.*

298

photolithography and metal deposition techniques in vacuum. After wafer separation the diodes were mounted in packages with use of sapphire isolators so that lead-to-lead resistance was not less than 10^{15} Ohm.

Current-voltage (I-V) measurements have been taken for the diodes with use of B7-45 Electrometer at room temperature. The resolution of the voltage and current meter was 100μV and 0.01 fA, respectively [3].

To establish a set of informative parameters for prediction GaP-diodes serviceability the analysis of forward and reverse I-V curves before and under aging the devices in the forced current mode (j = 40 A/cm^2) during 270 h at room temperature was made. The I-V curves were analysed for diodes of different batches (10 pieces) with currents not more than 10^{-14} A at U = 100 mV before aging [4].

3. Results and Discussion

Typical forward and reverse bias I-V curves for the GaP diodes shown in Figure 1 have been obtained from fitting data taken at 290 K between 0.01 and 0.6 V. These are representative of a large number (10 or more) of diodes.

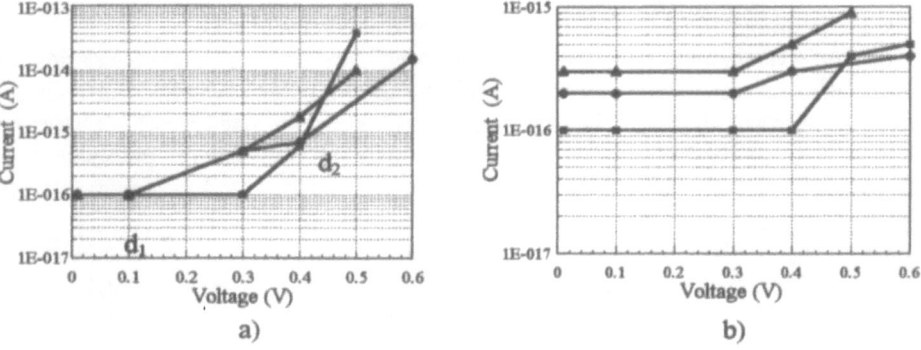

Figure 1. Forward (a) and reverse (b) bias1 current I -V curves for GaP p-n diodes:
$$N_d = 3.2 \cdot 10^{17} \text{ cm}^{-3}, T = 290 \text{ K. Diode area} = 0.0005 \text{ cm}^2.$$

The p-n junction total current consists of several components which have their origin in different processes. There are generation-recombination, diffusion, tunneling, and surface leakage currents [5, 6]

$$I = I_{gen} + I_{diff} + I_{tun} + I_{sur} \tag{1}$$

The last term in (1) is not a fundamental one and depends mainly on p-n junction passivation techniques.

Forward bias I-V curves at voltages less than 0.6 V consist of two regions. The first region (d_1) is due to linear tunneling current and described by $I_{tun} = V/R_{sh}$. Shunt resistance (R_{sh}) is identified as a combination of multistep tunneling and thermal trapping-detrapping of carriers through the defect states in the space-charge region, and V is a voltage applied to the p-n junction. The second region (d_2) is due to exponentional tunneling current of the form $I_{tun} = I_0 \exp(BV)$, where I_0 and B are empirical parameters [6 - 8].

It is well known that Igen dominates in reverse bias current for wide gap semiconductor based p-n junction and may be explained by the recombination of electrons and holes through the defect states within the junction space-charge region [5, 9].

To reveal the degradation mechanisms of GaP diodes the changes in I-V curves under action of the great forward current density have been studied. The Figure 2 represents the investigation results for diodes fabricated in the same technological run.

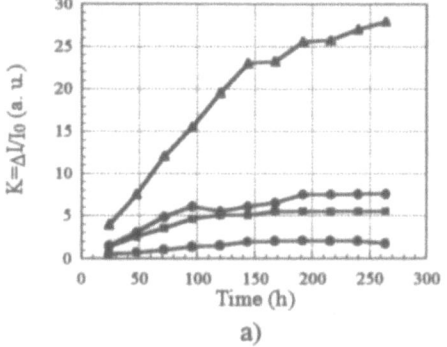
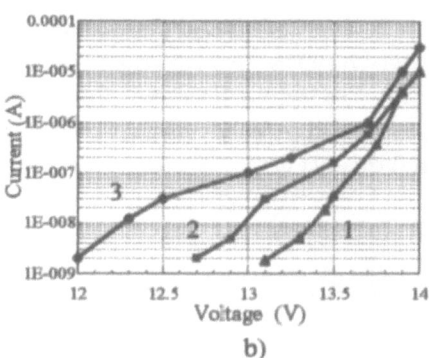

a)   b)

Figure 2.. GaP diodes forward current variation versus aging time (a) and
reverse bias I-V curves (b) before (1), after 90 h aging (2), after 270 h aging (3)

Figure 2(a) shows the ratio of forward current change ΔI to its value I_0 before aging (K = $\Delta I/I_0$) versus aging. As it follows from the curves, degradation process of the diodes consists of two stages: fast and slow ones. The most severe degradation takes place during the first 90 to 120 aging hours. It may be seen that the current change rates $\Delta K = K/\Delta t$ differ for various diodes on the fast degradation stage. Figure 2(b) shows the post-aging reverse bias I-V curves. The analysis of forward and reverse I-V curves before and after aging has shown that the forward current change rates ΔK and reverse bias voltage changes ΔU_{rev} at fixed reverse current $I_{rev} = 5 \times 10^{-9}$ A are substantially different for diodes studied. Comparison of these parameters give the following results: for diodes with $\Delta K < 3 \times 10^{-2}$ h^{-1} and ΔU_{rev} about 0.35 V the currents in the initial region of the I-V curves at U = 100

300

mV increase 1.5-3 times, and for diodes with $\Delta K > 5 \times 10^{-2}$ h^{-1} and ΔU_{rev} about 0.6 V the currents increase by an order of magnitude.

It can be noted that in III-V semiconductor structures degradation processes localize mainly in space charge region. It is due to strong electric field which stimulates drifting the charge point defects and trapping the carriers by deep centers in p-n junction depletion layer. As a rule, the forward and reverse currents are rising because of electron-hole recombination rate increase and electron tunneling through the p-n junction deep levels [10]. The studies carried out for many diodes allow us to conclude that ΔK and ΔU_{rev} can be taken as informative parameters for reliability prediction.

4. Conclusions

Thus, on the basis of the investigation results the following conclusions can be made:

1. Low and extra low current GaP diodes for application in input circuits of electrometric instrumentation have been obtained.

2. The GaP diodes parameters degradation process consists of two stages: fast and slow ones. The fast degradation stage lasts about first 90 to 120 hours of aging.

3. The forward current change rate ΔK in the fast degradation stage and reverse bias voltage change ΔU_{rev} at the fixed value of reverse current comprise a set of informative parameters to predict GaP diodes reliability.

5. References

1. Miles, R. (1979) Supersensitive measurement demands critical input design, *Electronics* **20**, 145-149.
2. Emelyanenko, Yu.S., Gushchinskaya, E.V., Malyshev, S.A., Privalov, V.I., and Zaitsev, I.I. (1989) Diffusion of zinc into III-V compounds in open gas flow system, *Crystal Properties & Preparation* **19&20**, 141-144.
3. Aslamov, P.P., Vedenin, A.S., Volodkevich, A.A., Vysotskii, K.S., Malyshev, S.A., Matskevich, T.P., Survilo, L.N., and Trofimov, Yu.V. (1990) GaP-based diodes with low leakage currents, *Elektronnaya Promyshlennost* **10**, 119-120.
4. Bocharova, I.A., Vedenin, A.S., Malyshev, S.A., and Nikonchuk, I.A. (1994) Degradation failure prediction for GaP-based protective diodes, *Vestsi Akademii Navuk Belarusi (Serya Fisiko-Matematychnykh Navuk)* **2**, 91-94.
5. Sze, S.M. (1981) *Physics of Semiconductor Devices*, A Wiley-Interscience Publication, John Wiley & Sons, New York.
6. Reinhardt, K.C., Yeo, Y.K., and Hengehold, R.L. (1995) Junction characteristics of Ga$_{0.5}$In$_{0.5}$Pn$^+$-p diodes and solar cells, *J. Appl. Phys.* **77(11)** 5763-5772.
7. Banerjee, S., and Anderson, W.A. (1986) Temperature dependence of shunt resistance in photovoltaic devices, *Appl. Phys. Lett.* **49(1)**, 38-40.
8. Riben, A.R., and Feucht, D.L. (1966) nGe-pGaAs heterojunctions, *Solid-State Electronics* **9**, 1055-1065.
9. Sah, C.-T., Noyce, R., and Shockley, W. (1957) Carrier generation and recombination in p-n junctions and p-n junction characteristics, *Proc. IRE* **45**, 1228-1243.
10. Ptashchenko, A.A. (1980) Light emitter diodes degradation, *Zhurnal Priklad noi Spectroskopii* **5**, 781-803

DESIGN OF InGaAs/InAlGaAs/InP RCE PIN PHOTODIODE

A. ŠATKA, D. W. E. ALLSOPP[*1], J. KOVÁČ, F. UHEREK,
B. RHEINLÄNDER[2], V. GOTTSCHALCH[*3]
*Department of Microelectronics, Slovak Technical University,
Ilkovičova 3, SK-812 19 Bratislava, Slovak Republic*
[*1]*Department of Electronics, University of York, Heslington,
YO1 5DD York, United Kingdom*
[*2]*Fakultät für Physik und Geowissenschaften, Universität Leipzig,
Linnéstrasse 3, D-07010 Leipzig, Federal Republic of Germany*
[*3]*Fakultät für Chemie und Mineralogie, Universität Leipzig,
Linnéstrasse 3, D-07010 Leipzig, Federal Republic of Germany*

High-sensitive and high-speed photodetectors are of great importance in high-bit-rate optical communication systems, information processing and ultrafast measurement techniques. PIN heterostructure photodiodes are widely used for these applications. Quantum efficiency η of a conventional PIN structure photodiodes is limited mainly by the surface reflection R and absorption in active layer. Thickness d_{abs} of absorption layer is usually made of twofold of absorption length $1/\alpha$. Along with antireflection coating of the semiconductor surface it ensures achievement of the photodiode quantum efficiency $\eta \cong 85\%$. For the high-speed applications, time photoresponse of PIN structure photodiodes is limited by the photocarrier sweep-out from the space charge region of thickness $w = d_{abs}$ in case of negligible influence of heterojunctions. The corresponding frequency bandwidth $B_v = 0.45w/v_s$ depends on saturation velocity v_s of slower carriers. B_v can be increased by the use of a thinner absorption layer in expense of decrease of η, increase of diode capacitance C_d and decrease of the frequency bandwidth to the value $B_{RC} = 1/(2\pi R_L C_d)$. As an example the bandwidth-efficiency product for InGaAs/InP PIN photodiode of 10x10 μm area and 700 nm thick absorption layer reaches the value $\eta B \cong 15$ GHz.

Resonant cavity-enhanced (RCE) PIN photodiode (Fig. 1a) with undoped absorption layer embedded in the cavity of the asymmetric optical resonator is one of the most promising solution. This allows decrease of the absorption layer thickness ($d_{abs} \ll 2/\alpha$), selective increase of the η at resonant wavelength λ_{design} and substantial increase of ηB. If absorption in exterior of the absorption layer is negligible and standing electric wave is developed in optical resonator cavity, then quantum efficiency η of the RCE PIN structure is [1]

$$\eta = \frac{\left(1 + R_2 e^{-\alpha d_{abs}}\right)}{\left(1 - \sqrt{R_1 R_2} e^{-\alpha d_{abs}}\right)^2} (1 - R_1)\left(1 - e^{-\alpha d_{abs}}\right). \qquad (1)$$

301

J. Novák and A. Schlachetzki (eds.), Heterostructure Epitaxy and Devices, 301–304.
© 1996 *Kluwer Academic Publishers.*

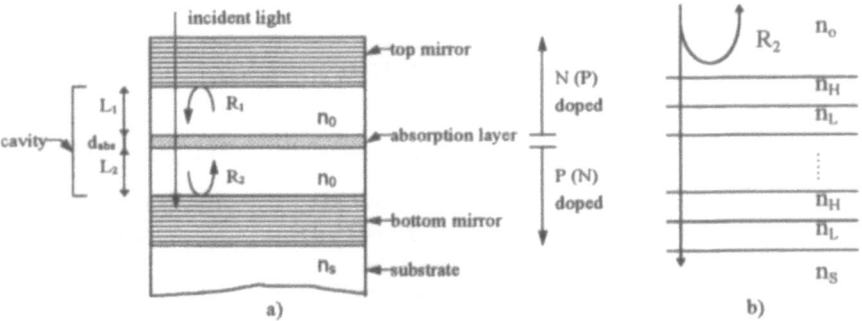

Figure 1. a) Schematic cross-section of RCE PIN photodiode, b) schematic cross-section of Bragg mirror.

As it follows from (1), reflectivity R_2 of the bottom mirror should be as high as possible, in practice it should be $R_2 \geq 0.9$. Despite of the other less important possibilities, it can be made by pairs of alternating semiconductor layers of high n_H and low n_L refractive index (Fig. 1b), each layer being $\lambda_{design}/4n$ thick. Its reflectivity R_2 increases with number N of $n_H n_L$ layer pairs [2]

$$R_2 = \left(\frac{1 - \dfrac{n_0}{n_s}\left(\dfrac{n_H}{n_L}\right)^{2N}}{1 + \dfrac{n_0}{n_s}\left(\dfrac{n_H}{n_L}\right)^{2N}} \right)^2 , \quad \text{where} \quad N = 0.5 \, \text{Int} \left| \ln\left[\frac{1 - \sqrt{R_2}}{\dfrac{n_0}{n_s}\left(1 + \sqrt{R_2}\right)}\right]^{\ln\left(\frac{n_H}{n_L}\right)} \right| . \quad (2)$$

In these equations, n_0 is refractive index of the semiconductor layer in the cavity and n_s refractive index of the substrate. The same type of mirror can be used for the top mirror. From technological point of view the use of the semiconductor/air interface of reflectivity $R_1 = (n-1)^2/(n+1)^2 \cong 0.3$ is a much convenient.

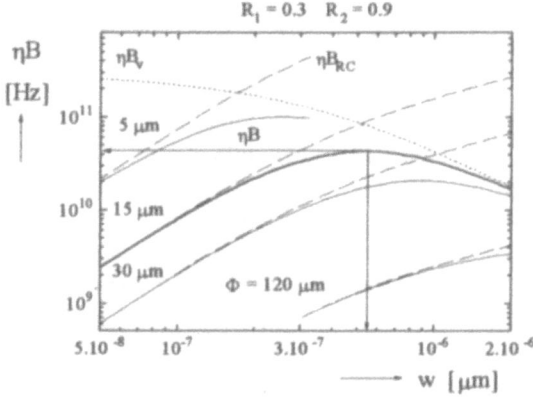

Figure 2. Calculated dependence of the bandwidth-efficiency product of RCE PIN photodiode on thickness of space charge region for different PN junction diameter Φ.

Dependence of the bandwidth-efficiency product ηB on thickness of the absorption layer $d_{abs} = w$, calculated using (1) along with expressions for B_v and B_{RC}, is shown in Fig. 2. Reflection delay in Bragg reflector has been neglected as it is of the order of a few femtoseconds [3]. The optimal thickness of the space charge region w for a given diameter Φ of the photodiode can be find from these dependencies. For monomode fibre applications appropriate photodiode diameter is $\Phi = 15$ μm. For this diameter ηB reaches maximum $\eta B = 43$ GHz for $w = 0.55$ μm and $\alpha = 10^6$ m^{-1}. As it follows from (1), quantum efficiency η reaches maximum for $R_2 = R_1 e^{2\alpha d_{abs}}$. For $R_1 = 0.3$ and $d_{abs} = 0.55$ μm the reflectivity of the bottom mirror should be $R_2 = 0.9$. This reflectivity also ensures efficiency higher than 90 % at optimal ηB product.

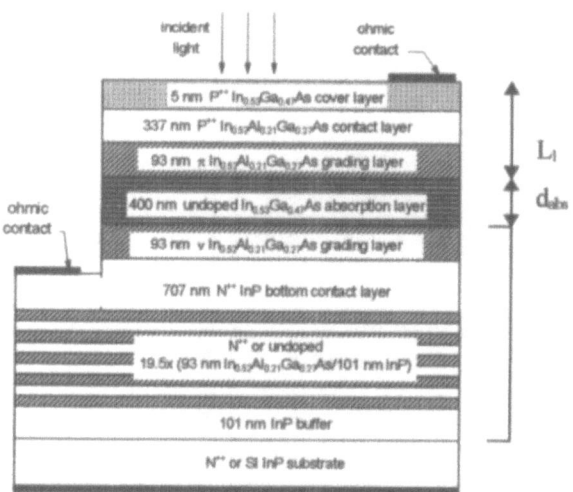

Figure 3. Schematic cross-section of the designed RCE PIN photodiode structure.

Schematic cross-sectional view of the designed photodiode is shown in Fig. 3. $In_xGa_xAs_yP_{1-y}/InP$ and $(In_{0.52}Al)_z(In_{0.53}Ga_{0.47})_{1-z}As/InP$ are two the most promising material systems for design of the RCE PIN photodiodes working at 1300 or 1550 nm wavelength. As it follows from equation (2), the number of the quarter-wavelength layer pairs N in bottom Bragg mirror decreases with increase of the ratio n_H/n_L. From the technological point of view this ratio should be as high as possible. Because the ratio $n_H/n_L = 3.4/3.22$ for InGaAsP/InP and $n_H/n_L = 3.5/3.22$ for InAlGaAs/InP at $\lambda_{design} = 1300$ nm, the last material system has been chosen for the designed RCE PIN photodiode. Reflectivity $R_2 \cong 0.9$ can be achieved for $N = 20$ of $In_{0.53}Al_{0.21}Ga_{0.26}As/InP$ quarter-wavelength layer pairs. Another important advantage of InAlGaAs/InP system is a transition from Type I to Type II of the heterojunction and disappearing of the conductance band offset for a total Al content of $x \cong 0.2 \div 0.25$ [4]. The N-type doping of $In_{0.53}Al_{0.21}Ga_{0.26}As/InP$ mirror therefore ensure low resistance of the bottom mirror layer stack.

$3\lambda_{design}/2n$ thick N^{++} doped InP contact layer, inserted into the bottom Bragg mirror has practically no effect to the optical properties. This layer enables to prepare "planar" structure photodiode and supresses the problems of N$^+$ doping of the bottom mirror. Thickness of the bottom part of the optical cavity is set to $L_2 = 0$. $In_{0.53}Ga_{0.47}As$ absorption layer on the top of the bottom mirror ensures a high absorption at 1300 nm and 1550 nm wavelengths. InAlGaAs of thickness L_2 is used for the upper part of the

304

optical cavity. It has higher reflectivity at the semiconductor/air interface in comparision with InP. The top, very thin $In_{0.53}Ga_{0.47}As$ layer of negligible total absorption prevents the oxidation of InAlGaAs layer and improves properties of the top ohmic contact. Thickness of the layers in optical cavity has been calculated by optical transfer matrix method [2] to obtain maximum reflectivity at working wavelength λ_{design} = 1300 nm. Following previous assumptions on ηB product, thickness of the space charge region and absorption layer has been taken as a parameter.

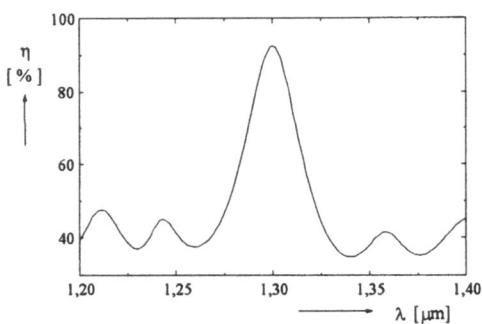

Figure 4. Quantum efficiency spectrum of the $In_{0.53}Ga_{0.47}As/In_{0.53}Al_{0.21}Ga_{0.26}As/InP$ RCE PIN photodiode designed for 1300 nm.

In summary, calculated spectral dependence of quantum efficiency of designed RCE PIN photodiode is shown in Fig. 4. Maximal efficiency is η = 92 % at working wavelength λ_{design} = 1300 nm. Expected bandwith-efficiency product of the photodiode is ηB = 43 GHz for photodiode diameter Φ = 15 μm.

The work has been supported by the EU program COPERNICUS, DEMACOMINT project #12 283, and by the Slovak Grant Agency project #1738/95. The MOCVD growth at Leipzig University was partially supported by the TEMPUS project IMG-94-SQ-1126.

References.

1. Kishino K., Ünlü M.S., Chyi J.I., Reed J., Arsenault L., and Morkoç H. (1991) Resonant Cavity-Enhanced (RCE) Photodetectors. *IEEE J. Quantum Electron.* **27** (8), 2025-2034.
2. Knitl Z. (1976) *Optics of thin films (An Optical Multilayer Theory).* John Wiley & Sons, London, 548 p.
3. Babic D.I. and Corzine, S.W. (1992) Analytic expressions of the reflection delay, penetration depth, and absorptance of quarter-wave dielectric mirrors. *IEEE J. Quantum Electron.* **28** (2), 514-524.
4. Sacilotti M., Motisuke F., Monteil Y, Abraham P., Iikawa F., Montes C., Furtado M., Horiuchi L., Landers R., Morais J., Cardoso L., Decobert J., and Waldman B. (1992) Growth and characterization of type-II / type-I AlGaInAs/InP interface. *J. Cryst. Growth* **124**, 589-595.

InP/GaInAs MSM PHOTODETECTOR FOR SIMPLE INTEGRATION IN HEMT CIRCUITS

M. HORSTMANN, M. MARSO, K. SCHIMPF, H. HARDTDEGEN,
M. HOLLFELDER, AND P. KORDOŠ
Institute of Thin Film and Ion Technology (ISI)
Research Centre Jülich GmbH (KFA)
D-52425 Jülich, Germany

Abstract. We report on the optoelectronic properties of a novel Metal-Semiconductor-Metal (MSM) photodetector (PD) at 1.3µm wavelength. The photodetector is based on a High Electron Mobility Transistor (HEMT) layer structure and can easily be integrated into HEMT receiver circuits. Optoelectronic measurements of photodetectors with interdigitated semitransparent Schottky-contacts show an 8.4GHz bandwidth and a responsivity of 0.26A/W. HEMTs processed on the same wafer have f_T and f_{max} values of 45GHz and 85GHz at 0.36µm gatelength.

1. Introduction

Different procedures exist for the growth of monolithically integrated photoreceivers. One approach is sequential growth with an insulating layer between the photodetector (MSM-PD or pin-PD) and amplifier (HEMT or FET) layers [1]. An advantage of this procedure is that only a single epitaxial growth is needed. Disadvantages are the non planar surface, the parasitic electronic coupling between doped layers and temperature degradation effects caused by propagation of dislocations during the growth process. To avoid the non planarity the epilayer growth is often initiated on substrates which have grooves etched to the appropiate depth so that the complete circuit will be planar. Another approach is two step or selective epitaxial growth. The photodetector and the amplifier layers are grown each for themselves [2]. Therefore parasitic coupling is avoided. One problem of this method is the very critical wafer preparation before the regrowth step. Besides that, another problem is the temperature degradation during the growth process.

In the last time considerable attention is given to the preparation of the devices on identical layer structures to avoid all these problems. This concept has been applied only to pin-PD/HBT counterpart but the best microwave performance has been shown on GaInAs-HEMTs [3]. We report on the investigation on photodetectors and HEMTs for long wave-length systems, based on a InP/GaInAs HEMT structure, which can be easily integrated to photoreceiver circuits due to the same layer structure of both devices [4].

2. Layer structure of MSM PD

The layer structure (Fig. 1) is prepared by LP-MOCVD on (100) oriented semi-insulating InP substrates. The 2DEG is formed in an 8nm thick strained $In_{77}Ga_{23}As$

305

J. Novák and A. Schlachetzki (eds.), Heterostructure Epitaxy and Devices, 305–308.
© 1996 *Kluwer Academic Publishers.*

layer with a carrier supply layer below the channel. Above the 2DEG layer sequence a lattice matched GaInAs layer with 125nm-175nm thickness is grown for the absorption of infrared light. To avoid all Al-related problems like ageing, lack of etch control and kink effects the conventional used InAlAs Schottky-barrier enhancement layer is replaced by a p-doped InP layer [5]. The 15nm i-InP barrier layer reduces diffusion of acceptor atoms to the layers below. On the top of the structure a 20nm thick n^+-doped

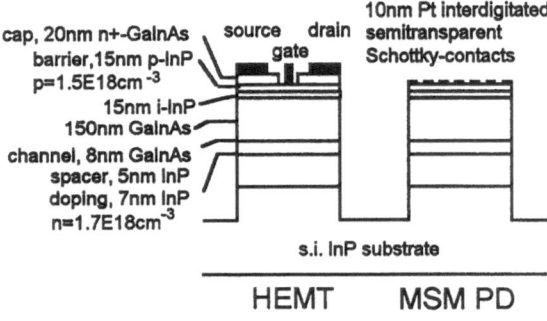

GaInAs layer, which is not used for the MSM PD, is grown to reduce the ohmic contact resistivity of the HEMT. The HEMT can be fabricated simultanously with the photo-detector. Hall measurements yield a 2DEG sheet carrier density of about 1.5 $*10^{12}$cm^{-3} and a mobility of about 12600cm^2/Vs at 300K. Device fabrication consists of standard optical and E-beam lithograpy , metal evaporation and lift-off processes. E-beam litho-graphy is used to define the gate of the HEMT and the finger structure of MSM PD. Due to the 2DEG the electric field lies mainly below the finger-shaped Schottky-contacts of the PD, perpendicular to the surface, leading to a lower electric field between the contact fingers. This lower electric field zone can be avoided by reducing the finger spacing down to subμm dimension. Because the 2DEG induces a strong electric field below the Schottky contacts, these contacts can be carried out semitransparent by evaporating 10nm Pt to enhance the optical active area without degradation of the bandwidth [6].

Fig. 1. Cross section of HEMT and MSM PD on the same layer sequence

3. Measurements

Optoelectronic on wafer measurements are performed by using an optoelectronic HP83420A lightwave test set in combination with an HP8510B network analyzer.

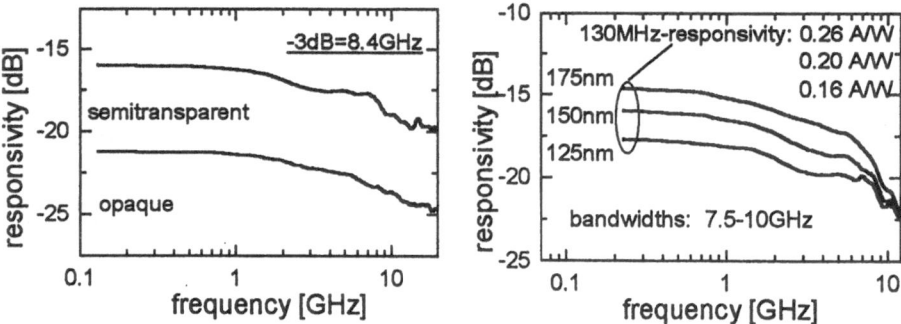

Fig. 2. 1.3μm frequency response of MSM PD above the 2DEG at 5V applied bias and with 0.5μm fingerspacing and width: (left) Comparison between opaque and semitransparent Schottky-contacts, (right) 130MHz responsivity for different GaInAs absorption thicknesses

The measurement frequency range lies between 130MHz and 20GHz. The light of 1.3μm wavelength is directed on the device by a glas fiber. The frequency domain response was measured by the network analyzer.

Electric and optoelectronic measurements are performed on finger-shaped MSM PD above the 2DEG, with the same dimension of finger spacing and -width. The devices have an area of about 50μm*50μm. A saturation current density of about 100pA/μm^2 and a breakdown voltage of about 30V are measured. The saturation capacitance at 5V bias has a value of about 60fF resulting in a small RC constant [7].

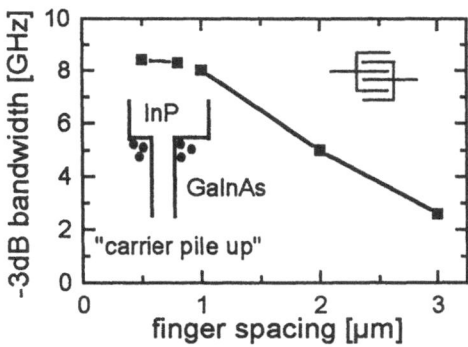

Fig. 3. -3dB bandwidth as function of finger spacing at 5V bias voltage

Fig. 2. (left) shows the frequency response of a device with 0.5μm fingerspacing. The PD has a -3dB bandwidth of 8.4GHz. Measurements on devices with opaque fingers show a similar frequency dependence, but a much smaller responsivity. This clearly demonstrates that the use of semitransparent Schottky-contacts enhances the responsivity of the MSM PD above the 2DEG without decreasing the bandwidth. The low frequency responsivity at 130MHz depends on the GaInAs absorption layer thickness and has a value of 0.16A/W up to 0.26A/W (Fig. 2. (right)). Fig. 3. shows the bandwidth of the device as function of finger spacing for a device with 150nm GaInAs absorption layer thickness. As predicted from theory the bandwidth increases with decreasing finger spacing, but at finger spacings below 1μm a saturation of bandwidth is observed [8], [9]. This saturation is caused by carrier pile up at the top InP/GaInAs interface and is subject of further investigation [10].

4. HEMT performance

InP/GaInAs HEMTs and MSM PDs above the 2DEG are fabricated on the same wafer [11]. The HEMTs are not expected to have record RF properties because the layer system is a compromise between photodetector responsivity and HEMT performance. RF measurements are performed on wafers with an Alessi probe station and a HP8510B network analyzer from 45MHz to 26GHz. Transistors have been fabricated with different gatelengths from 1.1μm down to 0.36μm and different GaInAs layer thicknesses (175nm, 150nm and 125nm, Fig. 4.). The values for

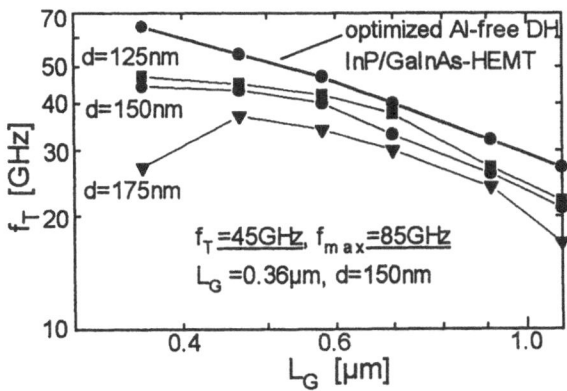

Fig. 4. Cutoff frequency f_T versus gatelengths L_G of HEMTs on detector suitable layers for different GaInAs absorption layer thicknesses

308

HEMTs in the same material system with optimized layer structure (double heterostructure) are included and are only 15% higher at gatelenghts above 0.5μm. With increasing GaInAs thickness the cutoff frequency decreases and at gatelengths below 0.5μm for every GaInAs thickness a saturation of cutoff frequency is observed. This behaviour is caused by short channel effects which dominate the rf characteristic of the HEMT at small gate to channel aspect ratios.

Summarizing best values of f_T=45GHz and f_{max} =85 GHz were measured at a gatelength of 0.36μm and with a GaInAs absorption thickness of d=150nm which is also a suitable choice for the MSM PD above the 2DEG.

5. Conclusion

We have demonstrated the capabilities of MSM photodetectors and HEMTs on an identical layer system. Therefore both devices can easily be integrated to complex receiver circuits. The photodetector has a responsivity up to 0.26A/W, without use of an antireflexion coating. The -3dB bandwidth of a device with 0.5μm finger spacing and -width and at 5V applied bias is 8.4GHz. The transistor cutoff frequency is 45GHz and f_{max}=85GHz at 0.36μm gatelength.

Further works will be the investigation of carrier pile up effects at the InP/GaInAs interface of the photodetector, responsivity enhancement by applying an anti-reflexion coating and fabrication of a monolithically integrated photoreceiver.

References

[1] P. Fay et al. (1995), Electron. Lett., 31, pp. 755-756

[2] R. Lai et al. (1991), Electron Lett., 27, pp. 364-366

[3] A.L. Gutierrez-Aitken et al. (1995), Proceedings of the seventh International Conference on InP and Related Materials, Hokkaido, Japan, p. 357

[4] M. Horstmann et al. (1995), Appl. Phys. Lett., 67, p. 106

[5] P. Kordoš et al. (1992), J. Appl. Phys., 72, pp. 2347-2355

[6] I. Adesida et al. (1994), Proceedings of the sixth International Conference on InP and Related Materials, Santa Barbara, California, USA, pp. 284-287

[7] P. Kordoš et al. (1992), Electron. Lett., 28, pp. 1689-1690

[8] I.S. Ashour et al. (1995), IEEE Trans. Electron. Devices, 42, p. 231

[9] J.B.D. Soole et al. (1990), IEEE Trans. Electron. Devices, 37, pp. 2285-2291

[10] J.H. Burroughes et al. (1991), IEEE Photonics Technol. Lett., 3, pp. 532-534

[11] M. Horstmann et al. (1995), Proceedings of the 25th European Solid State Device Research Conference, den Haag, the Netherlands, pp. 443-446

NATURE OF INTERNAL FEEDBACK IN THE SELF-ELECTRO-OPTIC EFFECT DEVICES

F. ŠROBÁR

Institute of Radio Engineering and Electronics, Czech. Acad. Sci.
Chaberská 57, CZ-182 51 Praha 8, Czech Republic

1. Introduction

The self-electro-optic effect device (*SEED*) [1] consists of a *p-i-n* structure connected via series resistor to *dc* voltage drive. With reverse bias and light of suitable wavelength impinging perpendicularly on the junction planes, this arrangement exhibits the attributes of both the detector and the electroabsorptive modulator. Under suitable conditions, this combination can lead to optical bistability, a feature valued in the context of the emerging parallel information processing technology. *SEED* has also other merits that make it a device of choice for optical computing.

The intrinsic (*i*) region is composed of multiple quantum well (*MQW*) stack of, usually, $A^{III}B^V$ layers. Due to possibility of formation of quasi-two-dimensional excitons the *MQW* layer absorption features are highly sensitive to changes in the electric field. This so-called quantum-confined Stark effect (*QCSE*) greatly improves the modulating properties of the *MQW* layer as compared with the bulk semiconductor material.

The presence of internal feedback in *SEED* has been qualitatively appreciated at the very outset [1]. Even though the feedback concept is implicit in any adequate treatment of the subject, the description is for the most part vague and only qualitative. We propose a topological (diagrammatic) representation that affords to locate the feedback relationships in a precise way. The method, which is a version of the signal flow graphs, is described, e.g., in [2]. (The basic convention is this: differential relation $\delta y = t_{xy}\delta x$ is represented by the diagram $x \rightarrow y$; t_{xy} is termed the *transmission function* of the diagram edge xy. Diagrams are composed of such elementary units. For instance, the serial concatenation of two diagrams, $x \rightarrow y \rightarrow z$, is equivalent to diagram $x \rightarrow z$ with the transmission function $t_{xz} = t_{xy}t_{yz}$.)

The development in this paper starts from equations pertaining to a simple quantitative model of *SEED*. Following standard procedures, diagrammatic representation of this model is found. This allows exact topological and analytical definition of the feedback relationshps in the *SEED* as well as expressing the differential equation of its input-output characteristic in terms of the transmission functions of diagram edges. Numerical evaluations are performed for realistic parameter values. The dependences thus obtained are discussed.

309

J. Novák and A. Schlachetzki (eds.), Heterostructure Epitaxy and Devices, 309–312.
© *1996 Kluwer Academic Publishers.*

310

2. The Model of *SEED* and its Diagrammatical Representation

Nara *et al.* [3] formulated a simple model of *SEED* that we follow. The voltage drop on the insulating layer V_{in}, the sum of the external source voltage and the built-in junction voltage V_0, and the voltage across the external resistor RI, obey the equation

$$V_{in} = V_0 - RI \tag{1}$$

The current I is supposed to be independent of the junction voltage, a good approximation for device with reverse-biased *p-n* junction. It is then dependent only on the photocarrier density n:

$$I = \eta n \tag{2}$$

where η is a constant.

It is further assumed that the exciton absorption spectrum is given by the Lorentzian expression

$$\frac{P_{in}\Omega_0}{(\omega - \xi_0)^2 + (\Omega_0/2)^2}$$

with P_{in} the input optical power, Ω_0 (equal to inverse exciton lifetime) the spectral halfwidth, ω the light frequency, and ξ_0 the central frequency of the exciton absorption spectrum. The latter quantity is dependent, via the *QCSE* which defines the spectral shift constant β, on V_{in}:

$$\xi_0 = \omega_0 - \beta V_{in} \tag{3}$$

Finally, we have the stationary version of the rate equation for photocarrier density which reads

$$0 = -\frac{n}{\tau} + \frac{\alpha P_{in}\Omega_0}{(\omega - \xi_0)^2 + (\Omega_0/2)^2} \tag{4}$$

where τ is the carrier relaxation time and α is the photocarrier generation rate (via exciton dissociation).

Next we follow the standard procedure of the diagrammatic method [2], i.e., we differentiate the equations (1) - (4) and identify the coefficients before the differentials as the transmission functions of a diagram:

$$\delta V_{in} = t_{IV_{in}} \delta I \tag{1'}$$

$$\delta I = t_{nI} \delta n \tag{2'}$$

$$\delta \xi_0 = t_{V_{in}\xi_0} \delta V_{in} \tag{3'}$$

$$\delta n = t_{P_{in}n} \delta P_{in} + t_{\xi_0 n} \delta \xi_0 \tag{4'}$$

Here we have used the substitutions

$$t_{IV_{in}} = -R, \qquad t_{nI} = \eta, \qquad t_{V_{in}\xi_0} = -\beta \tag{5}$$

$$t_{P_{in}n} = \frac{\alpha \Omega_0 \tau}{(\omega - \xi_0)^2 + (\Omega_0/2)^2} \tag{6}$$

$$t_{\xi_0 n} = \frac{2\alpha \Omega_0 \tau (\omega - \xi_0) P_{in}}{[(\omega - \xi_0)^2 + (\Omega_0/2)^2]^2} \tag{7}$$

Equation set (1') - (4') is represented by the following diagram

$$\begin{array}{ccc} \rightarrow P_{in} \rightarrow n \rightarrow I \rightarrow \\ \uparrow \qquad \downarrow \\ \xi_o \leftarrow V_{in} \end{array}$$

It is implied that the change in P_{in}, δP_{in}, is externally controlled (input) and the change in I, δI, is monitored (output). The diagram contains a feedback loop $loop \equiv IV_{in}\xi_0 nI$ with the transmission function t_{loop} equal to the product of transmission functions of the edges composing it. Transmission function of the whole diagram

$$t_{P_{in}I} \equiv \frac{dI}{dP_{in}} = \frac{t_{P_{in}n} t_{nI}}{1 - t_{loop}} \tag{8}$$

presents differential form of the input-output (P_{in}-versus-I) characteristic. It is known [2] that bistability occurs if the equation $t_{loop}(I) = 1$ has two real roots.

312

3. Numerical Evaluation and Discussion

We computed dependences of P_{in} and t_{loop} on I for two different values of R (0.5 and 3 MΩ, see Fig.1). Numerical values of other constants and variables were chosen in conformity with [3]: $\omega = 2.350\times10^{15}$ Hz, $\omega_o = 2.344\times10^{15}$ Hz, $\Omega_o = 2.22\times10^{12}$ Hz, $\beta = 4.688\times10^{12}$ Hz.V^{-1}, $\alpha\tau = 1.125\times10^{27}$ m^{-3}.J^{-1}, $\eta = 2\times10^{-16}$ A.m^3, and $V_o = 2.74$ V.

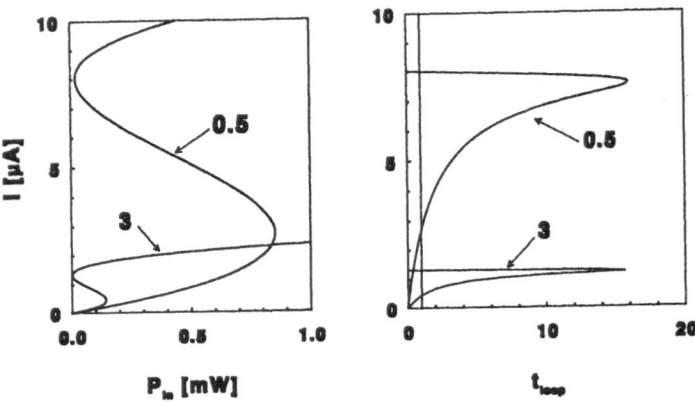

Figure 1. Dependences of the input optical power and the transmission function of the feedback loop on the photocurrent. Curve labels designate values of the external resistor R in MΩ. The vertical line on the right-hand panel indicates the critical level of unity.

One can see the correspondence between the singular points of the I-versus-P_{in} characteristics (i.e. points for which $dI/dP_{in} = \infty$) and points in which the $t_{loop}(I)$ curves cross the critical level of unity. Existence of two distinct critical points is a *sufficient* condition for the occurrence of bistability. (Existence of the feedback loop in the representative diagram forms the *necessary* condition.) Increasing the magnitude of external resistance shifts the maximum of the $t_{loop}(I)$ curve to lower I values; it also leads to a reduction of this originally broad asymmetric bell-shaped feature to a narrow peak. Accordingly, the bistability region of the input-output characteristic is displaced to lower current values and the two extrema shrink together, making the phenomenon difficult to observe.

Support of the Grant Agency of the Czech Republic is gratefully acknowledged (project No 102/93/0642).

4. References

1. Miller, D.A.B., Chemla, D.S., Damen, T.C., Gossard, A.C., Wiegmann, W., Wood, and T.H., Burrus, C.A. (1984) Novel hybrid optically bistable switch: The quantum self-electro-optic effect device, *Appl. Phys. Lett.* **45**, 13-15.
2. Šrobár, F. (1992) Feedback relationships - a neglected theme in physics, *Eur. J. Phys.* **13**, 1-8.
3. Nara, S., Tokuda, Y., Abe, Y., Yasukawa, M., Tsukada, N., and Totsuji, H. (1994) Multistable behaviour connected bistable devices, *J. Appl. Phys.* **75**, 3749-3755.

OPTOELECTRONIC INTEGRATED CIRCUIT A²B⁶-INSULATOR-A³B⁵ WITH POSITIVE FEEDBACK

S.A.MANEGO, A.S.POSEDKO, L.N.SURVILO,
Yu.V.TROFIMOV
Institute of Electronics, Belarus Academy of Sciences
22 Logoiski Trakt, 220841, Minsk-90, Republic of Belarus

1. Introduction

Optoelectronic designers are now paying much attention to the development of optoelectronic structures with expanded functional capabilities. In this context, an attempt has been made to develop and make a monolithic optoelectronic integrated circuit (OEIC) that consists of a photodetector based on a thin-film CdSSe photoresistor optically and electrically connected with a LED based on GaAsP material, and to reveal experimentally the presence of new functional characteristics that may be of practical significance. The semiconductor materials chosen for this optoelectronic structure may have both identical and different spectral ranges of photosensitivity and radiation, different temperature coefficients of photocurrent and radiation, as well as a number of other differences. Therefore, we have considered two main versions of the OEIC with the following set of optoelectronic characteristics:

1. Photoresistor and LED with matched spectral characteristics.
2. Photoresistor and LED with different spectral characteristics, or an optoelectronic structure where optical interaction between the above-mentioned elements is minimum.

It is known that in case of integrating photodetectors and emitters in an OEIC, i.e. in a small physical volume, serious problems are normally encountered in obtaining the required performance characteristics due to mutual thermal influence of structural elements [1]. Therefore, to make our study more detailed, we decided to consider this problem by creating both integrated and hybrid micro-opto-structures and by performing a comparative analysis of their characteristics.

2. Experimental

The experimentally obtained specimens of an OEIC have a monolithic planar structure. The structure consists of a semiconductor LED, a dielectric light-transmitting medium and a thin-film photoresistor. A cross-section of such an

313

J. Novák and A. Schlachetzki (eds.), Heterostructure Epitaxy and Devices, 313–316.
© 1996 *Kluwer Academic Publishers.*

element, an equivalent circuit and SEM photograph of this structure are shown in Figure 1.

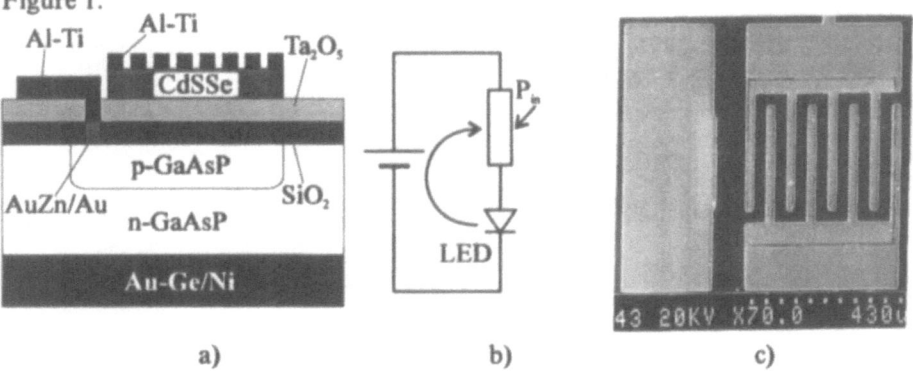

a)   b)   c)

Figure 1. Cross-section, equivalent circuit and SEM photograph of A^2B^6-insulator-A^3B^5 structure.

A masked zinc diffusion forms the p-electrode of the LED. The ohmic AuZn/Au contacts are defined by the lift-off process. Au-Ge/Ni contacts are deposited by thermal evaporation and alloyed in H_2 at 450 °C. Configuration of CVD-SiO_2 insulator region is formed by wet chemical etching. Deposited by electron beam evaporation Ta_2O_5-insulator regions are delineated by the lift-off process. CdSSe film is deposited by DC cathode three-electrode sputtering with RF-voltage applied to a cold-pressed target [2]. Activation of photosensitivity of the CdSSe film is made by diffusion of Cu, Cd, Cl impurities in open tube at 425°C. Al-Ti ohmic contacts to the photoconductive material are defined by the lift-off process. The photoresistor has a simple interdigitated electrode system with a 40 μm gap and about 0.1 mm² photosensitive (active) surface. The photoresistor proved to be most sensitive to wavelengths of about 675 nm which corresponded to the $CdS_{0.2}Se_{0.8}$ solid solution chosen for the experiment.

3. Results and Discussion

In the OEIC specimens produced, photoresistors and LEDs were seriesly connected to each other as shown in Figure 1(b). With voltage applied across the circuit and the photoresistor illuminated with external radiation, we had a light-controlled voltage divider. If the voltage magnitude on the LED reached 1.7 V, it started to illuminate light. It is quite natural that in this case the LED radiation intensity correlated with the values of the incident optical power (P_{in}) and applied voltage. Figure 2 shows the dependence of the total photocurrent (I_{ph}) flowing through a typical structure on the P_{in} for various voltages (2.5 V, 5 V, 10 V) applied across the OEIC. A GaAlAs LED was used as a radiation source. Let us consider the plots shown in Figure 2 in a greater detail. It should be noted,first of all, that since the voltage-current characteristic of the LED was nonlinear up to 1.7 V, its ohmic resistance was in the range 50-200 kOhm, and near 2.5 V it was 200-1000 Ohm.

In this connection, even for small levels of the P_{in} the voltage value on the LED exceeded 1.7 V, and the amount of I_{ph} flowing through the structure was limited by the photoresistor resistance value. The curve A corresponds to the case where 10 V is applied across the OEIC. Section (A-A') of the curve A corresponds to a linear relationship between the P_{in} and the OEIC's I_{ph}. However, after the point A' there occurs saturation of the photocurrent on the section A'-A" followed by its fall, the current magnitude being unstable in time.

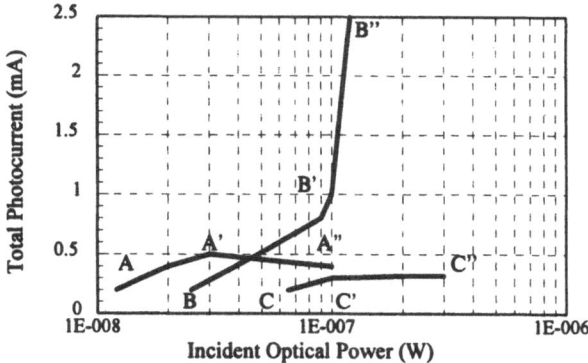

Figure 2. Dependence of the total photocurrent as a function of incident optical power.

This occurs due to heat evolution in an amount exceeding thermal limit of operation of thin-film A^2B^6 structures under conditions of natural cooling that is approximately equal to 1.6 W/cm². The curve B corresponds to the case where 5 V voltage is applied across the structure. This case is most interesting. Section (B-B') of the curve B reveals a behaviour close to a linear one with a small saturation in the region close to the point B'. After the P_{in} reaches 10^{-7} W, the I_{ph} magnitude rises sharply even though the subsequent P_{in} is practically unchanged, i.e. positive feedback effect is observed. The point B" on the section B'-B" reached various values in various structures, lying normally in the 2.5-4mA range. When optical radiation source was removed, the current value returned back to the point B' and remained at that level for a fairly long period of time (about 1 hour), i.e. memory function was realized. In our opinion, failure of the memory effect is mainly caused by temporal instability of the photocurrent (5% fall in photocurrent/1 hour). Some OEIC specimens showed the effect of periodic ON/OFF switching (blinking) of the LED near the section B'-B", the blinking frequency being largely determined by the applied voltage value. The curve C corresponds to the case where 2.5 V voltage is applied across the OEIC. Similarly to the behaviour of the section B-B' of the curve B, section C-C' of the curve C is close to a linear dependence of the I_{ph} on the P_{in}. However, as the P_{in} is further increased the I_{ph} flowing in the structure comes smoothly to saturation and stabilization due saturation of the photoresistor photosensitivity and LED resistivity. Much interest, in our opinion, is associated with the fact that the structure under consideration exhibits a property of changing the LED light intensity depending on the P_{in}. Manifestation of this property can be practically implemented in automatic control of LED luminance as a function of

316

external illumination conditions. In Figure 2, the property of automatic luminance control is characteristic of the sections A-A' and B-B' , the angle of inclination to the axis X being a function of the photoresistor photosensitivity.

In order to perform a more detailed study of the revealed relationships between the OEIC's elements and to find possible ways of improving design and fabrication process, a hybrid optoelectronic structure has been developed and made which is a functional analogue of the OEIC desribed above. Figure 3 shows a photograph and schematic cross-section of a hybrid optoelectronic element made of a thin-film CdSSe photoresistor and GaAlAs LED chip.

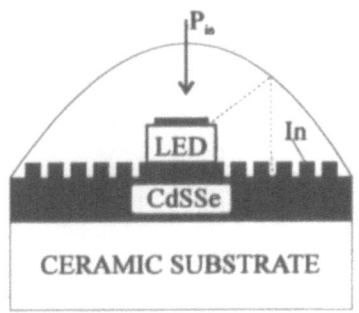 

Figure 3. Cross-section and SEM photograph of hybrid optoelectronic element.

The implemented hybrid structure made it quite simple for us to make the following purposeful changes: to use a considerably more effective light-emitting semiconductor material, namely, GaAlAs, instead of GaAsP; to choose photoconductive materials from CdS, through CdS-CdSe solid solutions, to CdSe in the wavelength range from 0.52 to 0.72 μm by using a standard process of A^2B^6 film deposition on more cheap ceramic substrates; to change the coefficient of optical coupling between the hybrid circuit elements by using compounds with different transmission coefficients and potting forms.

Investigations of the structures mentioned above have demonstrated that characteristics of the OEIC's also manifest themselves in hybrid elements in full measure. This fact enabled us to perform a comparative analysis of such structures with a view to define future research tasks aimed at creation of new functional optoelectronic elements with improved performance characteristics.

4. References

1. Trofimov, Yu.V., Vedenin, A.S., Posedko, A.S. (1993) A^2B^6-compound thin-film photoresistors for 3-D optical processors, Optoelectronic Instrumentation and Data Processing 3, 74-76.(Allerton Press Inc.).
2. Trofimov, Yu.V. (1994) Formation of thin CdSSe photoconductive films for optical information processing and measuring devices, in A.M. Girro, O.V. Goncharova et al. (eds.), New Materials for Thin-Film Functional Elements of Electronic Equipment, Nauka i Tekhnika Publishers, Minsk, pp.89-98 (in Russian).

CHARACTERISTICS OF MULTIPLE δ-DOPED GaAs STRUCTURES

T. LALINSKÝ, J. ŠAFRÁNKOVÁ, Ž. MOZOLOVÁ
Institute of Electrical Engineering, Slovak Academy of Sciences
Dúbravská cesta 9, 842 39 Bratislava, Slovakia
R. HARMAN, Š. NÉMETH, M. BUJDÁK
Faculty of Electrical Engineering and Information Technology,
Slovak Technical University
Ilkovičova 3, 812 19 Bratislava, Slovakia

1. Introduction

Recently, there has been a lot of interest in the physics and applications of the delta (δ) doping (or planar doping). With the advances in molecular beam epitaxy (MBE) and metalorganic chemical vapor deposition (MOCVD), it became possible to control the growth in such a way that the dopants could be put in a plane during growth interruption, resulting in a doped region confined to very narow space (plane). The carriers released from the dopants are confined by the potential well induced by the space charges, themeselves. Si-delta doping GaAs has received much attention as a means of obtaining very high two-dimensional electron gas (2DEG) density in different homomorphic or pseudomorphic field-effect transistor (FET) structures. If a δ-doped layer is used as an electron conduction channel in a GaAs FET structures [1-3], a significant increase of both the device performances and a potential applications could be achieved. In this letter we report on MBE grown multiple δ-doped GaAs layer test structures. The characteristics of fabricated vertical diode structures as well as a planar FET structures based on the multiple δ-doped layers are investigated.

2. Experiment

The multiple δ-doped layers were formed by Si in GaAs matrix. The growth was carried out in MBE system at Slovak Technical University in Bratislava. All structures consist of an undoped GaAs buffer layer grown on a highly Si-doped ($N_d = 3 \times 10^{18} \text{cm}^{-3}$) GaAs (001) substrate, the δ-doped GaAs planes, a separation layers from undoped GaAs, an undoped GaAs cap layer, and finally n^+-GaAs top layer. The eight δ-layers with both the three different plane doses (from $1 \times 10^{12} \text{cm}^{-2}$ till $6 \times 10^{12} \text{cm}^{-2}$) and different separation layers were designed.

J. Novák and A. Schlachetzki (eds.), Heterostructure Epitaxy and Devices, 317–320.
© 1996 *Kluwer Academic Publishers.*

The growth temperature of all GaAs layers was 595°C. The growth rate used was 0.5 μm/h. Si delta doping was achieved by depositing Si atoms while suspending the deposition of Ga atoms. The arsenic flux was not interrupted during Si deposition. The Si dose was estimated by the δ-doping time intervals using the relationship $\tau = N^{2D}/N_d^{3D} * v_{growth}$. The eightfold δ-doped layer structures were further processed into a vertical diodes as well as a planar FET structures schematically shown in Figure 1 and Figure 2, respectively. Nine circular Schottky contacts (Figure 1) with various diameters (d=40-800μm) were patterned using the lift-off technique. In order to suppress the influence of a surface leakage currents, a deep wet Mesa-etching of the layer structures using the Schottky contacts masking technique was additionaly introduced into the diode processing. A recess-gate FET technology design based on a Mesa-isolation technique (Figure 2) was applicated. In addition, a plasma deposited silicon nitride layer was used to improve the contact pads isolation.

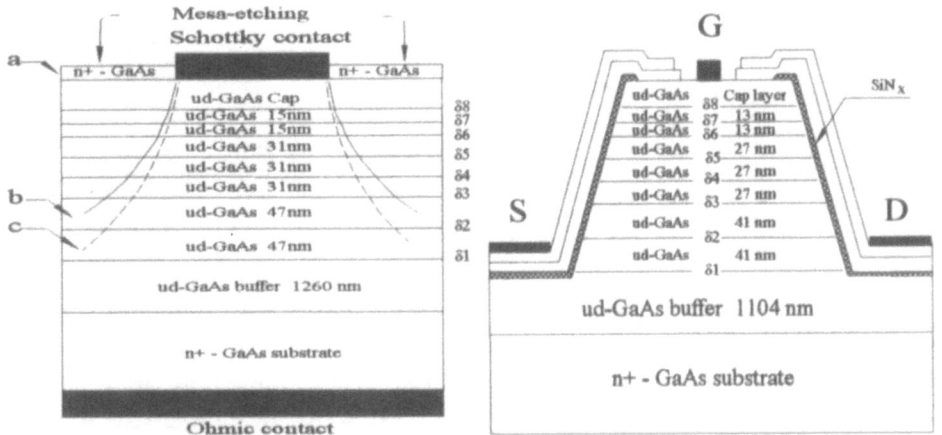

Figure 1. Cross section of diode structure. *Figure 2.* Cross section of FET structure.

3. Results and discussion

3.1. DIODE STRUCTURES

Figure 3 shows a typical behaviour of I-V characteristics of as formed diode structures type *a* (without any Mesa-etching) at various Schottky contact diameters. A typical Schottky diode-like behaviour can be seen. Likewise, the I-V characteristics follow the change of the Schottky contact diameters. The extracted barrier height and ideality factor values were 0.74V and 1.12, respectively. Thermionic emission was assumed to be the dominant mechanism of current transport in the diodes.

If Mesa-etching of layer structures is applicated step by step (Figure 1), the increase of the diode series resistance can be indicated. Figure 4 shows a typical I-V characteristics of the Mesa-etched diodes at two different etching depths (diodes b and c in Figure 1). A linear I-V relationship corresponding to the high series resistance value ($R_s = 2 \times 10^8 \ \Omega$) was observed for diodes c, when all δ-doped layers were etch-out around the Schottky contact periphery. The Schottky diode-like behaviour is suppressed by the diode series resistance, so the I-V characteristics exhibit a current symmetry.

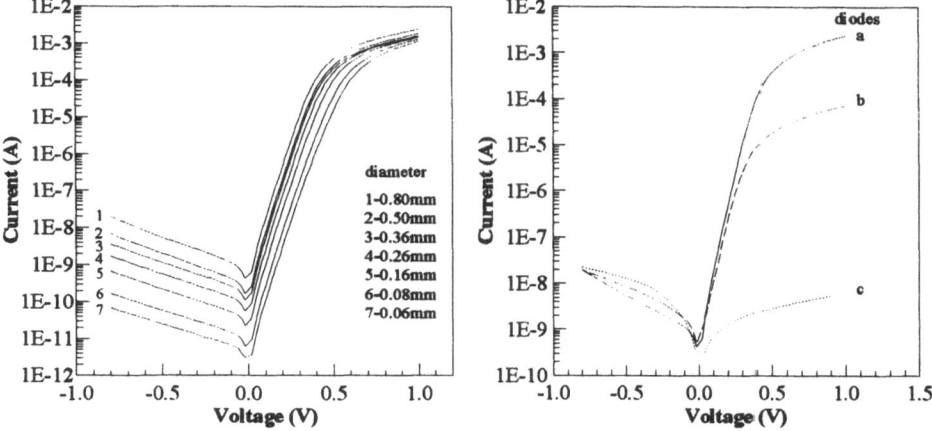

Figure 3. I-V characteristics of as formed diode structure (diode a).

Figure 4. Typical I-V characteristics of Mesa -etched diodes (diodes b and c).

3.2. FET STRUCTURES

The eightfold δ-doped layer test structures was designed to applicate also as a GaAs MESFET multiple-electron conductive channel (Figure 2). A significant increase of the 2DEG concentration in the channel can be expected. A typical output I-V characteristics of a 2µm gate length GaAs MESFET structures at two different temperatures (T=300K and 77K) are shown in Figure 5. Good saturation is achieved. There is seen the conductivity decreasing in the linear range (at 77K) probably due to by the decreasing of the electron mobility.

On the other hand some conductivity increasing in the saturation range was indicated which could be explained by the increase of the electron velocity saturation. A very interesting behaviour in the transconductance-gate bias curve (Figure 6) was observed. The transconductance is shown to exhibit the expected stepped form. In the transconductance-gate bias curve at 77K the eight separate peaks corresponding to the separated eight δ-channel layers were observed. Flat and broad transconductance profile is formed by using the multiple δ-doping structure.

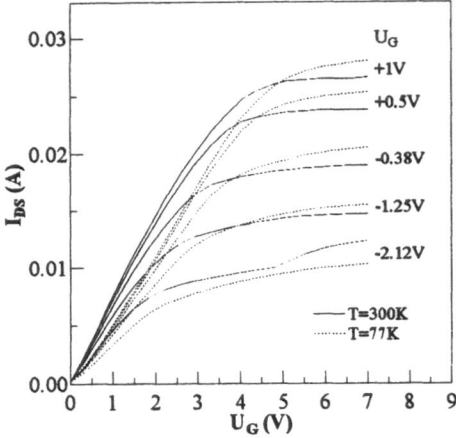
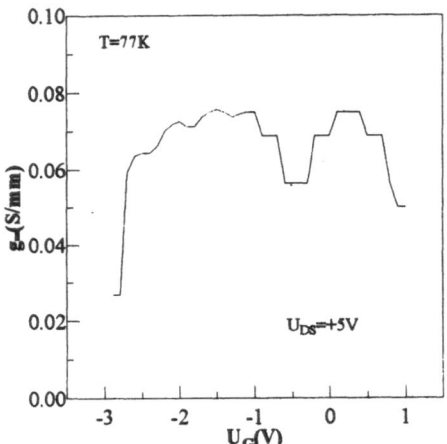

Figure 5. Typical output I-V characteristics of 2μm gate length GaAs MESFET structure at 300K and 77K.

Figure 6. Extrinsic transconductance-gate bias curve at 77K.

4. Conclusion

The characteristics of fabricated vertical diodes as well as a planar FET structures based on the eightfold δ-doped layers have been demonstrated to our knowledge for the first time. The Schottky diode-like behaviour was observed for as formed diode structures. If all the diode δ-layers were etch-out around the Schottky contact periphery, the increase of the diode series resistance value was observed, so the I-V characteristics exhibited a current symmetry. Furthermore, eight separate peaks in the transconductance-gate bias curve corresponding to the eight separated δ-channel layers were indicated. Flat and broad transconductance profile was formed by using the multiple δ-doping structure.

5. References

1. Kao, M.J., Hsu, W.C., Hsu, R.T., Wu, Y.H., Lin, T.Y., and Chang, C.Y. (1995) Characteristics of graded-like multiple-delta-doped GaAs field effect transistors, *Appl. Phys. Lett.* **66**, 2505-2506.
2. Board, K., and Nutt, H.C. (1992) Double delta-doped FETs in GaAs, *Electronics Letters* **28**, 469-471.
3. Wu, C.L., Hsu, W.C., Tsai, M.S., and Shieh, H.M. (1994) High performance three-terminal δ-doped GaAs negative resistance field-effect transistor based on real-space transfer, *Electronics Letters* **30**, 1537-1539.

SUBJECT INDEX